Methods in Enzymology

Volume 418
EMBRYONIC STEM CELLS

METHODS IN ENZYMOLOGY

EDITORS-IN-CHIEF

John N. Abelson   Melvin I. Simon

DIVISION OF BIOLOGY
CALIFORNIA INSTITUTE OF TECHNOLOGY
PASADENA, CALIFORNIA

FOUNDING EDITORS

Sidney P. Colowick and Nathan O. Kaplan

Methods in Enzymology

Volume 418

Embryonic Stem Cells

EDITED BY

Irina Klimanskaya

Robert Lanza

ADVANCED CELL TECHNOLOGY
WORCESTER, MASSACHUSETTS

ELSEVIER

AMSTERDAM • BOSTON • HEIDELBERG • LONDON
NEW YORK • OXFORD • PARIS • SAN DIEGO
SAN FRANCISCO • SINGAPORE • SYDNEY • TOKYO
Academic Press is an imprint of Elsevier

Academic Press is an imprint of Elsevier
525 B Street, Suite 1900, San Diego, California 92101-4495, USA
84 Theobald's Road, London WC1X 8RR, UK

This book is printed on acid-free paper. ∞

For information on all Elsevier Academic Press publications
visit our Web site at www.books.elsevier.com

ISBN-13: 978-0-12-373648-2
ISBN-10: 0-12-373648-X

PRINTED IN THE UNITED STATES OF AMERICA
06 07 08 09 9 8 7 6 5 4 3 2 1

To Richard Latsis, the Teacher
　　　　　　　　　　–Irina

Table of Contents

Section I. Derivation and Maintenance of Embryonic Stem Cells

Section II. Differentiation of Embryonic Stem Cells

Contributors to Volume 418

Article numbers are in parentheses following the names of contributors.
Affiliations listed are current.

HIDENORI AKUTSU (5), *Harvard University, Howard Hughes Medical Institute, Cambridge, Massachusetts*

TIZIANO BARBERI (12), *Division of Neurosciences, Beckman Research Institute of the City of Hope, Duarte, California*

GILAD BECK (15), *Stem Cell Center, Bruce Rappaport Faculty of Medicine, Technion, Haifa, Israel*

SANDY BECKER (7, 9), *Advanced Cell Technology, Biotech Five, Worcester, Massachusetts*

ALEX BENCHOUA (10), *Institute for Stem Cell Research, University of Edinburgh, Edinburgh, United Kingdom*

ANNE E. BISHOP (20), *Imperial College Faculty of Medicine, Tissue Engineering and Regenerative Medicine Centre, London, United Kingdom*

PRZEMYSLAW BLYSZCZUK (19), *Leibniz Institute of Plant Genetics and Crop Plant Research, In Vitro Differentiation Group, Gatersleben, Germany*

YOUNG CHUNG (7, 9), *Advanced Cell Technology, Biotech Five, Worcester, Massachusetts*

JOSE B. CIBELLI (8), *Departments of Animal Science and Physiology, Michigan State University, East Lansing, Michigan*

PAUL COLLODI (4), *Purdue University, Department of Animal Sciences, West Lafayette, Indiana*

CHAD A. COWAN (5), *Harvard University, Howard Hughes Medical Institute, Cambridge, Massachusetts*

KERRIANNE CUNNIFF (8), *Departments of Animal Science and Physiology, Michigan State University, East Lansing, Michigan*

GEORGE Q. DALEY (18), *Harvard Medical School, Children's Hospital Boston, Boston, Massachusetts*

LIANCHUN FAN (4), *Department of Animal Sciences, Purdue University, West Lafayette, Indiana*

NIELS GEIJSEN (18), *Harvard Stem Cell Institute, Center for Regenerative Medicine and Technology, Boston, Massachusetts*

ILANA GOLDBERG-COHEN (15), *Stem Cell Center, Bruce Rappaport Faculty of Medicine, Technion, Haifa, Israel*

XI-MIN GUO (16), *Institute of Basic Medical Sciences and Tissue Engineering Research Center, Academy of Military Medical Sciences, Beijing, People's Republic of China*

GEORGE R. HONIG (14), *Advanced Cell Technology, Biotech Five, Worcester, Massachusetts*

KARIN HÜBNER (17), *Max Planck Institute for Molecular Biomedicine, Cell and Developmental Biology, Muenster, Germany*

JOSEPH ITSKOVITZ-ELDOR (15), *Department of Obstetrics and Gynecology, Rambam Medical Center, Haifa, Israel*

GABRIELA KANIA (19), *Leibniz Institute of Plant Genetics and Crop Plant Research, In Vitro Differentiation Group, Gatersleben, Germany*

JAMES KEHLER (17), *Max Planck Institute for Molecular Biomedicine, Cell and Developmental Biology, Muenster, Germany*

IRINA KLIMANSKAYA (11), *Advanced Cell Technology, Worcester, Massachusetts*

GABSANG LEE (13), *Moore Laboratory, Cell Biology Program, Memorial Sloan-Kettering Cancer Center, New York, New York*

FEI LI (14), *Advanced Cell Technology, Biotech Five, Worcester, Massachusetts*

SALLY LOWELL (10), *Institute for Stem Cell Research, University of Edinburgh, Edinburgh, United Kingdom*

SHI-JIANG LU (14), *Advanced Cell Technology, Biotech Five, Worcester, Massachusetts*

CHRISTINE MATHER-LOVE (3), *Origen Therapeutics, Burlingame, California*

DOUGLAS MELTON (5), *Harvard University, Howard Hughes Medical Institute, Cambridge, Massachusetts*

MALCOLM A. S. MOORE (13), *Moore Laboratory, Cell Biology Program, Memorial Sloan-Kettering Cancer Center, New York, New York*

ANDRAS NAGY (1), *Mount Sinai Hospital, Samuel Lunenfeld Research Institute, Toronto, Ontario, Canada*

JULIA M. POLAK (20), *Imperial College Faculty of Medicine, Tissue Engineering and Regenerative Medicine Centre, London, United Kingdom*

STEVEN M. POLLARD (10), *Institute for Stem Cell Research, University of Edinburgh, Edinburgh, United Kingdom*

MARSHA ROACH (2), *Pfizer Global Research and Development, Genetically Modified Models CoE, Groton, Connecticut*

HANS R. SCHÖLER (17), *Max Planck Institute for Molecular Biomedicine, Cell and Developmental Biology, Muenster, Germany*

INSA S. SCHROEDER (19), *Leibniz Institute of Plant Genetics and Crop Plant Research, In Vitro Differentiation Group, Gatersleben, Germany*

JAE-HUNG SHIEH (13), *Moore Laboratory, Cell Biology Program, Memorial Sloan-Kettering Cancer Center, New York, New York*

NICK STRELCHENKO (6), *Reproductive Genetics Institute, Chicago, Illinois*

LORENZ STUDER (12), *Developmental Biology and Neurosurgery, Memorial Sloan-Kettering Cancer Center, New York, New York*

X. CINDY TIAN (2, 16), *Pfizer Global Research and Development, Genetically Modified Models CoE, Groton, Connecticut*

MARIE-CECILE VAN DE LAVOIR (3), *Origen Therapeutics, Burlingame, California*

YURY VERLINSKY (6), *Reproductive Genetics Institute, Chicago, Illinois*

KRISTINA VINTERSTEN (1), *Mount Sinai Hospital, Samuel Lunenfeld Research Institute, Toronto, Ontario, Canada*

KENT E. VRANA (8), *Penn State College of Medicine, Department of Pharmacology, Hershey, Pennsylvania*

CHANG-YONG WANG (16), *Institute of Basic Medical Sciences and Tissue Engineering Research Center, Academy of Military Medical Sciences, Beijing, People's Republic of China*

Li Wang (2), *Stem Cell Research Center, Department of Cell Biology, Beijing, China*

Anna M. Wobus (19), *Leibniz Institute of Plant Genetics and Crop Plant Research, In Vitro Differentiation Group, Gatersleben, Germany*

Xiangzhong Yang (2, 16), *Pfizer Global Research and Development, Genetically Modified Models CoE, Groton, Connecticut*

Anna Ziskind (15), *Stem Cell Center, Bruce Rappaport Faculty of Medicine, Technion, Haifa, Israel*

Preface

Stem cells are of great interest to scientists and clinicians due to their unique ability to differentiate into various tissues of the body. In addition to being a promising source of cells for transplantation and regenerative medicine, they also serve as an excellent model of vertebrate development. In the recent years, the interest in stem cell research has spread beyond the scientific community to the public at large as a result of heated political and ethical debate.

There are two broad categories of stem cells—"embryonic" and "adult." Embryonic stem cells—also known as *"pluripotent"* stem cells—are derived from preimplantation-stage embryos and retain the capacity to grow in culture indefinitely, as well as to differentiate into virtually all the tissues of the body. Adult stem cells are found in most tissues of the adult organism; scientists are beginning to learn how to isolate, culture, and differentiate them into a range of tissue-specific types (and are thus considered *multipotent*).

Growing stem cells in culture and differentiating them on demand requires specific skills and knowledge beyond basic cell culture techniques. We have tried to assemble the most robust and current techniques (including both conventional and novel methods) in the stem cell field and invited the world's leading scientists with hands-on expertise to write the chapters on methods they are experts in or even established themselves. Volume 418, "Embryonic Stem Cells," offers a variety of know-how from derivation to differentiation of embryonic stem cells, including such sought-after methods as human embryonic stem cell derivation and maintenance, morula- and single blastomere-derived ES cells, ES cells created via parthenogenesis and nuclear transfer, as well as techniques for derivation of ES cells from other species, including mouse, bovine, zebrafish, and avian. The second section of this volume covers the recent advances in differentiation and maintenance of ES cell derivatives from all three germ layers: cells of neural lineage, retinal pigment epithelium, cardiomyocytes, haematopoietic and vascular cells, oocytes and male germ cells, pulmonary and insulin-producing cells, among others.

Volume 419, "Adult Stem Cells," covers stem cells of all three germ layers and organ systems. The methods include isolation, maintenance, analysis, and differentiation of a wide range of adult stem cell types, including neural, retinal, epithelial cells, dental, skeletal, and haematopoietic cells, as well as ovarian, spermatogonial, lung, pancreatic, intestinal, trophoblast, germ, cord blood, amniotic fluid, and placental stem cells.

Volume 420, "Tools for Stem Cell Research and Tissue Engineering," has collected specific stem cells applications as well as a variety of techniques, including gene trapping, gene expression profiling, RNAi and gene delivery, embryo culture for human ES cell derivation, characterization and purification of stem cells, and cellular reprogramming. The second section of this volume addresses tissue engineering using derivatives of adult and embryonic stem cells, including important issues such as immunogenicity and clinical applications of stem cell derivatives.

Each chapter is written as a short review of the field followed by an easy-to-follow set of protocols that enables even the least experienced researchers to successfully establish the techniques in their laboratories.

We wish to thank the contributors to all three volumes for sharing their invaluable expertise in comprehensive and easy to follow step-by-step protocols. We also would like to acknowledge Cindy Minor at Elsevier for her invaluable assistance assembling this three-volume series.

IRINA KLIMANSKAYA
ROBERT LANZA

Foreword

As stem cell researchers, we are frequently asked by politicians, patients, reporters, and other non-scientists about the relative merits of studying embryonic stem cells *versus* adult stem cells, and when stem cells will provide novel therapies for human diseases. The persistence of these two questions and the passion with which they are asked reveals the extent to which stem cells have penetrated the vernacular, captured public attention, and become an icon for the scientific, social, and political circumstances of our times.

Focusing first on the biological context of stem cells, it is clear that the emergence of stem cells as a distinct research field is one of the most important scientific initiatives of the 'post-genomic' era. Stem cell research is the confluence between cell and developmental biology. It is shaped at every turn by the maturing knowledge base of genetics and biochemistry and is accelerated by the platform technologies of recombinant DNA, monoclonal antibodies, and other biotechnologies. Stem cells are interesting and useful because of their dual capacity to differentiate and to proliferate in an undifferentiated state. Thus, they are expected to yield insights not only into pluripotency and differentiation, but also into cell cycle regulation and other areas, thereby having an impact on fields ranging from cancer to aging.

This directs us to why it is necessary to study different types of stem cells, including those whose origins from early stages of development confers ethical complexity (embryonic stem cells) and those that are difficult to find, grow, or maintain as undifferentiated populations (most types of adult stem cells). The question itself veils a deeper purpose for studying the biology of stem cells, which is to gain a fundamental understanding of the nature of cell fate decisions during development. We still have a relatively shallow understanding of how stem cells maintain their undifferentiated state for prolonged periods and then 'choose' to specialize along the pathways they are competent to pursue. Achieving a precise understanding of such 'stemness' and of differentiation will require information from as wide a variety of sources as possible. This process of triangulation could be compared to how global positioning satellites enable us to locate ourselves: signal from a single satellite tells us relatively little, and precision is achieved only when we acquire signals from three or more. Similarly, it is necessary to study multiple types of stem cells and their progeny if we are to evaluate the outcome of cellular development *in vitro* in comparison with normal development.

An answer to the question of when stem cells will yield novel clinical outcomes requires us to define the likely therapeutic achievements. Of course, adult and neonatal blood stem cells have been used in transplantation for many years, and it is likely that new sources and applications for them will emerge from current studies. It is less likely, however, that transplantation will be the first application of research on other adult stem cell types or of research on the differentiated progeny of embryonic stem cells. This reflects in part the degree of characterization of such progeny that will be needed to ensure their long-term safety and efficacy when transplanted to humans. It is their use as *in vitro* cellular models that is more likely to pioneer novel clinical applications of the specialized human cell types that can be derived from stem cells. These cells, including cultured neurons, cardiomyocytes, kidney cells, lung cells, and numerous others, will imminently provide a novel platform technology for drug discovery and testing. The applications of such cellular models are likely to be extensive, leading to development of new medicines for a myriad of human health problems. The wide availability of these specialized human cells will also provide an opportunity to evaluate the stability and function of stem cell progeny in the Petri dish well before they are used in transplantation. Finally, we should not overlook the importance of stem cells and their progeny as models for understanding human developmental processes. While we cannot foresee the impact of a profound understanding of human cellular differentiation, it has the potential of transcending even the most remarkable applications that we can imagine involving transplantation.

Despite the links of stem cell research to other fields of biology and to established technologies, growing and differentiating stem cells systematically in culture requires specific skills and knowledge beyond basic cell and developmental biology techniques. These *Methods in Enzymology* volumes include the most current techniques in the stem cell field, written by leading scientists with hands-on expertise in methods they have developed or in which they are recognized as experts. Each chapter is written as a short review of outcomes from the particular method, with an easy-to-follow set of protocols that should enable less experienced researchers to successfully establish the method in their laboratories. Together, the three volumes cover the spectrum of both embryonic and adult stem cells and provide tools for extending the uses of stem cells to tissue engineering. It is hoped that the availability and wide dissemination of these methods will provide wider access to the stem cell field, thereby accelerating acquisition of the knowledge needed to apply stem cell research in novel ways to improve our understanding of human biology and health.

ROGER A. PEDERSEN, PH.D.
PROFESSOR OF REGENERATIVE MEDICINE
UNIVERSITY OF CAMBRIDGE

METHODS IN ENZYMOLOGY

Section I

Derivation and Maintenance of Embryonic Stem Cells

[1] Murine Embryonic Stem Cells

By ANDRAS NAGY and KRISTINA VINTERSTEN

Abstract

Embryonic stem (ES) cells are derived from preimplantation stage mouse embryos at the time when they have reached the blastocyst stage. It is at this point that the first steps of differentiation take place during mammalian embryonic development. The individual blastomeres now start to organize themselves into three distinct locations, each encompassing a different cell type: outside epithelial cells, trophectoderm; cells at the blastocele surface of the inner cell mass (ICM), the primitive endoderm; and inside cells of the ICM, the primitive ectoderm. ES cells originate from the third population, the primitive ectoderm, which is a transiently existing group of cells in the embryo. Primitive ectoderm cells diminish within a day as the embryo is entering into the next steps of differentiation. ES cells, however, while retaining the property of their origin in terms of developmental potential, also have the ability to self-renew. It is hence important to realize that ES cells do not exist *in vivo*; they should be regarded simply as tissue culture artifact. Nevertheless, these powerful cells have the potential to differentiate into all the cells of the embryo proper and postnatal animal. Furthermore, they retain the limitation of their origin through their inability to contribute to the trophectoderm lineage (the trophoblast of the placenta) and the lineages of the primitive endoderm, the visceral and parietal endoderm. Due to these unique features, we must admit that even if we regard ES cells as products of *in vitro* culture and should not compare them to true somatic stem cells found in the adult organism, they certainly offer us a fantastic tool for genetic, developmental, and disease studies.

Historical Overview

The year 2006 marks the 25th anniversary of two milestone publications reporting the establishment of embryonic stem (ES) cell lines from the mouse (Evans and Kaufman, 1981; Martin, 1981). By placing the cells in specific culture conditions, the authors could block the cells in their differentiation program and induce them to self-renew. Amazingly, however, once the cells were released from these conditions and placed into a differentiation-promoting environment (*in vitro* or *in vivo*), the cells proved capable of giving rise to a vast number of various cell types. The *in vivo* studies were done by

METHODS IN ENZYMOLOGY, VOL. 418   0076-6879/06 $35.00
DOI: 10.1016/S0076-6879(06)18001-5

injection of ES cells into a blastocyst stage host embryo. In such chimeras, ES cell derivatives could be found in all somatic cell lineages. Three years later, the ability of ES cells to contribute to the germ line was also demonstrated (Bradley *et al.*, 1984). The real breakthrough, however, took place 2 years later as two groups succeeded in passing along the genome of genetically altered ES cells through the germ line (Gossler *et al.*, 1986; Robertson *et al.*, 1986). In parallel to this development, Smithies *et al.* (1985) established the technique of homologous recombination in eukaryotic cells, which later became the means of targeting (mutating) any desired gene in the mouse genome. As a result, by the late 1980s, the stage was set for a revolution in mouse genetics (Koller *et al.*, 1989; Thomas and Capecchi, 1990).

Since the early 1990s, nearly 6000 genes have been knocked out in the mouse genome. These mutants have revealed a tremendous amount of information on the role of various genes in normal development as well as in disease processes. Today, the technology of gene targeting has developed to the level where high throughput generation of gene knockouts has become feasible. As a result, several international consortiums have been formed to generate a bank of targeted ES cells covering the entire mouse genome (Austin *et al.*, 2004). Once this goal has been achieved, will the golden era of mouse ES cells fade? In fact, why would we need to derive further lines in addition to those that already exist and have long proven their high quality? To answer these questions, we have to look more closely into the treasure chest and see what other riddles these cells could help us solve.

The germ line compatibility of mouse ES cells has been utilized almost exclusively in genetic studies, while not much attention has been devoted to their somatic abilities. It has been known for a long time that they are capable of supporting the entire embryonic (Nagy *et al.*, 1990) and adult (Nagy *et al.*, 1993) development of the mouse if some extraembryonic lineages (trophoblast, visceral, and parietal endoderm) are provided by tetraploid embryos. One of the first, and strongest, applications of ES cell <=> tetraploid embryo aggregation chimeras was analysis of the peculiar vascular endothelial growth factor-A (VEGF-A) knockout. This gene was identified as having a lethal heterozygous phenotype, hindering the "classical" germ line transmission-based gene targeting analysis (Carmeliet *et al.*, 1996). In order to overcome this obstacle, a homozygous VEGF-A null mutant ES cell line was created *in vitro* with high concentration G418 selection (Mortensen *et al.*, 1992) and then aggregated to wild-type tetraploid embryos. The embryos resulting from these experiments revealed the phenotypic consequence of VEGF-A deficiency, namely the lack of vessel formation (Carmeliet *et al.*, 1996). The very clear segregation of ES cell- and tetraploid embryo-derived compartments in the embryo proper and extraembryonic membranes, respectively, makes this type of chimera extremely useful also to rescue extraembryonic phenotypes of gene deficiency (Duncan *et al.*, 1997; Guillemot *et al.*, 1994). Due to this "complementing segregation," the method earned the name "tetraploid complementation assay" (TCA). In order to utilize the TCA, one must first establish mutant ES cell lines. The most commonly used method to generate *gain* of function mutations is by introducing a transgene into the ES cell genome. Loss of function can be achieved by "knock down" of a gene function though RNAi by a transgene expressing a small hairpin RNA (shRNA) (Kunath *et al.*, 2003) or by classical gene targeting. Creating a "knock out" with the latter method, however, requires elimination of the function of both alleles of the gene in order to visualize a recessive phenotype. This is usually done by targeting the two alleles of a gene of interest in a consecutive manner. In some cases, however, it may be more efficient to generate loss-of-function ES cell lines by deriving them from F2 generation embryos homozygous for the knockout allele (Bryja *et al.*, 2006). The huge advantage of this method becomes obvious when deficiency in two or three genes is required for studies (Ding *et al.*, 2004). In all these cases, TCA can provide very fast access to the deficient phenotypes (Duncan, 2005).

Factors Affecting the Efficiency of Mouse ES Cell Establishment

Despite our significantly increased knowledge in embryonic stem cell biology and experience with culturing these cells, the success rate of ES cell line establishment is still highly dependent on the genetic background of the source embryo (Gardner and Brook, 1997). The most permissive strains are the inbred 129 substrains. Consequently, ES cell lines with this genetic background have been used in the vast majority of ES cell-mediated mouse mutant generations. However, the 129 substrains come with the drawback of both minimal characterization and known anatomical (Livy and Wahlsten, 1997) and behavioral anomalies (Royce, 1972). Another relatively permissive inbred strain is the C57BL/6. This strain is strongly favored and considered to be somewhat of a "gold standard" in research using the mouse as a model. The germ line competence of the C57BL mES cell lines, however, falls behind that of 129 (Seong *et al.*, 2004), making the technology more expensive. As a result, the most common approach is to use 129 ES cells, and once mouse lines have been created, backcross these to the C57BL/6 background. The obvious drawback of this regimen lies in the extensive time required for breeding. Recent developments in establishment and maintenance conditions, however, have made the C57BL/6 lines more and more feasible to work with, which has led to the tendency to move the ES cell-mediated mouse genetics toward the C57BL/6.

Newly developed culture conditions have even affected the accessibility of so-called "nonpermissive" inbred strains, such as SVB, CBA (Roach *et al.*, 1995), and NOD (Chen *et al.*, 2005). Despite these advances, the genetic background still plays a vital role in the final potency of resulting mES cell lines. This fact is illustrated by the many attempts to derive ES cells from NOD embryos. There were no problems generating a large number of cell lines, but all of these lacked the ability to contribute to chimeras, obviously also including the germ line (Chen *et al.*, 2005). On the other extreme of this spectrum is the superior developmental potential of some F1 lines (Eggan *et al.*, 2001). ES cell-derived embryos produced with TCA using these hybrid cells not only developed to term at a very high frequency, but also survived after birth and developed into normal adults (Schwenk *et al.*, 2003). These new ES cell lines combined with TCA have tremendously improved the efficiency with which we can generate information on gene function (Nagy *et al.*, 2003).

Although mouse ES cells have been around since the early 1980s, the culture conditions supporting their self-renewal and inhibiting differentiation are still not fully understood. The success rate of establishment varies from laboratory to laboratory due to different levels of expertise and different batches of undefined reagents, most notably fetal bovine serum (FBS). In an attempt to reduce these variations, efforts have been made to move toward developing defined culture conditions, for example, through the use of chemically defined Serum Replacement (SR)(Invitrogen knockout Serum Replacement). There are signs, however, that SR also has lot-to-lot variations and may not contain all the necessary reagents for ES cell establishment. Therefore, alternation of SR and FBS has been suggested as a solution (Bryja *et al.*, 2006). The authors reported an impressive 70% success rate of mouse ES cell establishment compared to the about 25% that can be expected from standard derivation attempts. The ability to give rise to ES cell lines from a certain genetic background is, however, only a precondition—there are many more hurdles to overcome along the road to an ES cell line suitable for genetic studies.

Factors Affecting the Contribution of Mouse ES Cell to
    Chimeric Embryos

Apart from the genetic background of ES cells, other factors also influence their capacity to contribute to the somatic tissue and, most importantly, to the germ line of chimeras. As is the case in all tissue culture, cells accumulate random genetic and epigenetic changes with time. These changes range from large chromosomal aberrations to small methylation changes in the DNA affecting critical gene expressions. If these alterations increase the proliferation rate under the given culture conditions, the population of abnormal cells could take over the culture in a short time (Liu *et al.*, 1997). An ES cell culture plate always represents a competitive field favoring the "speedy." Therefore, it is essential to keep the time cells spend in this competition as short as possible (low passage number) in order to keep the accumulation of genome damage low to maintain high developmental potential. It is important to keep in mind that suboptimal culture conditions allow for a larger competitive field than optimal settings. Consequently, extreme care is necessary to provide the best possible physical environment during both establishment and maintenance of the cells.

Once mES cells are introduced back *in vivo*, another stage of competition is initiated: that between cells originating from the host embryo versus the ES cell. These two groups of cells are competing for colonization of the different lineages. Different genetic backgrounds have different strengths in these pairwise competitions. Taking the germ line competence as the ultimate measure, many years of worldwide experience have yielded a few, well-documented successful combinations of ES cell–host embryo genetic backgrounds. 129 ES cells are generally injected into C57BL/6 blastocysts and C57BL/6 ES cells are injected into BALB/c (Kontgen *et al.*, 1993; Ledermann and Burki, 1991; Lemckert *et al.*, 1997) or coat color coisogenic (albino) C57BL/6-Tyr(c)-2J (c2J) (Schuster-Gossler *et al.*, 2001) embryos. However, in cases where 129 ES cells are aggregated with morula stage embryos, the outbred CD1 or ICR is the preferred host (Nagy *et al.*, 2003; Wood *et al.*, 1993).

Critical Events During Mouse ES Cell Establishment

Source Embryos

Embryos for the derivation of ES cells can be obtained from either naturally mated or superovulated female mice. The former option generally yields higher-quality embryos, whereas the latter results in higher numbers. The choice ultimately depends on the age and genetic background of the donor animals. Strains that respond well to superovulation are worth placing under this regimen, whereas strains that are poor responders are better naturally mated. Embryos are usually collected from the uterus of 3.5-dpc animals. At this time, they have reached the blastocyst stage and are ready for ES cell establishment without the need for further *in vitro* culture. However, all embryos from a single mouse are not always at exactly the same developmental stage at a given time. This may result in the isolation of late morula stage as well as expanded and perhaps even a few hatched blastocysts from the same female donor. In this case, morula stage embryos should be given a few hours further culture in embryo culture media before proceeding to the next step.

Before recovery of the embryos is attempted, all necessary reagents, media, and instruments should be prepared in such a way that the procedure can be carried out speedily. The time the embryos spend outside the *in vivo* environment or the incubator influences their quality greatly. If a large cohort of embryos with the same genotype is available, it is worthwhile selecting only embryos of good quality (Fig. 1A). However, if embryos are isolated from mutant crosses, it is crucial to include every single embryo, as the variation of genotype may cause a slightly varied phenotype or developmental rate.

Placement of Embryos into ES Cell Conditions and Subsequent Hatching

Derivation of ES cells directly from live animals always encompasses the risk of transmitting pathogens from the animal to the tissue culture. For this reason, great care should be taken to use separate media, reagents, hoods, and incubators as far as at all possible until established cell lines have been screened and declared free of pathogens.

At the time of recovery, the embryo is still surrounded by a thick protective glycoprotein layer called the zona pellucida (ZP). *In vivo*, the ZP gradually thins and cracks, and the blastocyst hatches at around 4.5 dpc as it arrives into the uterine cavity. This hatching process also occurs *in vitro*, provided that the culture conditions are optimal (Fig. 1B). Once out of its protective shell, the blastocyst becomes very fragile, collapses, and immediately starts to attach to the uterine wall. *In vitro*, it is possible to lure the blastocyst to attach to the tissue culture plate or a layer of mitotically inactivated mouse embryonic fibroblasts (MEFs). Needless to say, the quality of these feeder cells has to be optimal, they should be prepared fresh (no longer than 5 days prior to use), and be of optimal density (see later) and low passage number (no higher than p3). Also, the culture media and the physical environment in which the further *in vitro* culture will take place have a fundamental impact on the success rate of derivation. Culture media and reagents should be prepared fresh and be of highest possible quality. Incubators should be checked for their temperature and CO_2 reading accuracy, and it should be ensured that the humidity remains as high as possible at all times.

Attachment

The following few days are very exciting: the blastocyst emerges from the zona pellucida and starts to attach to the feeder layer (Fig. 1C). The temptation is great to peek in for a quick look on this process from time

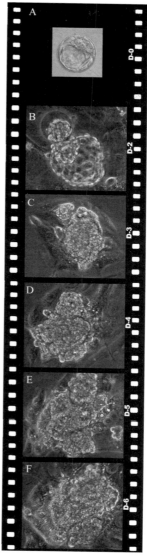

FIG. 1. Phases of blastocyst outgrowth development during preparation for ES cell establishment. All the pictures are of the same scale. (A) High-quality blastocyst ready to be plated on MEFs for ES cell derivation (day 0). (B) Embryo at the final stage of hatching (day 2). (C) Attaching embryo (day 3). Attached trophoblast cells are clearly visible under the outgrowth. (D–F) The outgrowth is increasing in size (days 4 to 6). Areas with ES cell-like cells become visible (E) and grow larger. This outgrowth is now ready for disaggregation (F). (See color insert.)

to time. However, attachment is best achieved if the plates are not disturbed. As a general rule, the door of the incubator should remain shut for 48 h. Two days after plating (if the embryos were well-expanded blastocysts, 3 days if they were smaller), the cultures should be examined carefully and the media replaced.

Formation of Outgrowth

Once the blastocyst has attached, cells will start to grow out on top of the feeder layer very soon. During the coming few days, the recommendation about checking on the culture is reversed: they should be investigated carefully every day in order to determine the optimal time point for the first disaggregation (Fig. 1D–1F). However, it is still important to keep the time the cultures spend outside the incubator to a minimum. Media should be replaced every other day.

First Disaggregation

Timing the first disaggregation right is perhaps the most crucial determinant for the success of ES cell establishment. Done too early, the outgrowth will not contain enough cells and the culture will die. Done too late, the outgrowth will have already started to differentiate into other, more specialized cell types, and the culture will not result in ES-like colonies. Each individual outgrowth has to be assessed carefully and the optimal time point determined individually. Figure 1 illustrates the stages the outgrowth goes through, and the optimal size/time when it should be disaggregated is illustrated in Fig. 1F.

The next critical step is the dissociation procedure. The outgrowth can, at this point, not be compared to a simple cell colony, which after the seeding of a single-cell preparation will form new colonies readily. In fact, harsh enzymatic dissociation will inevitably result in the death of the cells. Instead, a gentle process has to be adopted where the outgrowth is divided into small cell clumps of approximately 5 to 10 cells each.

Expansion of First ES Cell-Like Colonies

After the first dissociation of the outgrowth, the culture usually grows slowly. Occasionally, it may appear as if all cells (except the feeders) have died. It is important to again practice patience and leave the cultures alone until small colonies become visible. Once cell growth can be identified, however, the cultures will again require daily attention. Three possible scenarios might occur: (1) growth of non–ES-like colonies/cell types (Fig. 2B),

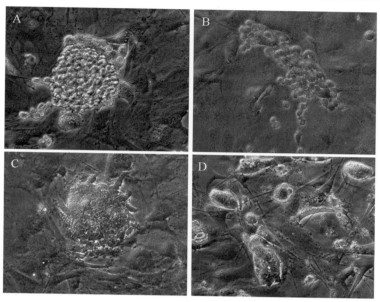

Fig. 2. Cell colonies in the early phase of mES cell establishment. (A) Colony of mixed cell types as a result of improper disaggregation of the initial outgrowth. (B) Differentiated non-ES-like cell. (C) Three days after disaggregation of the outgrowth (Fig. 1F). ES cell-like colonies can be recognized easily by the characteristic morphology. The colonies may, at this stage, still contain a few differentiated cells, but these usually diminish after a few passages. (D) Small colonies of pure mouse ES cells. (See color insert.)

(2) growth of both ES-like colonies and other cell types (i.e., Fig. 2C and A, respectively), and (3) mainly ES type colonies (Fig. 2D). In case of no visible growth for more than 6 days, the culture can be discarded. Cultures with only non-ES-like cells can also be terminated. In case both ES-like and other cell types are present, single nice colonies can be picked and transferred after gentle disaggregation to a well with fresh feeders. Once typical ES-like colonies have become visible and grown to an appropriate size (Fig. 2D), they can be passaged according to standard protocols for mouse ES cell culture. Each line should be expanded enough to freeze a small but safe number of vials for future characterization steps. It is important to start keeping track of the passage number right from the beginning. The most widely used method is to start counting passage #1 when the cells are plated for the first time onto a larger surface area than what they were derived on (this usually is either two wells of a four-well dish or a 35-mm plate).

Freezing of ES Cell Lines

As mentioned earlier, keeping the passage number low is of vital importance for the quality of ES cell lines. Randomly acquired chromosomal or epigenetic changes that give individual cells a growth advantage will inevitably result in the accumulation of abnormal cells with increased passage numbers. For this reason, it is important to cryopreserve each line as soon as possible. Early passage vials can be used later to expand the line. Expansion should be done in a way such that vials are frozen from each consecutive passage, and a sufficiently large pool of vials is created for future use. This way, in case the cells in the final passage would have acquired suboptimal characteristics, one can fall back on the earlier passages for renewed expansion. Although time-consuming, this approach is well worth all the invested efforts. Failing to establish a "ladder" of vials from earlier to later passage numbers may result in the loss of the entire line. One last word of caution: it is easy to lose the most precious early passage vials. These few valuable aliquots should not be wasted on anything other than initial characterization and expansion.

Characterization

Even if all possible precautions are taken and protocols are followed to the letter, far from all resulting ES cell lines will display the desired pluripotency. Derivation also carries the innate risk of unintentional contamination with pathogens. Hence, the careful characterization of candidate lines should be given due consideration.

The first (and easiest) screening strategy for identifying potentially "good" lines is based on morphology, homogeneity of the culture, and speed of growth. These initial steps can be undertaken during the actual derivation process, before the lines are frozen. Good morphology is depicted in Fig. 2D; the aim should be to achieve cultures with predominantly these kinds of colonies. Optimally, established cultures should grow at a rate at which they become subconfluent in 2 days if passaged at a rate of 1:6 (the initial passages during establishment, however, should be kept at a lower expansion rate [1:2 or 1:3] until the cells have gained growth momentum). However, morphology and growth speed alone are not enough criteria for distinguishing good ES cell lines. Further characterization steps could include the following.

1. Pathogen testing. An aliquot of cells should be cultured for a minimum of three passages in media without antibiotics. The cell supernatant and/or cell suspension should then be screened for a panel of mouse pathogens. This step is not only important if future use of the ES cells is aimed at creating animals in a specific pathogen-free facility, but also as a general precaution

for avoiding transfer of contamination to other cell cultures. One of the most common pathogens found in mouse ES cell cultures is *Mycoplasma* species.

2. Karyotyping. Karyotype analysis can be used to determine the sex of the line and to detect possible chromosomal abnormalities. A complete spectral karyotype painting ultimately gives the most information, but involves considerable costs. A more economical alternative is G-banding or simple chromosome counting. In either case, a minimum of 20 metaphase spreads should be analyzed in order to correctly pinpoint the overall euploidy of the line.

3. *In vitro* differentiation. Placing ES cells in *in vitro* differentiation assays can provide some information about potency. A large number of assays have been established, allowing induction of a vast number of cell types. However, all these assays are time, labor, and cost intensive, and information gain is limited to the *in vitro* potential of the cells.

4. Teratomas. The classical method of determining the ability of ES cells to contribute to all three germ layers is by teratoma assay. This is done by injecting ES cells under the skin, kidney capsule, or testicle of immunologically compatible (or compromised) mice. The tumor formation ability and composition of the teratoma give a good indication of the developmental potency of the ES cells. However, this assay is also time and cost intensive.

5. Chimera formation. The most widely used and very informative method of determining the quality of ES cell lines is to introduce them back into an embryonic environment through morula aggregation or blastocyst injection (Nagy *et al.*, 2003). ES cells of high quality will contribute to all somatic tissues and the germ line of resulting chimeras.

6. Tetraploid complementation assay. The ultimate test of mES cell potency, however, can be seen when they are forced to form the entire embryo proper. This can be achieved by combining ES cells with tetraploid host embryos, as tetraploid embryos do not contribute well to the embryo proper, but they do form normal placentas.

Protocols

Mouse Embryonic Fibroblast Feeder Layer Preparation

   Materials and Equipment
   Sterile horizontal flow hood
   Sterile incubator 37°, 5% CO_2, 100% humidity
   Centrifuge
   70% EtOH

Tissue culture plates (four-well, 35, 60, 1000 mm)

10-ml sterile plastic tubes

Phosphate-buffered saline (PBS) without Ca^{2+} and Mg^{2+}

0.05 to 0.1% trypsin in saline/EDTA

Mitomycin C (1 mg/ml Sigma)

MEFs, early passage, frozen vial

MEF culture media: KO-DMEM (GIBCO) supplemented with 10% fetal bovine serum (FBS), 100 μM nonessential amino acids (100× stock, GIBCO), 100 μM β-mercaptoethanol (100 × stock Sigma), 2 mM GlutaMax (Invitrogen), and penicillin/streptomycin (final concentration 50 μg/ml) (100× stock GIBCO). *Note:* if standard DMEM is used instead of KO-DMEM, the media should be supplemented with 1 mM sodium pyruvate as well (100× stock, GIBCO). (If only KO-DMEM is recommended it already has sodium pyruvate so there is no need to supplement.)

Method

1. Thaw the vial of MEFs quickly at 37°. Clean the outside of the vial by wiping with 70% EtOH.

2. Add the contents of the vial to 5 ml culture medium in a 10-ml sterile plastic tube.

3. Centrifuge for 3 min at 200*g*.

4. Remove the supernatant. Flick the tube to loosen the cell pellet.

5. Add the appropriate amount of culture media (depending on the cell number present in the vial) and plate the cells on tissue culture dishes.

6. Replace the media the next day and thereafter every other day.

7. Inspect the cultures daily to determine the optimal density for inactivation. Initially the cultures will display a typical thin elongated fibroblast morphology. As the culture grows in density and space becomes sparse, cells start to take on a "cobblestone" appearance. It is at this point they should be passaged. If the culture is left to grow longer, the fibroblasts will start growing on top of each other, a phenomenon that should be avoided.

8. Add 10 μl mitomycin C (1 mg/ml) per milliliter culture media directly to the cultures. Rock the plates gently to mix the mitomycin C with the media. Incubate at 37° for 2 h.

9. Remove the media and wash cells three times with PBS.

10. Add 0.1 ml of trypsin per 10- to 15-mm diameter of plate surface.

11. Incubate for 3 to 5 min at 37°. Periodically check the plates under a microscope and stop the trypsin reaction when cells start to lift off the surface.

12. Add 1 ml of culture media per 10- to 15-mm diameter of plate surface to stop the trypsin reaction (serum contained in the medium will inhibit the trypsin immediately).

13. Resuspend cells by pipetting up and down several times.

14. Add the cell suspension to a sterile plastic tube and centrifuge for 3 min at 200g.

15. Remove the supernatant. Flick the tube to loosen the cell pellet.

16. Dilute the cell suspension in a small amount of culture media. Count the cell concentration and make the appropriate dilution in such a way that 40,000 to 50,000 cells/cm^2 are plated.

17. Incubate overnight to allow the MEFs to adhere properly to the tissue culture plate.

18. Inactivated MEFs can be used as feeder layers for mES cells no later than 5 days after plating.

Establishment

Material and Equipment

Sterile horizontal flow hood

Sterile incubator 37°, 5% CO_2, 100% humidity

Centrifuge

3.5 dpc pregnant mice

Dissecting microscope

HEPES-buffered embryo culture medium, for example, M2 (Specialty Media/Chemicon)

KSOM-AA (Specialty Media/Chemicon)

Pulled Pasteur pipettes

Pipette P200 with sterile tips

5-ml syringe with 27-gauge needle

Tissue culture plates (four-well, 35, 60, 1000 mm) with mitotically inactivated MEFs

10-ml sterile plastic tubes

PBS without Ca^{2+} and Mg^{2+}

0.05 to 0.1% trypsin in saline/EDTA

ES culture media: KO-DMEM (GIBCO) supplemented with 15% mES cell-qualified FBS, 100 μM nonessential amino acids (100× stock, GIBCO), 100 μM β-mercaptoethanol (100× stock, −20°, Sigma), 2 mM L-glutamine (100× stock, −20°, GIBCO), penicillin/streptomycin (final concentration 50 μg/ml) (100× stock GIBCO), and 2000 U/ml leukemia inhibitory factor (LIF). *Note:* if standard DMEM is used instead of KO-DMEM, media should be supplemented with 1 mM sodium pyruvate as well (100× stock, GIBCO). (If only KO-DMEM is recommended it already has sodium pyruvate so there is no need to supplement.)

Method

PLATING

1. One day prior to the experiment, remove media in the appropriate number of four-well tissue culture plates with MEFs (one well per embryo). Add freshly prepared ES culture media.

2. Sacrifice pregnant mice at 3.5 dpc in a humane way following local animal welfare practices. Dissect uteri.

3. Isolate embryos from the uterine horns by inserting a 27-gauge needle (with a 5-ml syringe filled with M2 medium attached) in each end of the uterus close to the ovaries and flush with approximately 0.5 ml medium.

4. Using a finely pulled Pasteur pipette, locate the blastocysts and rinse them several times through M2 medium (for more details on this step, see Nagy *et al.* [2003]).

5. Using a pulled Pasteur pipette, place one blastocyst in the center of each well. These steps can be performed using a dissecting microscope placed in a laminar flow hood. All consecutive procedures should be carried out under strictly sterile conditions.

6. Culture the blastocysts undisturbed at 37°, 5% CO_2 for 48 h.

DISAGGREGATION OF OUTGROWTH. After 48 h of undisturbed culture, outgrowths should be inspected daily to determine the right stage at which to perform the first disaggregation (usually the 4th to 6th day after plating). Due to variability between different embryos, it might be necessary to perform the disaggregation on different days. The ICM outgrowth ready for disaggregation should be as large as possible but not yet differentiated. The evolving morphology of outgrowths is illustrated in Fig. 1. During this time, media should be replaced on the cultures every other day.

1. One day prior to the planned disaggregation, replace media on four-well plates with freshly inactivated MEFs using ES culture media.

2. On the day of disaggregation, remove media in wells with outgrowths. Add 0.5 ml PBS per well.

3. Place 25-μl drops of trypsin in a 96-well tissue culture plate without MEFs.

4. Using a finely pulled Pasteur pipette, gently circle the ICM clump, remove it from the surrounding trophoblast cells, and place it into the trypsin in one well of the 96-well plate. Repeat the process with up to 10 to 20 outgrowths (depending on experience). If a larger number is ready for disaggregation on the same day, these should be done in a separate round in order to avoid the initially picked cells spending too long in the enzyme.

5. Incubate at 37° for 3 to 5 min.

6. Using a P200 pipette and yellow tips, break up the outgrowth into smaller clumps of 5 to 10 cells. Watch the process under a microscope, as some clumps might need repeated pipetting. Take care not to pipette too much, as single cells will not survive!

7. Add 30 μl of media to each well to stop the trypsin reaction.

8. Transfer the cell suspension into one well of the four-well MEFs plate. Make sure that media in the well is ES cell culture media and not MEF media. Media should have been replaced in the wells 1 day in advance.

9. Change media after overnight incubation.

CULTURE OF INITIAL COLONIES. A few days after disaggregation, small colonies may become visible in the cultures. However, because the initial cell growth may be very slow, colonies may not appear for several days. During this time, media on the cultures are best changed every other day.

1. Observe the cultures every day and keep a log on each well. As soon as cell growth can be seen, start changing media every day.

2. Wells that do not show sign of cell proliferation within 10 days can be discarded. Also, wells in which solely cells of non-ES-like morphology are present can be terminated.

3. In wells in which only a few mES-like colonies are present among other cells of varying morphology, renewed picking can be performed: mES-like colonies are disaggregated individually as described earlier and placed into a new well with MEFs.

4. Slow-growing colonies can be trypsinized (see later) and replated back in the same well to prevent differentiation.

5. When ES cells have reached near confluency in a well, they should be passaged. Near confluency means that the colonies cover approximately 75% of the surface area, but are not yet so large that they have come in contact with each other. Passaging should be done at a rate of 1:2 (into two wells of a four-well plate) or 1:3 into a 35-mm feeder plate. This is considered passage 1.

BEYOND THE BASIC DERIVATION PROTOCOL. As with all techniques that have been utilized for a long time in many laboratories over the world, a number of alternate approaches have been developed. Depending on the individual experiment and genetic background of donor embryos, the following variations may prove useful for increasing efficiency:

1. Increasing the FBS concentration in the culture media to 25% for the initial plating.

2. Using DMEM with low glucose instead of KO-DMEM for culture media.

3. Supplementing culture media with knockout Serum Replacement (GIBCO) instead of FBS (*note:* MEFs do not attach to the tissue culture plates when grown in SR. Culture media for feeders should always be supplemented with FBS) or using an alternating approach (Bryja *et al.*, 2006).

4. Adding nucleosides (Specialty Media mES-008D) to culture media.

5. Isolating delayed blastocysts, prevented from implantation by ovariectomy, and administration of progesterone (Brook and Gardner, 1997; Nagy *et al.*, 2003).

6. Removing trophoblast cells from the blastocyst by immunosurgery (Knowles *et al.*, 1977) prior to plating.

7. Using the proprietary conditioned medium ResGro (Chemicon) (Schoonjans *et al.*, 2003) instead of standard ES cell culture media.

Maintenance

Method

Once an ES cell culture has been successfully initialized from an ICM outgrowth, it should be maintained at a density allowing for optimal growth. This means that the culture should be passaged every other day at a rate of approximately 1:6. However, this rule of thumb is only a guidance. Each individual culture plate should be inspected daily and passaged as soon it has reached subconfluency. A few additional important points to remember include the following.

> Media should always be kept at 4° but warmed to room temperature before use.
>
> Make sure to always create a single cell suspension during passaging. It is better to slightly overtrypsinize cells than to leave cell clumps.
>
> Always keep track of the passage number of each culture dish.
>
> Always grow ES cells on a freshly inactivated feeder layer (ideally no older than 5 days).

PASSAGING MOUSE ES CELL CULTURES

1. Aspirate media.

2. Rinse with 1 ml of PBS per 10- to 15-mm diameter of plate surface (2 ml for 30-mm plates, 5 ml for 60-mm plates, etc.).

3. Add 0.1 ml trypsin per 10- to 15-mm diameter of plate surface.

4. Incubate for 3 to 5 min at 37°. Periodically check the plates under a microscope and stop the trypsin reaction when colonies start to lift off the surface.

5. Add 1 ml of culture media per 10- to 15-mm diameter of plate surface to stop the trypsin reaction (serum contained in the medium will inhibit the trypsin immediately).

6. Resuspend cells by pipetting up and down several times until a single cell suspension has been achieved, but not so excessively that cell damage is caused. Until experience has been gained with this step, periodically check the suspension under a microscope.

7. Add the cell suspension to a sterile plastic tube and centrifuge for 3 min at 200g.

8. Remove the supernatant. Flick the tube to loosen the cell pellet.

9. Dilute the cell suspension in an appropriate amount of ES media and add to a six times larger growing area than before passaging.

Acknowledgments

We gratefully acknowledge Marina Gertsenstein for providing her expert view on the manuscript and for preparing the embryos photographed for the figures.

References

Austin, C. P., Battey, J. F., Bradley, A., Bucan, M., Capecchi, M., Collins, F. S., Dove, W. F., Duyk, G., Dymecki, S., Eppig, J. T., Grieder, F. B., Heintz, N., Hicks, G., Insel, T. R., Joyner, A., Koller, B. H., Lloyd, K. C., Magnuson, T., Moore, M. W., Nagy, A., Pollock, J. D., Roses, A. D., Sands, A. T., Seed, B., Skarnes, W. C., Snoddy, J., Soriano, P., Stewart, D. J., Stewart, F., Stillman, B., Varmus, H., Varticovski, L., Verma, I. M., Vogt, T. F., von Melchner, H., Witkowski, J., Woychik, R. P., Wurst, W., Yancopoulos, G. D., Young, S. G., and Zambrowicz, B. (2004). The knockout mouse project. *Nature Genet.* **36,** 921–924.

Bradley, A., Evans, M., Kaufman, M. H., and Robertson, E. (1984). Formation of germ-line chimaeras from embryo-derived teratocarcinoma cell lines. *Nature* **309,** 255–256.

Brook, F. A., and Gardner, R. L. (1997). The origin and efficient derivation of embryonic stem cells in the mouse. *Proc. Natl. Acad. Sci. USA* **94,** 5709–5712.

Bryja, V., Bonilla, S., Cajanek, L., Parish, C. L., Schwartz, C. M., Luo, Y., Rao, M. S., and Arenas, E. (2006). An efficient method for the derivation of mouse embryonic stem cells. *Stem Cells* **24,** 844–849.

Carmeliet, P., Ferreira, V., Breier, G., Pollefeyt, S., Kieckens, L., Gertsenstein, M., Fahrig, M., Vandenhoeck, A., Harpal, K., Eberhardt, C., Declercq, C., Pawling, J., Moons, L., Collen, D., Risau, W., and Nagy, A. (1996). Abnormal blood vessel development and lethality in embryos lacking a single VEGF allele. *Nature* **380,** 435–439.

Chen, J., Reifsnyder, P. C., Scheuplein, F., Schott, W. H., Mileikovsky, M., Soodeen-Karamath, S., Nagy, A., Dosch, M. H., Ellis, J., Koch-Nolte, F., and Leiter, E. H. (2005). "Agouti NOD": Identification of a CBA-derived Idd locus on chromosome 7 and its use for chimera production with NOD embryonic stem cells. *Mamm. Genome* **16,** 775–783.

Ding, H., Wu, X., Bostrom, H., Kim, I., Wong, N., Tsoi, B., O'Rourke, M., Koh, G. Y., Soriano, P., Betsholtz, C., Hart, T. C., Marazita, M. L., Field, L. L., Tam, P. P., and Nagy, A. (2004). A specific requirement for PDGF-C in palate formation and PDGFR-alpha signaling. *Nature Genet.* **36,** 1111–1116.

Duncan, S. A. (2005). Generation of embryos directly from embryonic stem cells by tetraploid embryo complementation reveals a role for GATA factors in organogenesis. *Biochem. Soc. Trans.* **33,** 1534–1536.

Duncan, S. A., Nagy, A., and Chan, W. (1997). Murine gastrulation requires HNF-4 regulated gene expression in the visceral endoderm: Tetraploid rescue of Hnf-4$^{-/-}$ embryos. *Development* **124,** 279–287.

Eggan, K., Akutsu, H., Loring, J., Jackson-Grusby, L., Klemm, M., Rideout, W. M., 3rd, Yanagimachi, R., and Jaenisch, R. (2001). Hybrid vigor, fetal overgrowth, and viability of mice derived by nuclear cloning and tetraploid embryo complementation. *Proc. Natl. Acad. Sci. USA* **98,** 6209–6214.

Evans, M. J., and Kaufman, M. H. (1981). Establishment in culture of pluripotential cells from mouse embryos. *Nature* **292,** 154–156.

Gardner, R. L., and Brook, F. A. (1997). Reflections on the biology of embryonic stem (ES) cells. *Int. J. Dev. Biol.* **41,** 235–243.

Gossler, A., Doetschman, T., Korn, R., Serfling, E., and Kemler, R. (1986). Transgenesis by means of blastocyst-derived embryonic stem cell lines. *Proc. Natl. Acad. Sci. USA* **83,** 9065–9069.

Guillemot, F., Nagy, A., Auerbach, A., Rossant, J., and Joyner, A. L. (1994). Essential role of Mash-2 in extraembryonic development. *Nature* **371,** 333–336.

Knowles, B. B., Solter, D., Trinchieri, G., Maloney, K. M., Ford, S. R., and Aden, D. P. (1977). Complement-mediated antiserum cytotoxic reactions to human chromosome 7 coded antigen(s): Immunoselection of rearranged human chromosome 7 in human-mouse somatic cell hybrids. *J. Exp. Med.* **145,** 314–326.

Koller, B. H., Hagemann, L. J., Doetschman, T., Hagaman, J. R., Huang, S., Williams, P. J., First, N. L., Maeda, N., and Smithies, O. (1989). Germ-line transmission of a planned alteration made in a hypoxanthine phosphoribosyltransferase gene by homologous recombination in embryonic stem cells. *Proc. Natl. Acad. Sci. USA* **86,** 8927–8931.

Kontgen, F., Suss, G., Stewart, C., Steinmetz, M., and Bluethmann, H. (1993). Targeted disruption of the MHC class II Aa gene in C57BL/6 mice. *Int. Immunol.* **5,** 957–964.

Kunath, T., Gish, G., Lickert, H., Jones, N., Pawson, T., and Rossant, J. (2003). Transgenic RNA interference in ES cell-derived embryos recapitulates a genetic null phenotype. *Nature Biotechnol.* **21,** 559–561.

Ledermann, B., and Burki, K. (1991). Establishment of a germ-line competent C57BL/6 embryonic stem cell line. *Exp. Cell Res.* **197,** 254–258.

Lemckert, F. A., Sedgwick, J. D., and Korner, H. (1997). Gene targeting in C57BL/6 ES cells: Successful germ line transmission using recipient BALB/c blastocysts developmentally matured *in vitro*. *Nucleic Acids Res.* **25,** 917–918.

Liu, X., Wu, H., Loring, J., Hormuzdi, S., Disteche, C. M., Bornstein, P., and Jaenisch, R. (1997). Trisomy eight in ES cells is a common potential problem in gene targeting and interferes with germ line transmission. *Dev. Dyn.* **209,** 85–91.

Livy, D. J., and Wahlsten, D. (1997). Retarded formation of the hippocampal commissure in embryos from mouse strains lacking a corpus callosum. *Hippocampus* **7,** 2–14.

Martin, G. R. (1981). Isolation of a pluripotent cell line from early mouse embryos cultured in medium conditioned by teratocarcinoma stem cells. *Proc. Natl. Acad. Sci. USA* **78,** 7634–7638.

Mortensen, R. M., Conner, D. A., Chao, S., Geisterfer-Lowrance, A. A., and Seidman, J. G. (1992). Production of homozygous mutant ES cells with a single targeting construct. *Mol. Cell. Biol.* **12,** 2391–2395.

Nagy, A., Gertsenstein, M., Vintersten, K., and Behringer, R. (2003). "Manipulating the Mouse Embryo, a Laboratory Manual." Cold Spring Harbor Press, Cold Spring Harbor, NY.

Nagy, A., Gocza, E., Diaz, E. M., Prideaux, V. R., Ivanyi, E., Markkula, M., and Rossant, J. (1990). Embryonic stem cells alone are able to support fetal development in the mouse. *Development* **110,** 815–821.

Nagy, A., Rossant, J., Nagy, R., Abramow-Newerly, W., and Roder, J. C. (1993). Derivation of completely cell culture-derived mice from early-passage embryonic stem cells. *Proc. Natl. Acad. Sci. USA* **90,** 8424–8428.

Roach, M. L., Stock, J. L., Byrum, R., Koller, B. H., and McNeish, J. D. (1995). A new embryonic stem cell line from DBA/1lacJ mice allows genetic modification in a murine model of human inflammation. *Exp. Cell Res.* **221**, 520–525.

Robertson, E., Bradley, A., Kuehn, M., and Evans, M. (1986). Germ-line transmission of genes introduced into cultured pluripotential cells by retroviral vector. *Nature* **323**, 445–448.

Royce, J. R. (1972). Avoidance conditioning in nine strains of inbred mice using optimal stimulus parameters. *Behav. Genet.* **2**, 107–110.

Schoonjans, L., Kreemers, V., Danloy, S., Moreadith, R. W., Laroche, Y., and Collen, D. (2003). Improved generation of germline-competent embryonic stem cell lines from inbred mouse strains. *Stem Cells* **21**, 90–97.

Schuster-Gossler, K., Lee, A. W., Lerner, C. P., Parker, H. J., Dyer, V. W., Scott, V. E., Gossler, A., and Conover, J. C. (2001). Use of coisogenic host blastocysts for efficient establishment of germline chimeras with C57BL/6J ES cell lines. *Biotechniques* **31**, 1022–1024, 1026.

Schwenk, F., Zevnik, B., Bruning, J., Rohl, M., Willuweit, A., Rode, A., Hennek, T., Kauselmann, G., Jaenisch, R., and Kuhn, R. (2003). Hybrid embryonic stem cell-derived tetraploid mice show apparently normal morphological, physiological, and neurological characteristics. *Mol. Cell. Biol.* **23**, 3982–3989.

Seong, E., Saunders, T. L., Stewart, C. L., and Burmeister, M. (2004). To knockout in 129 or in C57BL/6: That is the question. *Trends Genet.* **20**, 59–62.

Smithies, O., Gregg, R. G., Boggs, S. S., Koralewski, M. A., and Kucherlapati, R. S. (1985). Insertion of DNA sequences into the human chromosomal beta-globin locus by homologous recombination. *Nature* **317**, 230–234.

Thomas, K. R., and Capecchi, M. R. (1990). Targeted disruption of the murine int-1 proto-oncogene resulting in severe abnormalities in midbrain and cerebellar development. *Nature* **346**, 847–850.

Wood, S. A., Allen, N. D., Rossant, J., Auerbach, A., and Nagy, A. (1993). Non-injection methods for the production of embryonic stem cell-embryo chimaeras. *Nature* **365**, 87–89.

[2] Bovine Embryonic Stem Cells

By Marsha Roach, Li Wang,
Xiangzhong Yang, and X. Cindy Tian

Abstract

Bovine embryonic stem (bES) cell lines reported to date vary in morphology and marker expression, such as alkaline phosphatase (ALPL), stage-specific embryonic antigen 4 (SSEA4), and octamer-binding transcription factor-4 (OCT4), that normally are associated with the undifferentiated, pluripotent state. This chapter introduces the methods of isolating and maintaining bovine ES cells. These bovine ES cells grow in large, multicellular colonies resembling the mouse ES and embryonic germ (EG) cells, as well as human EG cells. Throughout the culture period, most of the cells within the colonies stain positive for ALPL and cell surface markers SSEA4

METHODS IN ENZYMOLOGY, VOL. 418   0076-6879/06 $35.00
Copyright 2006, Elsevier Inc. All rights reserved.   DOI: 10.1016/S0076-6879(06)18002-7

and OCT4. The staining patterns of the bES cells are identical to those of the blastocysts fertilized *in vitro* (IVF), yet different from most previously reported bovine ES cell lines, which are either negative or not detected. After undifferentiated culture for more than 1 year, these cells maintained the ability to differentiate into embryoid bodies and derivatives of all three EG layers, thus demonstrating their pluripotency. In addition to bES from IVF, this chapter introduces two methods of generating blastocyst stage embryos other than *in vitro* fertilization, which are parthenogenetic activation and somatic cell nuclear transfer for the potential application of generation "patient-specific" ES cells.

Introduction

Bovine embryonic stem (bES) cells had been isolated from *in vitro* fertilized and nuclear transfer embryos (Cibelli *et al.*, 1998; Mitalipova *et al.*, 2001; Saito *et al.*, 2003; Stice *et al.*, 1996; Wang *et al.*, 2005). These cells offer the possibilities of making large numbers of offspring by nuclear transfer and performing genetic manipulations followed by chimeric production or cloning (First *et al.*, 1994; Yang *et al.*, 2000). Although several ES-like cell lines established by various research teams have been reported to exhibit pluripotency both *in vitro* and *in vivo*, they vary in morphology and marker expression that normally are associated with the undifferentiated, pluripotent state of ES cells in other species (Table I). For instance, OCT4, also known as POU-domain transcription factor POU5F1, is found to be associated with the pluripotency of ES cells in many species (Reubinoff *et al.*, 2000; Suzuki *et al.*, 1990; Vrana *et al.*, 2003). It was only detected on two of the reported bovine ES cell lines (Saito *et al.*, 2003; Wang *et al.*, 2005), including the one established in our laboratory.

Commonly, ES cells are propagated by enzymatically dissociating colonies and plating individual cells for new colony formation (Cibelli *et al.*, 2002; Evans and Kaufman, 1981; Thomson *et al.*, 1995, 1998). Unlike in other species, bovine ES cells reported to date are sensitive to enzymes, such as trypsin, type IV collagenase, and protease E. Bovine ES cells, therefore, failed to form colonies after enzymatic disassociation and replating (Mitalipova *et al.*, 2001; Stice *et al.*, 1996; Wang *et al.*, 2005). However, this chapter reports on the generation of bovine ES cells that can be dissociated with 0.05% trypsin EDTA, attach, and continue to proliferate following dissociation and replating on murine embryonic fibroblast (MEF) feeder layers.

Although the methods still need optimizing, we have developed bovine ES cell lines with morphology similar to those of established ES cells in humans and mice, as well as marker-staining patterns identical to those of bovine blastocysts. This chapter presents methods developed in our

TABLE I

SUMMARY OF CELLULAR AND MOLECULAR CHARACTERIZATIONS OF SELECTED BOVINE ES-LIKE CELLS AND COMPARISON TO THOSE OF ES OR ES-LIKE CELLS FROM OTHER SPECIES

Marker	Mouse	Cattle				Monkey	Human
		Wang (NT)	Cibbeli (NT)	Mitalipova (IVF)	Saito (IVF)		
SSEA1	+	−	Undetermined	+	+	−	−
SSEA3	−	Undetermined	Undetermined	+	−	+	+
SSEA4	−	+	Undetermined	+	−	+	+
TRA1-60	−	−	Undetermined	Undetermined	Undetermined	+	+
TRA1-81	−	−	Undetermined	Undetermined	Undetermined	+	+
ALPL	+	+	−	Undetermined	+	+	+
OCT4	+	+	Undetermined	Undetermined	+	+	+
Growth characteristics in vitro	Compact, round, multilayer clumps	Compact, round multilayer clumps	Monolayer cell sheet	Monolayer cell sheet	Colonies, with compact cells	Flat, loose aggregates; can form EBs	Flat, loose aggregates; can form EBs

laboratory for the generation of bovine ES cell lines and maintenance of their pluripotency.

Materials

Bovine Oocytes and Semen for Blastocyst Formation, Cumulus Cells for Somatic Cell Nuclear Transfer (SCNT), and Mouse Fetuses for Primary Embryonic Fibroblast (PEFS) Preparation

1. Frozen semen (0.25 ml/straw) is from Cooperative Resources International (Shawano, WI)
2. Cumulus–oocyte complexes (COC) are from slaughterhouse ovaries
3. CD1 mice are from Jackson Laboratories for the isolation of primary MEFs as feeder cells

Tissue Culture Plasticware and Glassware (see Table II)

Reagents (see Table III)

Methods

Media Recipes and Solutions

1. Brackett and Oliphant medium (BO medium; see Table IV) (Brackett and Oliphant, 1975)
2. CR1aa

    a. Make CR1 stock (Table V; prepare in a 100-ml volumetric flask): Add the first nine ingredients to volumetric flask. Add water (∼90 ml). Thoroughly dissolve constituents and then add hemicalcium lactate. Add remaining water. Adjust pH to 7.4 before bringing to final volume.

TABLE II
TISSUE CULTURE PLASTICWARE AND GLASSWARE

Supplies	Vendor
35-mm Petri dish	Falcon
35-mm tissue culture dish	Falcon
100-mm tissue culture dish	Falcon
Four-well multiwell tissue culture dish	Nunc
T-25 flask	Falcon
T-75 flask	Falcon
Cryogenic vial	Corning

TABLE III
REAGENTS AND SUPPLIERS

Reagent	Vendor
0.05% trypsin/EDTA	GIBCO/Invitrogen
β-Mercaptoethanol	Sigma
6-Dimethylaminopurine	Sigma
A23187	Sigma
Antiactin, α-smooth muscle antibody	Sigma
Anticytokeratin	Sigma
Anti-β-tubulin III antibody	Sigma
Biotinylated antimouse IgG	Vector Laboratories
Biotinylated antimouse IgM	Vector Laboratories
Biotinylated antirabbit IgG	Vector Laboratories
Bovine serum albumin	Sigma
Ca^{2+}/Mg^{2+}-free phosphate-buffered saline (PBS)	GIBCO/Invitrogen
Caffeine	Sigma
Calcium chloride dehydrate	Sigma
Cycloheximide	Sigma
DAB substrate kit	Vector Laboratories
Dimethyl sulfoxide	Sigma
Dulbecco's PBS	GIBCO/Invitrogen
ES cell characterization kit	Vector Laboratories
Fibroblast growth factor basic bovine	R&D Systems
Heparin	Sigma
Iscove's modified Dulbecco's medium (MDM)	GIBCO/Invitrogen
Knockout Dulbecco's modified Eagle's medium (DMEM)	GIBCO/Invitrogen
Knockout fetal bovine serum	GIBCO/Invitrogen
KaryoMAX	GIBCO
Leukemia inhibitory factor (human)	Cheimcon
L-Glutamine	GIBCO/Invitrogen
Light mineral oil	Chemicaon
Magnesium chloride hexahydrate	Sigma
Mitomycin C	Sigma
Nitro-blue tetrazolium chloride/5-bromo-4-chloro-3-indolylphosphate toluidine	GIBCO/Invitrogen
Nonessential amino acids	GIBCO/Invitrogen
PBS	GIBCO/Invitrogen
Penicillin/streptomycin	GIBCO/Invitrogen
Phenol red	Sigma
Potassium chloride	Sigma
Pyruvic acid (sodium salt)	Sigma
Sodium bicarbonate	Sigma
Vectastain ABC kit (mouse IgG)	Vector Laboratories
Vectastain ABC kit (mouse IgM)	Vector Laboratories

TABLE IV
COMPONENTS OF BRACKETT AND OLIPHANT (BO) MEDIUM

	Formula weight	Final concentration (mM)	g/100 ml
Sodium chloride	58.44	11.2	0.6545
Potassium chloride	74.56	0.402	0.03
Sodium phosphate monobasic monohydrate	137.99	0.083	0.0115
Magnesium chloride hexahydrate	203.30	0.052	0.0106
Calcium chloride dihydrate	147.02	0.225	0.0331
Glucose	181.16	1.39	0.2518
Penicillin/streptomycin	—	—	1 ml
0.2% phenol red		0.2%	0.4 ml
Sodium bicarbonate	84.01	37.0	0.3108
Pyruvic acid (sodium salt)	110.0	1.25	0.0138

TABLE V
COMPONENTS OF CR1 MEDIUM STOCK

Common name	mg/100 ml	Maker
Sodium chloride	670.3	Sigma
Potassium chloride	23.1	Sigma
Sodium bicarbonate	220.1	Sigma
Phenol red (0.5%)	200 μl	Sigma
L-Glutamine	14.6	Sigma
EAA (50×)	2 ml	Sigma
NEA (100×)	1 ml	Sigma
Antibiotic (100×)	1 ml	Invitrogen
Pyruvic acid (sodium salt)	4.4	Sigma
Hemicalcium lactate	55	Sigma

Check osmolarity (255–270 mOsm). Start over if significantly outside range. Filter into sterile bottles. Store for up to 1 week at 4°. Supplement with pyruvate, lactate, and bovine serum albumin (BSA) on the day of use. *Note:* constituents of this medium are known to precipitate out of solution. To minimize the chances of this occurring, make sure all constituents are dissolved before adding hemicalcium lactate and use immediately after making. If the medium appears white and cloudy, discard and prepare fresh medium.

b. To prepare CR1aa, add 3.00 mg/ml EFAF BSA to CR1 stock. Sterile filter medium through a 0.22-μm syringe filter. Use immediately.

TABLE VI
COMPONENTS OF I-SCML MEDIUM

Component	Amount	Concentration	Vender
Iscove's MDM	500 ml	—	Invitrogen
KO-FBS	90 ml	—	Invitrogen
L-Glutamine	6 ml	2 mM	Invitrogen
NEAA	6 ml	—	Invitrogen
Human leukemia inhibitory factor	500 μl	10 ng/ml	Chemicon
Bovine bFGF	5 μg	10 ng/ml	R&D System
β-Mercaptoethanol	1.75 μl	5×10^{-5} M	Sigma
Gentamicin	1.5 ml		Invitrogen

3. MEF sDMEM medium: KO-DMEM, 10% FBS, and Gentamicin

4. bES cell I-SCML medium (also see Table VI): Iscove's MDM, 15% ES Qualified FBS, 10 ng/ml bovine recombinant bFGF, 0.2 mM L-glutamine (100×), 1% nonessential amino acids, 0.1 mM β-mercaptoethanol, 10 ng/ml human recombinant LIF, 50 units/ml penicillin, and 50 μg/ml streptomycin

5. ES cell differentiation medium: KO-DMEM, 15% fetal bovine serum, 1 mM glutamine, 0.1 mM β-mercaptoethanol, 1% nonessential amino acids, 50 units/ml penicillin, and 50 μg/ml streptomycin

6. Mitomycin C: Dissolve 2 mg mitomycin C in 200 ml sDMEM medium to make a 10-μg/ml working stock. Store protected from light at 4° for up to 6 weeks or frozen at $-20°$ for longer storage.

Preparation of Feeder Cells

Isolation of Primary Mouse Embryonic Fibroblasts

1. Isolate. MEFs from pregnant (12.5 to 13.5 days) CD1 mice. Detailed information is available in Hogan (1994) and Robertson (1987).

2. Freeze nearly confluent MEFs in medium that contains 40% sDMEM, 10% dimethyl sulfoxide (DMSO), and 50% FBS at a concentration of 5×10^{6} MEFs/ml. Aliquot 1 ml cell suspension per cryovial.

3. Transfer the cryovials into a $-80°$ freezer overnight and place in liquid nitrogen ($-196°$) for long-term storage.

MEFs Treatment

1. Pretreat four-well dishes with 0.1% gelatin.

2. Thaw frozen vials of MEFs by agitation in a 37° water bath, and seed 1×10^{5} cells per well of four-well dishes.

3. When the MEFs culture is subconfluent (i.e., growing actively), remove old medium, add 500 μl mitomycin C medium to each well, and return to the incubator for 2 to 5 h.

4. Remove mitomycin C medium and wash with phosphate-buffered saline (PBS) three times.

5. Add 500 μl sDMEM to each well and return to incubator until ready to use. Primary MEFs can be used for 7 to 10 days following mitomycin C treatment.

Oocyte Collection and Maturation

1. Aspirate cumulus oocyte complexes (COCs) from antral follicles (2- to 5-mm diameter) of slaughterhouse ovaries using an 18-gauge needle attached to a 10-ml syringe.

2. Place these oocytes in sterile cryovials containing pregassed culture medium and ship to the laboratory in a portable incubator at 38.5° by overnight express.

3. Select COCs with uniform cytoplasm and intact or at least four layers of cumulus cell for maturation.

4. Mature oocytes for 22 h in M199 medium containing 10% fetal calf serum and hormones (0.5 μg/ml FSH, 5.0 μg/ml LH, and 1.0 μg/ml β-estradiol) at 39° in 5% CO_2 and 95% air using the BOMED standard procedure (Sirard et al., 1988).

Generation of Bovine Blastocysts

In Vitro Fertilization

1. An insemination dish is made by droplets of 60 μl BO medium containing 6 mg/ml of BSA and 10 mg/ml of heparin overlaid with 3 ml mineral oil. Preequilibrate the insemination dish for 2 h at 39° in 5% CO_2 in humidified air.

2. After maturation in vitro, wash bovine COCs twice in the insemination medium and transfer into droplets in the insemination dish (20 to 25 oocytes/drop).

3. Gently shake semen straw for 10 s in air (room temperature) and then thaw it for 10 s in a 37° water bath.

4. Wash spermatozoa twice by centrifugation (1000 rpm) for 8 min in 10 ml BO medium containing 3 mg/ml of BSA supplemented with 10 mM caffeine. Resuspend the washed spermatozoa pellet in BO sperm-wash solution at a concentration of 1×10^6 sperm/ml for subsequent fertilization.

5. Add 50 μl of sperm suspension to each droplet. Incubate oocytes with sperm for 6 h at 39° in 5% CO_2 in humidified air.

6. Culture the zygotes in potassium simplex-optimized medium (KSOM) plus 0.1% (w/v) BSA for the first 4 days. Switch the medium to KSOM containing 1% BSA for the remaining 3 days required to reach the blastocyst stage. Change the medium every 2 days during the course of the *in vitro* culture.

Parthenogenetic Activation

1. The parthenogenetic activation procedure has been described previously (Liu *et al.*, 1998).

2. Strip matured slaughterhouse bovine oocytes of cumulus cells and select those with a polar body and subject to activation.

3. Expose oocytes to 5 mM A23187 for 5 min, followed by incubation in KSOM containing 2.5 mM 6-dimethylaminutesopurine and 0.1% BSA under mineral oil for 3.5 h at 39° in a 5% CO_2 atmosphere in humidified air.

4. Following the activation treatment, wash oocytes in KSOM, cultured in KSOM plus 0.1% BSA for the first 4 days.

5. From day 5, switch the medium to KSOM containing 1% BSA for the remaining 3 days required to reach the blastocyst stage.

6. Change the medium every 2 days during the course of *in vitro* culture.

Somatic Cell Nuclear Transfer or Cloning

1. Conduct the SCNT according to the method reported by Kubota *et al.* (2000).

2. Briefly, after 22 h of maturation, enucleate oocytes by micromanipulation. Successful enucleation is confirmed by Hoechst 33342 staining before transfer of the somatic cell.

3. For donor cells, culture adult bovine cumulus cells for six passages and subject to serum starvation for 5 days after reaching confluency.

4. Immediately before NT, trypsinize, wash, and resuspend donor cells in PBS supplemented with 0.5% fetal bovine serum.

5. Transfer cells with an approximate diameter of 10 to 15 μm to the perivitelline space of the recipient cytoplast.

6. After transfer, induce the cell–cytoplast complexes to fuse with two pulses of direct current at 2.5 kV/cm for 10 ms each with an Electrocell Manipulator 200 (BTX, San Diego, CA). These electrical pulses also simultaneously induce the initial oocyte activation.

7. Confirm fusion by microscopic examination.

8. All fused cell–cytoplast complexes are activated further with 10 μg/ml of cycloheximide in CR1aa medium for 5 h.

9. Culture the cloned embryos in KSOM plus 0.1% (w/v) BSA for the first 4 days.

10. Switch the medium to KSOM containing 1% BSA for the remaining 3 days required to reach the blastocyst stage.

11. Change the medium every 2 days during the course of *in vitro* culture.

Mechanical Isolation of the Inner Cell Mass

1. On day 7 of embryo culture, the blastocysts expand (days 7.5 to 8 for parthenogenetic blastocysts, Fig. 1) and the inner cell mass (ICM) becomes apparent.

2. Prepare four-well MEF dishes by removing old medium and add 500 μl I-SCML medium to each well. Preequilibrate for 2 h at 37° in 6% CO_2 in humidified air.

3. Transfer one blastocyst per well and return dish to incubator. Most blastocysts hatch within 24 h and attach to feeder layer within 48 h postplating.

4. Prepare 24-well MEF dishes as described in step 2 for four-well dishes. A deeper well will make it easier to dissociate the cells later.

5. Once blastocysts have attached, using a 30-gauge needle, cut ICM away from trophoblasts. Using a drawn pipette, transfer the isolated ICM to a single well in a 24-well dish that contains MEF feeders.

6. Return dish to incubator. ICM will attach overnight. Observe daily for outgrowth.

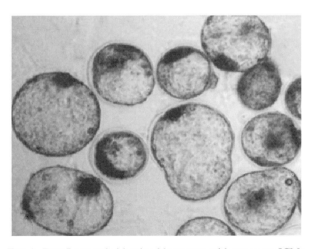

FIG. 1. Day 7 expanded bovine blastocysts with apparent ICMs.

Expansion of Putative ES Cells from Trophectoderm Colonies

1. On days 2 to 4 following ICM isolation, ICM outgrowth is apparent.

2. When the ICM outgrowth expands to a size of approximately 400 μm in diameter, remove the old medium, wash the cell layer with PBS (calcium and magnesium free), and add 100 μl trypsin EDTA (0.05%). After a 30-s exposure to trypsin, tap dish against palm of hand to dislodge cells. Repeat this every 15 s until cells come off the dish. Once cells float freely, neutralize trypsin with 2 ml I-SCML media and pipette up and down vigorously to separate cells into smaller clumps. Bovine cells do not dissociate into a single cell suspension. Return dish to incubator and record this as passage one.

3. The next day, remove the old medium and replace with fresh I-SCML medium. Complete media changes daily between passages.

4. The interval between passages ranges from 3 to 6 days. Each time the cells are exposed to trypsin is recorded as a passage. Keep accurate records of passage number as this is the way the age of a cell line is determined.

5. The first two to three passages occur in the 24-well dish as a split 1:1. Passages three and four move the cells into a six-well dish followed by passages four and five moving into a 100-mm dish. Each dish size increase will contain fresh mitomycin C-treated MEF feeder layers.

Maintenance of Bovine Embryonic Stem Cells

1. Between passages, bovine ES cells should be fed daily by complete medium change.

2. On the day cells are ready for passage, remove old medium and replace with fresh I-SCML medium 2 h prior to dissociation.

   a. Undifferentiated colonies exhibit the following morphological characteristics. Undifferentiated bES cells are large and tightly packed and have indistinguishable cell–cell boundaries. The cells reside in multicellular colonies with a smooth surface and a distinct colony boundary from the surrounding feeder layers (Fig. 2).

   b. Signs of differentiation include the colony boundary becoming unclear. The surface of the colony is not smooth, and cell–cell boundaries within the colony become obvious. Only a few cells are alkaline phosphatase (ALPL) positive (Figs. 3 and 4).

3. Following the feed, remove the old medium, wash cell layer with PBS, and add 0.05% trypsin EDTA. After a 30-s exposure to trypsin, tap dish against palm of hand to dislodge cells. Repeat this every 15 s until cells come off the dish.

4. Once cells float freely, neutralize trypsin with I-SCML (0.5 ml for six-well dish and 2 ml for 100-mm dish). Pipette vigorously up and down to break ES cell colonies into smaller clumps.

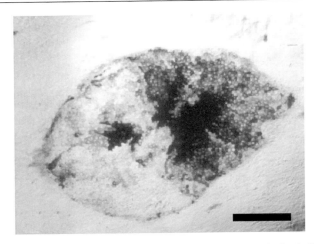

Fıg. 2. Multiple cell colonies derived growing on a feeder layer of mitotically inactivated mouse fetal fibroblasts. Note that the colonies are large and have clear boundaries with the feeder layer. Cells in the colonies are compact, with indistinguishable cell–cell boundaries. The morphology of these colonies is very similar to that of established ES/EG cell lines of the mouse/human. Scale bar: 100 mm.

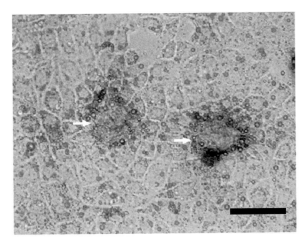

Fıg. 3. A group of differentiated cells (arrow) in an undifferentiated colony. Scale bar: 100 mm.

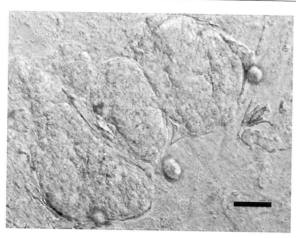

FIG. 4. A badly differentiated colony. Signs of differentiation are visible. The colony boundary becomes unclear. The surface of the colony is not smooth, and cell–cell boundaries within the colony become obvious. Only few cells are ALPL positive (arrows). Scale bar: 100 mm.

5. Transfer cell suspension into a 15-ml conical tube and pellet by centrifugation at 1000 rpm for 5 min.

6. Discard supernatant and resuspend in 10 ml I-SCML. Transfer 2 ml of cell suspension into a 100-mm dish containing equilibrated I-SCML on a new MEF. Record the passage number and the split ratio of 1:5. Cells are clumped so it is difficult to count.

7. The remaining cells suspension can either be plated for expansion or be prepared for cryopreservation.

8. For cryopreservation, pellet the cells as before, discard supernatant, and resuspend in 4 ml freezing medium that contains 40% I-SCML, 50% KO-FBS, and 10% DMSO.

9. Transfer 1 ml cell suspension/cryovial. Put cryovials into a controlled rate freezer and put in –80° freezer. The next day, transfer vials into liquid nitrogen for long-term storage.

Characterization of Bovine Embryonic Stem Cells

Detection of Alkaline Phosphatase

1. Once the ICM-derived cells have been expanded into 100-mm dishes, at the next passage (usually between five and seven), plate 12 wells in a 24-well dish for characterization of the cells.

2. Feed daily for 3 days and then remove the old medium, wash with PBS, and add 500 μl fixative (4% paraformaldehyde in PBS).

3. Fix the cells for 10 min and then wash with PBS. Leave PBS in all the wells except two for ALPL staining.

4. Use two wells to incubate the cells with an ALPL substrate, nitroblue tetrazolium chloride/5-bromo-4-chloro-3-indolylphosphate toluidine (toxic!) at room temperature in the dark for 20 min to 1 h.

5. Remove ALPL stain and rinse twice with PBS, 5 min each. Wrap dish in aluminum foil to protect from light. ALPL-stained cells are ready for viewing under a microscope.

6. The trophectoderm colonies contain flattened cells in the monolayer as well as multilayer clumps. Clumps stained positive for ALPL are presumably undifferentiated (Fig. 5). Most of the flattened cells in the monolayer stain negative for ALPL and may represent residual trophoblast cells of the embryos or differentiated ES cells (Fig. 6).

Immunostaining of Surface Markers

1. The remaining wells in the 24-well dish are ready for blocking solution.

2. Remove the PBS and add 500 μl blocking solution (0.1% Triton X-100/PBS with 5% donkey serum and 5% BSA fraction V). Block overnight at room temperature on a "belly dancer" at slow speed.

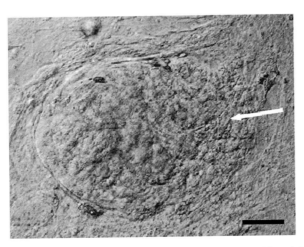

FIG. 5. A representative primary trophectoderm colony. After ALPL staining, part of the colony is multilayer clumps and stains positive (dark), whereas the rest is a monolayer, contains flattened cells, and stains negative. Scale bar: 1 mm.

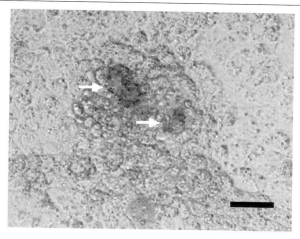

FIG. 6. Colonies selected for replating from passage-1 ICM cultures. The colonies are surrounded by cells growing in monolayer, which may be residual trophoblast cells. Scale bar: 100 mm.

3. The next morning, dilute primary antibodies in blocking solution: antibodies against SSEA1 (1:30), SSEA4 (1:30), TRA1-60 (1:20), TRA1-81 (1:20), and OCT4 (1:500). Add each antibody to two wells at room temperature and incubate for 2 h.

4. Following primary antibody incubation, wash three times (5 to 10 min each) with 0.1% Triton X-100/PBS.

5. Remove the last rinse and add diluted biotinylated secondary antibody (1:200 horse antimouse IgG and 1:200 goat antimouse IgM). Incubate for 1 h at room temperature.

6. Detection of specific binding is performed using the Elite ABC peroxidase staining kit with 3,39-diaminutesobenzidine as the substrate. Positive staining is gray-black in color.

Immunocytochemical Analysis of Differentiated Cultures

1. Place ES cell colonies onto a tissue culture dish in differentiation medium for 2 to 7 weeks for random differentiation.

2. Perform indirect immunocytochemistry using markers for all three embryonic layers. These markers are cytokeratin for endoderm (Shamblott *et al.*, 1998), actin $\alpha 2$ smooth muscle aorta (ACTA2) for mesoderm (Shamblott *et al.*, 1998), and tubulin $\beta 3$ (TUBB3) for ectoderm (Vrana *et al.*, 2003).

3. Fix differentiated ES cells in 4% paraformaldehyde/PBS for 15 to 20 min at room temperature.

4. Wash twice (5 to 10 min each) with PBS.

5. Permeabilize cells with 0.1% Triton X-100/PBS for 10 min at room temperature.

6. Wash twice (5 to 10 min each) with PBS.

7. Apply a blocking solution (e.g., 4% serum/PBS) for 30 min at room temperature.

8. Remove excess blocking solution.

9. Incubate the cells with primary antibodies—ACTA2 (1:100), TUBB3 (1:200), and cytokeratin (1:400)—diluted with PBS at room temperature for 1 h.

10. Wash three times (5 to 10 min each) with PBS.

11. Incubate sections for 30 min with diluted biotinylated secondary antibody solution (1:200 horse antimouse IgG).

12. Perform detection of specific binding using the Elite ABC peroxidase staining kit with 3,39-diaminutesobenzidine as the substrate. Positive staining is gray-black in color.

13. Mount slides.

Karyotyping

1. The karyotyping procedure has been described previously (Mitalipova *et al.*, 2001; Shamblott *et al.*, 1998).

2. Cells prepared for cytogenetic analysis are incubated in growth medium supplemented with 0.08 mg/ml of KaryoMAX for 4 to 6 h at 37°.

3. Trypsinize and treat cells with hypotonic KCl (0.57%) for 25 min at 36.5° and fix in acetic methanol (1:3, v/v); spread drops of cell suspension on clean microscopic slides.

4. Stain the chromosomes with 5% Giemsa for 40 min. Examine the chromosomes at 1000× magnification under oil.

5. Count at least 100 spreads to determine the percentage of cells with the correct karyotype.

References

Brackett, B. G., and Oliphant, G. (1975). Capacitation of rabbit spermatozoa *in vitro*. *Biol. Reprod.* **1975**, 260–274.

Cibelli, J. B., Grant, K. A., Chapman, K. B., Cunniff, K., Worst, T., Green, H. L., Walker, S. J., Gutin, P. H., Vilner, L., Tabar, V., Dominko, T., Kane, J., Wettstein, P. J., Lanza, R. P., Studer, L., Vrana, K. E., and West, M. D. (2002). Parthenogenetic stem cells in nonhuman primates. *Science* **295**, 819.

Cibelli, J. B., Stice, S. L., Golueke, P. J., Kane, J. J., Jerry, J., Blackwell, C., Ponce de Leon, F. A., and Robl, J. M. (1998). Transgenic bovine chimeric offspring produced from somatic cell-derived stem-like cells. *Nature Biotechnol.* **16**, 642–646.

Evans, M. J., and Kaufman, M. H. (1981). Establishment in culture of pluripotential cells from mouse embryos. *Nature* **292,** 154–156.

First, N. L., Sims, M. M., Park, S. P., and Kent-First, M. J. (1994). Systems for production of calves from cultured bovine embryonic cells. *Reprod. Fertil. Dev.* **6,** 553–562.

Hogan, B. (1994). "Manipulating the Mouse Embryo: A Laboratory Manual." Cold Spring Harbor Laboratory Press, Plainview, NY.

Kubota, C., Yamakuchi, H., Todoroki, J., Mizoshita, K., Tabara, N., Barber, M., and Yang, X. (2000). Six cloned calves produced from adult fibroblast cells after long-term culture. *Proc. Natl. Acad. Sci. USA* **97,** 990–995.

Liu, L., Ju, J. C., and Yang, X. (1998). Parthenogenetic development and protein patterns of newly matured bovine oocytes after chemical activation. *Mol. Reprod. Dev.* **49,** 298–307.

Mitalipova, M., Beyhan, Z., and First, N. L. (2001). Pluripotency of bovine embryonic cell line derived from precompacting embryos. *Cloning* **3,** 59–67.

Reubinoff, B. E., Pera, M. F., Fong, C. Y., Trounson, A., and Bongso, A. (2000). Embryonic stem cell lines from human blastocysts: Somatic differentiation *in vitro. Nature Biotechnol.* **18,** 399–404.

Robertson, E. J. (1987). "Teratocarcinomas and Embryonic Stem Cells: A Practical Approach." IRL, Oxford.

Saito, S., Sawai, K., Ugai, H., Moriyasu, S., Minamihashi, A., Yamamoto, Y., Hirayama, H., Kageyama, S., Pan, J., Murata, T., Kobayashi, Y., Obata, Y., and Yokoyama, K. K. (2003). Generation of cloned calves and transgenic chimeric embryos from bovine embryonic stem-like cells. *Biochem. Biophys. Res. Commun.* **309,** 104–113.

Shamblott, M. J., Axelman, J., Wang, S., Bugg, E. M., Littlefield, J. W., Donovan, P. J., Blumenthal, P. D., Huggins, G. R., and Gearhart, J. D. (1998). Derivation of pluripotent stem cells from cultured human primordial germ cells. *Proc. Natl. Acad. Sci. USA* **95,** 13726–13731.

Sirard, M. A., Parrish, J. J., Ware, C. B., Leibfried-Rutledge, M. L., and First, N. L. (1988). The culture of bovine oocytes to obtain developmentally competent embryos. *Biol. Reprod.* **39,** 546–552.

Stice, S. L., Strelchenko, N. S., Keefer, C. L., and Matthews, L. (1996). Pluripotent bovine embryonic cell lines direct embryonic development following nuclear transfer. *Biol. Reprod.* **54,** 100–110.

Suzuki, N., Rohdewohld, H., Neuman, T., Gruss, P., and Schöler, H. R. (1990). Oct-6: A POU transcription factor expressed in embryonal stem cells and in the developing brain. *EMBO J.* **9,** 3723–3732.

Thomson, J. A., Itskovitz-Eldor, J., Shapiro, S. S., Waknitz, M. A., Swiergiel, J. J., Marshall, V. S., and Jones, J. M. (1998). Embryonic stem cell lines derived from human blastocysts. *Science* **282,** 1145–1147.

Thomson, J. A., Kalishman, J., Golos, T. G., Durning, M., Harris, C. P., Becker, R. A., and Hearn, J. P. (1995). Isolation of a primate embryonic stem cell line. *Proc. Natl. Acad. Sci. USA* **92,** 7844–7848.

Vrana, K. E., Hipp, J. D., Goss, A. M., McCool, B. A., Riddle, D. R., Walker, S. J., Wettstein, P. J., Studer, L. P., Tabar, V., Cunniff, K., Chapman, K., Vilner, L., West, M. D., Grant, K. A., and Cibelli, J. B. (2003). Nonhuman primate parthenogenetic stem cells. *Proc. Natl. Acad. Sci. USA* **100**(Suppl. 1), 11911–11916.

Wang, L., Duan, E., Sung, L. Y., Jeong, B. S., Yang, X., and Tian, X. C. (2005). Generation and characterization of pluripotent stem cells from cloned bovine embryos. *Biol. Reprod.* **73,** 149–155.

Yang, X., Tian, X. C., Dai, Y., and Wang, B. (2000). Transgenic farm animals: Applications in agriculture and biomedicine. *Biotechnol. Annu. Rev.* **5,** 269–292.

[3] Avian Embryonic Stem Cells

By Marie-Cecile van de Lavoir and Christine Mather-Love

Abstract

Blastodermal cells derived from the area pellucida of a stage X (EG&K) embryo have the potential to contribute to the somatic tissues and the germ line when reintroduced into a stage X (EG&K) recipient embryo. This chapter describes a method to culture chicken embryonic stem (cES) cells derived from blastodermal cells. Within the first week of culture, the cells change their morphology; they become smaller with a large nucleus and a prominent nucleolus. The cES cells remain chromosomally normal and can be cultured for extended periods. They can be modified genetically using standard electroporation procedures and, after injection into a recipient embryo, can contribute to all somatic tissues. Using a surrogate shell culture system, the injected embryos can be manipulated and visualized easily throughout incubation. We have generated high-grade chimeras by compromising the recipient embryos and maintaining the ES cells in stage X (EG&K) recipients for a few days at 15° before incubating them at 37.5°. The cES system provides a novel experimental paradigm for the investigation of developmental and physiological mechanisms in the chicken.

Introduction

The chicken embryo is an excellent model for the study of developmental biology; it is accessible, easy to study in the early stages of development, and there is a wealth of anatomical data describing the morphological changes that occur (Stern, 2005). Because of these attributes, chicken embryos have been used extensively in teaching and studying early developmental processes. However, the absence of methods to introduce genes into the chicken has limited the appeal of the chicken embryo in the search for a molecular understanding of development. The ease of genetic modification of murine embryonic stem (ES) cells and their potential to contribute to both somatic tissues and the germ line after reintroduction to a recipient embryo have opened avenues for study that have made it the model of choice among vertebrate animals. Although transgenic chickens can be obtained using lentiviral vectors (McGrew *et al.*, 2004), this technology precludes targeted insertions and the use of large transgenes.

METHODS IN ENZYMOLOGY, VOL. 418
Copyright 2006, Elsevier Inc. All rights reserved.

0076-6879/06 $35.00
DOI: 10.1016/S0076-6879(06)18003-9

In 1990 Petitte *et al.* (1990) showed that blastodermal cells retrieved from the area pellucida of a stage X (EG&K) (Eyal-Giladi and Kochav, 1976) chicken embryo were able to contribute to both the somatic tissues and the germ line of the recipient embryo. Since that time, there has been extensive activity to derive long-term chicken embryonic stem (cES) cell lines from blastodermal cells (Etches *et al.*, 1996; Pain *et al.*, 1996, 1999; Petitte, 2004; Petitte and Yang, 1993; Petitte *et al.*, 2004). Although cells could be cultured for several weeks, they were incapable of contributing to somatic tissues after more than 3 weeks in culture. Furthermore, they could not be transfected using conventional protocols, and only promoterless transgenes could be inserted into the genome. In addition, the technology lacked the robustness essential for widespread use. This chapter describes a method of deriving, maintaining, and transfecting chicken embryonic stem cells. We show how these cell lines can be maintained indefinitely, can be transfected reliably, and can retain their potential to contribute to somatic tissues when injected into recipient embryos.

The term *cES cell* has been coined since the first derivation of cell lines from the area pellucida (Pain *et al.*, 1996; Petitte, 2004). Although germ line transmission has not yet been obtained with these cells, we have retained the term *ES cell* because the cells are of embryonic origin, are self-renewing in culture, and give rise to ectodermal, mesodermal, and endodermal derivatives *in vivo*.

To inject and grow the embryos, we use a surrogate shell system to maximize access to the embryos with injection pipettes. Although this system is labor-intensive, the ease of access that the system provides creates the opportunity to combine genetic modification with other techniques, such as time-lapse photography, to gain new insight into the earliest stages of vertebrate development.

Selection of Culture Media and Routine Procedures

Chicken ES Cell Culture Medium

While Dulbecco's modified Eagle's medium (DMEM) can support the growth of cES cells, KnockOut DMEM (KO-DMEM) (Invitrogen) increases the growth rate substantially, making it easier to obtain sufficient cells for transfection. Although the composition of KO-DMEM is proprietary, the lower osmolarity (270 mOsm vs 320–350 mOsm for DMEM), optimized to approximate mouse embryonic tissue, might be important, as osmotic stress is implicated in the inhibition of embryo development *in vitro* (Lawitts and Biggers, 1992). Before use, the KO-DMEM is conditioned on buffalo rat liver (BRL) cells. Although a combination of growth factors

(Pain *et al.*, 1996) has been reported to sustain the development of cES cells, we have obtained very consistent results with this conditioned medium, eliminating the need to supplement the medium with growth factors. We have established efficient procedures for producing large batches of conditioned media and freeze aliquots at $-20°$ for several months. Fetal bovine serum (FBS) is an important component of the medium to support the growth of cES cells and needs to be evaluated before use. Serum that supports the growth of mouse ES cells may not sustain the growth of chicken ES cells (van de Lavoir, unpublished observations). FBS cannot be replaced by chicken serum. For manipulation of cells outside the incubator, for example, injection and cryopreservation, a manipulation medium based on CO_2 independent medium (Invitrogen) is used.

Media Formulations

> Chicken ES cell culture medium: KO-DMEM (Invitrogen) containing 80% BRL-conditioned medium supplemented with 10% FBS (characterized serum, Hyclone), 2 mM glutamine (Glutamax, Invitrogen), 1 mM pyruvate (Mediatech, Inc.), 1× nucleosides (Specialty Media), 1× nonessential amino acids (Mediatech, Inc.), and 0.1 mM β-mercaptoethanol (Invitrogen)
> Conditioning medium: KO-DMEM supplemented with 2 mM glutamine, 5% FBS, and 0.5% penicillin/streptomycin (Mediatech, Inc.)
> Manipulation medium: CO_2-independent medium (Invitrogen) supplemented with 10% FBS, 2 mM glutamine, and 1% penicillin/streptomycin

Prepare BRL-conditioned medium as follows.

1. Grow up buffalo rat liver cells (BRL; ATCC, CRL-1442) in either DMEM or KO-DMEM supplemented with 10% serum, 2 mM glutamine, and 1% penicillin/streptomycin. When confluent, split the cells 1:18 into a two-stack CellSTACK (1272 cm^2; Corning).

2. After 3 days, when the cells are confluent, remove the medium and add 221 ml of conditioning medium.

3. Condition the medium for 3 days; remove 218 ml of the BRL-conditioned medium and place in a sterile bottle.

4. Replenish the cell stack with 218 ml of fresh conditioning medium. Repeat collection two more times for a total of three collections.

5. Place collection bottle at $-20°$ between collections. Use the same bottle for all three collections, ending with a total volume of 654 ml. If smaller aliquots are desired, adjust the volume that is added to each collection bottle accordingly but make sure to have equal contributions from all three

collection days. Alternatively, the batch can be scaled down by using T175 (or similar) flasks and a similar medium:cm^2 ratio (Hooper, 1987; Petitte, 2004).

6. Store the BRL-conditioned medium at $-20°$ and filter it when the cES cell medium is made.

Selection and Preparation of Feeder Cells

We have derived cES lines on cell culture inserts (Transwell, Corning) (van de Lavoir *et al.*, 2006) but do not use them for routine culture because proliferation on the inserts is much slower than on a feeder layer. Primary chicken and mouse fetal fibroblasts do not support the growth of cES cells (Yang and Petitte, 1994; van de Lavoir, unpublished observations). Although the cES cells can be grown on BRL cells, the cES cells attach more firmly to BRL compared to Sandoz inbred mouse-derived thioguanine-resistant and ouabain-resistant (STO) cells (ATCC; CRL-1503), making passaging of the cES cells difficult. The only cell line identified that consistently supports the growth of cES cells is STO cells. However, a high concentration of STO cells will impede the growth of cES cells and therefore it is essential to seed the cells at a low concentration (10^4 cells/cm^2) that does not produce a confluent feeder. Due to this minimal STO concentration, care must be taken to ensure that the STO cells are plated evenly with few concentration differences throughout the well. When exposed to the cES cell culture medium, the irradiated STO cells take on an elongated morphology that at times can be confused with growing STO cells. We grow the STO cells in large batches, irradiate the cells with 12,000 rad from a cesium source, and cryopreserve aliquots, making plating of the irradiated STO cells predictable and efficient. The irradiated STO cells are seeded on dishes, gelatinized with 0.1% gelatin, a minimum of 1 day before use. The feeder layers should be used preferably within 5 days. Although mitomycin C can be used to inactivate the STO cells, we prefer not to use it because occasionally a few cells continue to grow following treatment with mitomycin C. Chicken ES cells do not proliferate fast enough to outgrow the STO cells and, eventually, STO cells become the predominant cell type in the culture.

Derivation and Maintenance of Chicken Embryonic Stem Cell Lines

Selection of Breeds

Conventionally, cES cells are derived from a chicken strain that differs in feather color from that of the recipient embryo used to make the chimera. This enables evaluation of the cES cell contribution to the chimera as early as the 10th day of incubation, at which time the melanocytes have begun to secrete melanin and the rudimentary feathers are visible. We derive our cES cells from one of two crosses, Barred Plymouth Rock (BR) × BR or Rhode Island Red × BR. These breeds have pigmented feathers that provide a marker when the cES cells are injected into recipient embryos from a white-feathered breed. White Leghorns, which are homozygous at the dominant white locus, provide excellent recipient embryos. A chimeric chicken will display dark feathers from the cES cells and white feathers from the recipient embryo.

Isolation of Blastodermal Cells

When fertile chicken eggs are laid, most of the embryos will be at stage X (EG&K), but factors such as flock age (Mather and Laughlin, 1979), egg handling, and storage conditions can influence the stage of development of the embryo. At stage X (EG&K), the embryo can be identified *in situ* on the yolk as a white disk, approximately 3 mm in diameter. The embryo consists of approximately 40,000 to 60,000 cells with two distinct morpho-logical areas: the area pellucida in the center, which will form the embryo proper, and the area opaca on the periphery, which will form the extra-embryonic membranes (Figs. 1A, 1B, and 2). Blastodermal cells isolated from the area pellucida can contribute to somatic tissues and the germ line when injected into a recipient embryo (Petitte *et al.*, 1990), making the area pellucida the tissue of choice for the derivation of cES cells.

Previous publications on cES cells describe the derivation of cES cells from pooled embryos (Pain *et al.*, 1996; Petitte, 2004), and although these cES cell lines are useful, there are advantages in deriving cES cell lines from single embryos. We have observed distinct differences in growth characteristics between blastodermal cells from individual embryos; some cultures develop into healthy, long-term cES cultures, whereas others self-terminate after a few days or differentiate easily and often. Furthermore, when producing live chimeras it is important to know the sex of the cell line. Male cell lines will induce male secondary sex characteristics in female birds, precluding the laying of eggs in high-grade (>65% feather pigmentation) chimeras.

Blastodermal cells disperse very easily when the embryo is triturated and enzymatic treatments are not needed to obtain a single-cell suspension. Blastodermal cells retrieved from single embryos are plated into 48 wells, seeded previously with irradiated STO cells. The cells are large and loaded with glycogen (Fig. 1C), which is lost within the first week of culture as cES cells become visible. At the time of seeding a smaller well size might seem more appropriate, but due to potentially rapid growth and the fragile nature

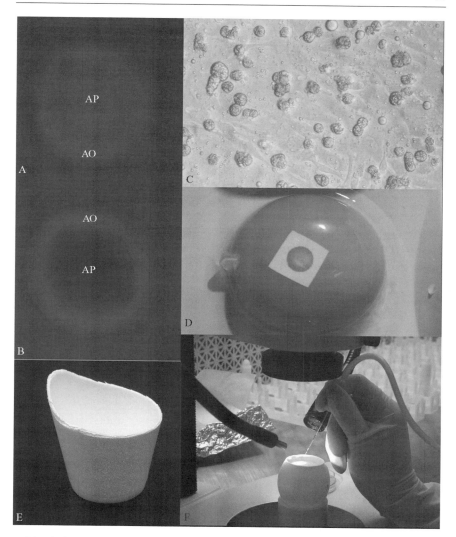

FIG. 1. (A and B) Stage X (EG&K) embryos showing the peripheral area opaca (AO) and the central area pellucida (AP). (B) After injection of 2 μl of medium. Note the expansion of the medium throughout the subgerminal cavity. Injections were done and pictures taken using a blue light source. (C) Blastodermal cells collected from the AP from a stage X (EG&K) embryo. (D) Recovery of stage X (EG&K) embryo using a paper filter square. (E) Flexible pouring cup for transferring contents of fertile egg into first surrogate shell. (F) Injection of a cell suspension into a stage X (EG&K) embryo. (See color insert.)

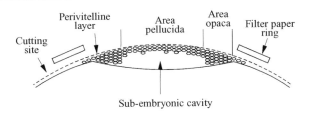

FIG. 2. Schematic showing a sagittal section through a stage X (EG&K) embryo; placement of the paper ring/square is indicated. cES cells are injected into the subgerminal cavity of the intact embryo. From Watt *et al.* (1993).

of the blastodermal cells, it is best not to passage the cultures for at least 1 week. If the cells are seeded into a smaller well, the culture can become confluent and differentiate before it can be passaged reliably. Chicken ES cell lines can also be derived on Transwell inserts that contain a polyester membrane.

Recover the blastodermal cells with the following procedure.

1. In a laminar flow hood, using sterile equipment, crack the egg open and separate the egg white from the egg yolk. Identify the location of the embryo on the yolk and slide the yolk onto a Petri dish (or weigh boat) with the embryo positioned on top. Remove albumen from the perivitelline membrane by laying a Kim-wipe gently on top of the embryo and peeling it back slowly.

2. To retrieve the embryo, take a small ring or square of filter paper (cut from Whatman paper) with a hole punched in the center and lay it on the yolk so that the embryo is left exposed in the middle of the hole (Figs. 1D and 2). While holding onto a corner of the filter paper with forceps, cut through the yolk membrane around the filter paper and lift it gently away from the yolk.

3. Slowly submerge the filter paper with the attached embryo perpendicularly into phosphate-buffered saline (PBS). This is a crucial step to eliminate yolk from the embryo. Rotate the filter paper so that the embryo is placed ventral side up. Under the microscope, while the embryo is submerged in PBS, remove the remaining yolk by using either a hair loop or a gentle jet of PBS. A hair loop is made by attaching a circle of fine hair to the end of a wooden applicator stick with adhesive material.

4. Isolate the area pellucida by cutting away the area opaca using the hair loop.

5. Pick up the area pellucida with a Pasteur pipette and place it in a small Eppendorf tube containing 200 μl manipulation medium. Let it settle to the bottom and wash it twice by replacing the medium (do not centrifuge).

6. Add 200 μl of culture medium and triturate with a P200 Pipetman until a single-cell suspension is obtained. Plate the cells in a 48-well containing irradiated STO cells.

Derivation of Chicken Embryonic Stem Cells

Deriving cES cells is the most challenging aspect of cES culture. It is therefore advisable to acquire experience growing established cES cell lines before attempting to derive a new line.

After the blastodermal cells are seeded, it takes approximately 1 week of culture before cES cells become visible. Occasionally, blastodermal cells grow so rapidly that it is easy to passage and expand them. More frequently, cES cells and differentiated cells are mixed and it can be a challenge to retrieve the population of cES cells. One option is to incubate the wells with Ca^{2+}/Mg^{2+}-free PBS before passage (as described under *Maintenance of Chicken Embryonic Stem Cells*), which will allow the cES cells to be passaged while the differentiated cells are left behind. Alternatively, the differentiated cells can be removed by aspiration.

Always add fresh medium to the old well after passage because some cES cells will inevitably be left behind and healthy colonies can be retrieved. We have developed this method as a backup strategy because the finicky nature of cES cells can cause them to differentiate and/or stop growing after passaging.

Maintenance of Chicken Embryonic Stem Cells

Maintaining cES cell cultures requires diligent attention to the details of the growth of the culture, as well as judgment derived from experience when making decisions about passaging and/or feeding the cells. Each culture has its own idiosyncrasies, and constant monitoring and vigilance are required for maintaining healthy cultures. Our culture methods have evolved since 2000 and our experience has taught us that chicken ES cells are best maintained when seeded at high densities (>40% confluency) and passaged in small clumps. Under these conditions the cES cells double every 24 h. We grow the cells to 80 to 90% confluency before passaging them in a 1:2 ratio, which implies that cultures are passaged daily. The passaging ratio can vary: in unevenly plated cultures and cultures with differentiation we passage 1:1; when the cultures are too dense to be left until the next day but too sparse to passage 1:2, we passage 2:3; and occasionally, when the cES cells are very dense, we passage 1:3. To maintain optimum growth and prevent differentiation, the cells are passaged by transferring 30 to 50% of the medium covering the cells into the new wells.

Although 0.25% trypsin can be used for dispersion, repeated use of trypsin is not recommended as it induces differentiation. Due to the loose attachment of cES cells to the feeder and to each other, a 1- to 2-min incubation (this can be shorter or longer depending on the culture) in Ca^{2+}/Mg^{2+}-free PBS followed by trituration in medium with a P1000 Pipetman is sufficient to disperse the cES cells into small clumps. This method of passaging precludes the growth of cES cells in flasks and we routinely grow the cells in six-well plates.

Cultures are passaged as follows.

1. Collect medium from well that is to be passaged and deposit into two new wells an amount that takes up 30 to 50% of the final volume.

2. Carefully wash once with Ca^{2+}/Mg^{2+}-free PBS. Add PBS carefully so as not to loosen and remove cES cells prematurely.

3. Add fresh Ca^{2+}/Mg^{2+}-free PBS and incubate cells for 1 to 2 min. When looking under a microscope, the cells appear to round up slightly; the edges of the cells may take on a whitish tinge, appear rougher, and visibly loosen their attachment.

4. Remove PBS.

5. Add medium (1 ml/six well) and pipette to collect all the cells, triturate moderately to obtain small clumps and divide the cells into two new wells.

6. Add fresh medium to obtain desired volume.

When growing up clones and deriving new lines, we recommend addition of medium to the wells that were passaged; cES cells that have remained in the well can be the source of very nice cultures. In contrast to mouse embryonic stem (mES) cells, where medium changes need to be frequent to prevent differentiation, cES cells are very sensitive to medium changes and will differentiate in response to a premature medium change. If the confluency of a culture on the day after passage is <50%, do not change the medium (see Figs. 3C and 3D). Instead, add some fresh medium if there is concern about nutrient depletion.

Morphology of Chicken Embryonic Stem Cells

Individual cES cells are morphologically similar to mES cells; the cells are small with a large nucleus and a pronounced nucleolus (Figs. 3A and 3B). In contrast to mES cells, which grow in smooth-edged, tight, round colonies where individual cells are difficult to distinguish, cES cells grow in single-layer colonies with clearly visible individual cells (Figs. 3A and 3B). In general, the morphology of ES cultures is dependent on a variety of factors. If the cells are grown on a denser STO feeder layer, the colonies take on a defined shape that is more reminiscent of mouse ES cells (Petitte

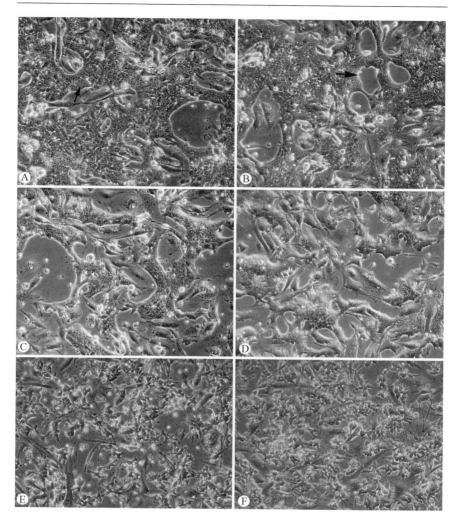

FIG. 3. (A and B) Chicken ES cell cultures with desired morphology. Note the single layer of cells. The cells have a large nucleus and a pronounced nucleolus (arrow in A). These panels represent the cell density minimally necessary for successful passaging. (B) The large openings between the cells (arrow) are indicative of good growing cES cells. These plates were seeded with STO cells, which have been pushed aside by the cES cells and are not visible in the openings. (C and D) These panels are representative of good cultures that are too sparse for passaging. Medium changes on these cultures will have a detrimental effect. Be aware that these cultures can grow very fast and therefore need to be checked first thing the next morning. (E and F) cES cultures that have taken on a pronounced fibroblast morphology. Although these cells do not have the distinct characteristics of cES cells, they can still contribute to the somatic tissues when injected into a recipient embryo. (See color insert.)

et al., 2004). However, the denser STO feeder layers impair the growth of the cES cells and the colonies will die off. Because we use a sparse feeder layer to enhance the growth of the cES cells, the feeder layer does not physically impede the cells and the colonies will spread out and change their morphology. This can result in an ES cell morphology that appears more fibroblast like; the cells are not closely attached but have small spaces between them (Figs. 3E and 3F). This can be pronounced in cultures that are very sparse (<40% confluency). This spacing between cells is different from the formation of large, round openings that can be seen developing in good growing cultures (Fig. 3B).

In contrast to mouse ES cells, cES cells stop growing when they differentiate (Fig. 4D). Sometimes a fibroblast culture is derived and occasionally

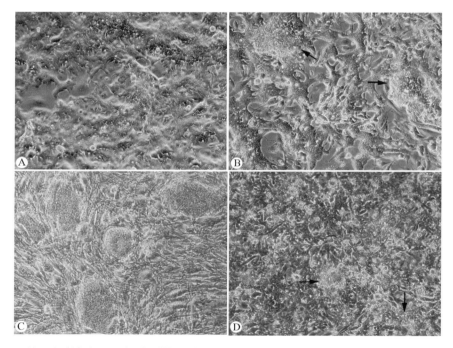

FIG. 4. (A) A completely differentiated cES culture. The culture in B displays areas of differentiation (indicated by arrows) that can be removed by passaging gently in Ca^{2+}/Mg^{2+}-free PBS in a 1:1 ratio. The differentiated areas will be left behind and a nice culture can be obtained. Generally a few passages are necessary to eliminate all the differentiation. (C) cES cells that have compacted into colonies reminiscent of mouse ES cells. The STO cells are dense and elongated due to their exposure to cES medium. (D) A culture that has become too dense. Areas of differentiation are apparent (arrows). (See color insert.)

neurons are seen but differentiation into muscle cells has not been observed. When differentiation develops in a culture, cES cells can be recovered and cultures reestablished with no differentiation (Fig. 4B). We take advantage of the fact that differentiated cells attach more firmly to the feeder layer compared to cES cells. By passaging the culture with Ca^{2+}/Mg^{2+}-free PBS, the differentiated cells can be left behind. Alternatively, when the differentiated cells are passaged, add medium to the plate that has been passaged and wait to see if more cES cells appear. If the differentiation is confined to only part of a well, aspiration of the differentiated area before passaging will help prevent its spread.

When cES cells develop vacuoles or the cES colonies compact, the cultures are lost, even if this problem initially affects only a small part of the culture. The compacted cES cell colonies are deceiving because they initially look like mouse ES cell colonies (Fig. 4C).

Cryopreservation

Chicken ES cells are cryopreserved using standard procedures in 10% dimethyl sulfoxide (DMSO) in manipulation medium (Robertson, 1987). To obtain single cells and small clumps, the cells are incubated for 10 to 20 s in 0.25% trypsin (no EDTA), dispersed in manipulation medium, and pelleted. The cells are resuspended in manipulation medium, and an equal volume of manipulation medium containing 20% DMSO is added. The cells are aliquoted to freezing vials, frozen in a cryocontainer at −80°, and transferred to LN2 the next day. To obtain good viability after thawing it is helpful to thaw the cells into 50% cES-conditioned medium. This can be collected from cultures that are growing well, filtered, and stored at 4° for up to 1 week. The medium is then mixed with fresh cES medium and equilibrated in the incubator before the cells are thawed and plated.

Genetic Modification of cES Cells

Widespread utilization of cES cell technology requires a simple method for genetic modification. Chicken ES cells, like mouse ES cells, can be modified genetically using standard electroporation procedures. Transfection efficiency is influenced by three important factors: quality of starting cell population, choice of promoter, and culture conditions of cells post-transfection.

In general, healthy, undifferentiated cultures with a nice morphology are the best choice for transfection. While one can still obtain colonies using cES cultures with some differentiation, choosing a culture with a doubling time of roughly 24 h or less and little or no differentiation will give the best clones. Harvesting the cells for transfection should be done at a time when the cells are ready to be passaged 1:2 (or approximately 80 to 90% confluency). cES cells are sensitive to passaging if not sufficiently confluent and will differentiate when they are overly confluent (Fig. 4D).

Providing a promoter strong enough to drive selection resistance is another important factor in cES cell modification. Although MC1 and PGK promoters are widely used in mouse transgenesis, we found them unsuccessful in cES cells. However, consistently good expression in cES cells is obtained using the CAG promoter (also known as the cx promoter), a combination of the chicken β-actin promoter and a cytomegolo virus (CMV) enhancer (Hadjantonakis *et al.*, 1998; Ikawa *et al.*, 1998; Niwa *et al.*, 1991).

The culture of cells after electroporation is an important part of the genetic modification procedure. Recovery of cells is maximized when the medium in which the cells were growing before transfection is used in a 1:1 mixture in the replating medium for the cells after transfection. Because cES cells spread out in a single, loosely connected layer, individual cells can dislodge easily during medium changes and even by movements in the medium induced when handling a culture dish. Seeding the cES cells back into large wells or plates and picking colonies after 10 days is, therefore, not an option. To prevent colonies from mixing, the electroporated cES cells are seeded into 1-cm^2 wells to obtain one colony per well. For this reason, the amount of DNA added to the electroporated cells is limited to 2 to 5 μg for constructs that are less than 10 kb. If additional colonies appear in the well, they are removed by aspiration. Resistant clones will become visible 8 to 12 days after transfection and are passaged to new 48 wells when they cover a minimum of 20% of the surface area of the well. The colonies are not "picked" but passaged using the Ca^{2+}/Mg^{2+}-free incubation (see *Maintenance of Chicken Embryonic Stem Cells*). The number of colonies depends on the size of the construct, the amount of DNA added, and the quality of the cells. Using 2 μg of an <10-kb construct, we passage, on average, 10 to 30 colonies per 10^7 cells transfected. Keep medium changes to a minimum when colonies are small; a colony can differentiate in response to a medium change. When a medium change is necessary and the colonies are still very small, growth can be promoted by including a portion of filtered conditioned medium from other cES cell cultures. If the STO feeder layer becomes too sparse, extra STO cells can be seeded into the wells. For selection using neomycin, neomycin-resistant STO cells need to be used as a feeder. For selection using puromycin, the effective concentration is low and will not impair STO viability. Hence, puromycin-resistant STO cells are not necessary.

Chicken ES cells are modified genetically using the following procedure.

1. Select cells to be transfected and remove and collect conditioned media in a sterile tube.

2. Wash the cells with Ca^{2+}/Mg^{2+}-free PBS, add 0.25% trypsin (no EDTA) to wells, and remove immediately. After 15 s add manipulation media to each well and pipette up/down to obtain a cell suspension. Transfer the cell suspension to a sterile tube and count cells. The transfection is done with approximately 10^7 cells.

3. Centrifuge cells for 2 min at 200g, resuspend the pellet in electro-poration buffer (specialty media), and add DNA. The final volume should be 800 μl. DNA should be linearized, precipitated, and resuspended in PBS.

4. Transfer the cell/DNA suspension to a 4-mm gap cuvette, cap cuvette, and expose cells to 8 square wave pulses of 2.8 kV/cm and 100-μs duration (BTX ECM830 electro square porator). Let cuvette stand for 10 min at room temperature after electroporation.

5. Gently resuspend cells in enough medium (use 50% conditioned medium set aside in step 1) to plate into all the wells of a single 48-well plate that has been seeded with irradiated STO cells the previous day.

6. Initiate antibiotic selection when the wells are confluent, approximately 2 days after transfection with either 0.5 μg/ml of puromycin or 100μg/ml neomycin, depending on selection cassette. The puromycin concentration is increased 1 to 2 days later to 1.0 μg/ml.

7. Change medium initially every day, followed by every other day until the majority of the cells are dead and then change the medium every 3 to 4 days until actively growing colonies are visible. The cells can be grown indefinitely in selection medium but this is not necessary.

Production of Chimeric Chickens

Maximizing the Level of Chimerism

To produce chimeric chickens, freshly laid eggs containing stage X (EG&K) embryos are the recipient of choice, principally because they are the developmental equivalents of the source of the cES cells, are readily available, and have the potential for generating high-grade chimeras (Fig. 5E). Chimeras can also be made, with some success, using recipient embryos at earlier stages of development (van de Lavoir et al., 2006). Injection of cES cells into stage X recipient embryos will generate somatic chimeras, but the frequency and extent of cES cell incorporation will be enhanced if the embryo is first compromised by exposure to γ-irradiation (Carsience et al., 1993). Exposure to 660 rad from a cesium source will retard subsequent development of the recipient by approximately 24 h without

FIG. 5. (A) System II cultures showing the window sealed with plastic film held in place by plastic rings and rubber bands. (B) System II eggs inside an Octagon 250 incubator. Note their horizontal position in the tray. (C) A Stage 17 (H&H) embryo immediately before transfer from system II to system III. (D) A tray of system III cultures containing 8-day-old embryos. Note how the surrogate shell system provides easy accessibility and visibility of the embryo and the extraembryonic vasculature. (E) Group of two BR and two chimeric chicks. Chimeras are indistinguishable from BR chicks, indicating high levels of chimerism. Reproduced from van de Lavoir *et al.* (2006). (See color insert.)

severely increasing mortality (Carsience *et al.*, 1993). Coring the recipient embryos (removing the center of the area pellucida) (Kagami *et al.*, 1997) can also increase the contribution of the cES cells to the embryo but is time-consuming and technically demanding. We have developed a method to increase the incorporation of cES cells to the embryo, exploiting embryonic diapause that occurs when the freshly laid egg is cooled to around 15 to 18°. When cES cells are injected into the irradiated stage X embryo at diapause and the injected embryo is maintained for 4 more days at diapause, the frequency and quality of somatic chimera production are increased (van de Lavoir *et al.*, 2006).

Culturing Chimeras in Surrogate Shells

To meet the physiological requirements of the recipient embryos, a two-phase culture system is required. The requirements of the embryo change over time and we meet these requirements by changing the shells and modifying the culture and incubation conditions (Perry and Mather, 1991). These two periods of culture are designated as systems II and III and are approximately equivalent to those in Perry's complete culture system (Borwornpinyo *et al.*, 2005; Perry, 1988, 1991). System II lasts from cES cell injection to day 4 of incubation and system III from day 4 to hatching. Perry's system I, which lasts from 2 h postfertilization to oviposition, that is, the oviductal phase of development, is not relevant to chimera production, as the stage X (EG&K) recipient we use is from the laid egg. If the injected embryos are evaluated before day 7 of incubation, a second surrogate shell is not required; the embryos can remain in the first surrogate shell.

Although more labor-intensive, we routinely culture the manipulated embryos in surrogate shells due to the unimpeded visibility for accurate ES cell injection and observation of embryo development. Embryos can be evaluated easily, and viability and progress of development can be recorded daily.

Preparation and Use of Surrogate Shells

Each fertile recipient egg needs two surrogate shells. Large openings, or windows, must be made in the surrogate shells to permit transfer of the fertile recipient egg between these shells, that is, before cES cell injection and as the chimera develops. Cutting the windows reduces the volume of the surrogate shells. In order to accommodate the fertile egg accurately, the egg supplying the first surrogate shell (for system II) must be 3 to 4 g heavier than the fertile egg. The second surrogate shell (for system III) must be large enough to provide an artificial air space into which the embryo will pip and begin lung ventilation prior to hatching. Consequently, the egg supplying this shell needs to be 30 to 40 g heavier than the fertile egg. For chimera production, sets of three eggs are matched by weight, that is, the fertile recipient egg (say 55 g) is matched with the egg that will supply the first surrogate shell (58 to 59 g) and with the egg that will supply the second surrogate shell (85 to 90 g).

As it develops, the embryo will elicit no immune response to a foreign shell or shell membranes (Ratcliffe, 1989) so surrogate shells from different individuals, chicken strains, or bird species such as turkeys and ducks can be used (Borwornpinyo *et al.*, 2005; Love *et al.*, 1994; Miura *et al.*, 1991; Zhu *et al.*, 2005). We prefer chicken eggs for the shells used in system II because they are readily available, and turkey eggs for system III because they give a hatchability of around 75% when used for unmanipulated embryos (Borwornpinyo *et al.*, 2005) (C. Mather-Love, unpublished data). In each surrogate shell, the window must be cut cleanly and accurately to avoid albumen leakage when the egg is resealed with a small square of plastic film. A template is used to draw a circle on the shell as a guide for cutting with a high-speed rotary tool (Dremel MultiPro, Model 395) fitted with a 220-mm diameter, two-sided, diamond-coated disc (DiamondMade, Global Enterprise Marketing Ltd.).

For the first shell, a window is cut at the pointed end of the egg to preserve the natural air space of the surrogate shell, which forms at the opposite (blunt) end as the egg cools after oviposition. The air space plays an important role during incubation, enlarging to compensate for water lost from the egg as water vapor through the porous shell. In the absence of an air space, that is, if the window is cut in the blunt end, the "sealed" window provides the path of least resistance for air to enter as the egg loses weight and bubbles will form in the albumen, causing abnormal development in some cases.

For the second shell, a larger window is made to facilitate transfer of the entire contents of the first shell into the second shell at the end of system II. In contrast to the first shell, a window is cut in the blunt end of the egg, removing the natural air space. Because the second shell has a much larger volume than the first shell, an artificial air space is generated above the embryo after transfer from system II to system III. In system III, air is not detrimental to development. However, if the embryo is submerged in albumen, it will fail to develop normally.

Preparation and Injection of cES Cells

The best chimeras are obtained with cells that have been growing well in the laboratory for at least several days, although it is possible to obtain somatic chimeras with cells thawed the day of injection. Preparing cES cells for injection is the only time we use trypsin with EDTA for dissociation, as a population of single cells is needed to prevent clumps from blocking the injection pipette. The viability of the recipient embryo is influenced greatly by the number of cells injected. Too many cells (>10,000/embryo) will result in few hatched chimeras; too few cells (1000 cells/embryo) will lead to large hatches but few high-grade chimeras. To obtain the greatest number of high-grade chimeras we inject 5000 cells per embryo. We use an injection volume of 1 to 2 μl; larger volumes tend to induce embryo twinning (J. Diamond, Origen Therapeutics, unpublished data).

Embryos can be injected without magnification, but for precise cES cell injection a microscope is recommended. We use a stereomicroscope (MZ6, Leica Microsystems Inc.) illuminated with a KL1500 LCD (Leica Microsystems Inc.) light source fitted with a blue filter. The blue filter facilitates accurate injections by giving better contrast of the white embryo on the yellow yolk (Figs. 1A and 1B).

The cES cells are injected into the subembryonic cavity (Fig. 2). This shallow, fluid-filled space lies immediately below the area pellucida, which is approximately two cells thick, so it is important not to insert the injection pipette too deeply or the cES cells will disperse into the yolk and will not contribute to the embryo. The cells should be injected slowly so that they spread out through the subembryonic cavity. If the injection has been performed effectively, the area pellucida will show up as a darker circle bordered by a clear white ring, the area opaca (Figs. 1A and 1B).

The procedure is summarized as follows.

To avoid contamination, work under laminar flow and with autoclaved equipment.

1. Prepare the first surrogate shells: cut 32-mm-diameter windows at the pointed end of the eggs; rinse inside and out with deionized water to remove albumen and shell dust; drain with the window down on absorbent paper; and store in a cool place until needed. This step can be done the day before step 2.

2. Irradiate the fertile recipient eggs with 660 rad within a few hours before the cES cells are to be injected.

3. Set up injection hood.

    a. Collect albumen from fresh eggs by separating it from the yolks and storing in a sealed container. Collect approximately 3 ml for each egg to be injected.

    b. Prepare a flexible pouring cup, for example, from an 8-oz. foam drinking cup, for transferring the recipient into the surrogate shell. The rim should be trimmed to provide a lip (Fig. 1E).

    c. Cut 55-mm squares of plastic film, interleaved between clean papers to prevent sticking. Use plastic that has low permeability to oxygen, carbon dioxide, and water vapor in order to minimize water loss and pH changes in the albumen (Borwornpinyo et al., 2005), for example, Saran Premium food wrap.

d. Have ready enough rings (two per recipient) and rubber bands (four per recipient); see step 9. Rings can be cut from PVC tubing and embedded with four screws (Borwornpinyo *et al.*, 2005) or custom cut from plastic sheeting.

e. Packs of autoclaved tissues are required for wiping excess albumen from the surrogate shells.

f. Have ready transfer pipettes for aspirating bubbles from albumen and 10-ml syringes for topping shells up with albumen.

4. About 30 min before injection of cES cells, crack the side of the irradiated egg, lever open the shell with your thumbs, and empty the contents carefully into the pouring cup. Pour into the first surrogate shell and cover with a small plastic lid until injection. Care should be taken at all times not to break or stretch the albumen capsule that surrounds the yolk. The animal pole, where the stage X (EG&K) embryo is located, has a lower density than the rest of the yolk and will normally rotate in the shell so that the embryo lies on top. Damage to the albumen capsule will hinder this rotation, rendering the embryo difficult to see and inject, and may even cause the yolk to float free and expose the embryo to dehydration.

5. Prepare cells for injection.

a. Remove medium from the wells and wash cells once with Ca^{2+}/Mg^{2+}-free PBS.

b. Add 0.25% trypsin/0.02% EDTA to the wells, remove, and let incubate for 20 to 30 s.

c. Add manipulation medium and disperse the cells using a P1000.

d. Collect cell suspension and centrifuge cells for 2 min at 200g.

e. Resuspend cells in a small volume and count cells.

f. Dilute cell suspension to 5000 cells/μl.

6. Attach a 35°-beveled glass microinjection pipette with an inner diameter of 60 μm (Humagen) to a microcapillary pipette (Sigma-Aldrich Co.) fitted with a 0.2-μm Luer filter (Sigma-Aldrich Co.). Calibrate the pipette by drawing up a known volume of buffer and marking the meniscus with a Sharpie pen. Alternatively, use a custom microinjector (Tritech Research Inc.).

7. Draw cES cells up to the loading mark and inject the cell suspension into the subembryonic cavity of the recipient embryo, inserting the pipette centrally through the perivitelline layer and the area pellucida (Figs. 1F and 2).

8. Top up excess space in the first surrogate shell with albumen collected from the fresh eggs and aspirate bubbles trapped in the albumen with a pipette.

9. Carefully lay a square of plastic film over the window. If air is trapped under the plastic film, remove it, top up again with albumen, and cover with

a fresh square. Hold the film in place with a pair of plastic rings secured by rubber bands (Fig. 5A)

10. Place the sealed system II cultures horizontally in incubator trays.

Storage of Recipient Embryos

Wrap the trays of recipient embryos in plastic to minimize evaporation from the shells and store them for 4 days at 15°. An incubator can be placed inside a cold room or, alternatively, the trays can be placed in a temperature-controlled cabinet and rocked manually approximately three to five times per day. To prevent the yolk sticking to the inside of the shell during storage, rock the eggs through 90°, with the long axis of the eggs parallel to the angle of rotation.

Incubation of Recipient Embryos

Incubation Days 0 to 4. After 4 days of preincubation storage, unwrap the trays and move them to incubators for incubation at 37.5° and 60% relative humidity (Fig. 5B). We use Octagon 250 (Brinsea Products Ltd.) incubators. Rock the system II cultures five times per hour through 90° to prevent the yolk from sticking to the inside of the shell.

Incubation Days 4 to 19. To allow normal development to continue past day 7, the entire contents of the first shell must be transferred to a second surrogate shell between stages 14 and 17 (H&H) (Hamburger and Hamilton, 1951) (Fig. 5C). Embryos compromised by irradiation will reach these stages on day 4 of incubation. By this time, the layer of albumen above the embryo is depleted and the yolk is very fragile. If the embryo is left to develop to stage 19 or 20 (H&H), the perivitelline layer will have ruptured and the yolk will be too fragile for successful transfer.

Transfer the viable embryos to the second shell as follows.

1. Prepare the second surrogate shells: cut 38-mm windows at the blunt end of the eggs; rinse inside and out with deionized water to remove albumen and shell dust; drain with the window down on absorbent paper; and store in a cool place until needed.

2. Take each egg from the incubator and, maintaining its horizontal orientation, remove the rings and bands without dislodging the plastic film.

3. With the prepared second shell close at hand, rotate the first shell so that the window is on top. The viscosity of the albumen slows the concomitant rotation of the yolk and allows time for the plastic film to be removed. Pour the contents of the first shell smoothly into the second shell without damaging the yolk. While pouring, the second shell should be held

closely against the lower edge of the window in the first shell. The final position of the yolk should be in the center of the egg with the embryo centered on top of the yolk.

4. Add 1 ml of tissue culture grade penicillin/streptomycin.

5. Seal the window with a 55-mm square of Saran Cling Plus. Any plastic film can be used providing that it is similar to the shell in permeability to water vapor, oxygen, and carbon dioxide. Stick the plastic film to the top edge of the shell with a thin layer of albumen containing 20% tissue culture grade penicillin/streptomycin.

6. Incubate at 37.5° and 60% relative humidity with hourly rocking through a 60° angle. The system III culture has an artificial air space of approximately 30 ml (see *Preparation and Use of Surrogate Shells* and Fig. 5D) and the shallower angle prevents extraembryonic membranes from touching the plastic film as the egg rocks.

7. The critical period for egg rocking is from day 3 to day 7 (Deeming, 1989), but, for system III cultures, continue rocking until day 10 to allow for retardation caused by irradiation, storage, and other manipulations.

8. Evaluate embryos as often as necessary (we recommend daily) by removing them from the incubator (briefly) and looking through the window (Fig. 5D). Remove dead embryos and/or low-grade chimeras.

Incubation Day 19 to Hatching. Irradiating the recipient embryo delays development for 24 h so the chimeras will hatch on day 22 instead of the normal day 21. Hatching in surrogate shells requires extra attention, as the chimera does not pip into the air space and then through the shell as an embryo does in the intact egg. Neither can the chimera push off the plastic film, so close monitoring is required over the period during which respiration is transferred from the chorioallantoic membrane (CAM) to the lungs and the chimera is ready to hatch.

1. On day 19, transfer the system III cultures to a hatcher that permits easy access and viewing so that the progress of the chimera can be monitored.

2. Stand the eggs vertically in racks or on individual egg rings. Maintain a temperature of 37.5° and 60% relative humidity. This relative humidity is lower than recommended for intact eggs but is adequate, as the embryo does not have to pip through a shell membrane in order to hatch.

3. When the blood has begun to drain from the vessels in the CAM and the chimera has pipped into the air space and is breathing consistently, perforate the plastic film by making two small holes with a sterile hypodermic needle. At this time the air space has a high concentration of CO_2 and a low concentration of O_2. Ambient air will enter through the perforations and gradually equilibrate the air space with the surrounding air.

4. Replace the plastic film with a small Petri dish when the embryo has made a large hole in the CAM and begins to move its head and beak

backward vigorously. These are instinctive hatching movements and are performed by the chimera despite it being in a nonconfining surrogate shell. The embryo will push off the lid as it "hatches."

5. Once the head of the chimera is up and it is looking around, lay the shell on the floor of the hatcher to allow the chimera to push itself free. This energetic process aids the final stages of yolk sac retraction and closing of the navel sphincter.

The correct hatching position is with the head and beak up and the legs tucked down underneath the body. Some chimeras that do not orientate themselves correctly can be rescued. For example, if the CAM has drained and the chimera is breathing well but the beak is pointing downward, the CAM can be cut carefully and the head lifted to free the chimera for hatching. Also, if there is just a little residual albumen it can be removed with a sterile pipette to prevent the chimera sticking to the shell and failing to hatch.

Acquisition, Storage, and Quality Control of Eggs

Only the highest-quality eggs should be used for each step in the process of embryonic stem cell derivation and chimera production. If access to flocks of laying birds is not possible, eggs of known production date should be purchased from a reputable supplier of eggs. The important criteria are that (a) there are quality control data on the health of the flock to ensure that no vertically transmissible diseases are present, (b) eggs are free from shell defects, for example, cracks, crazing, or damaged cuticle, and (c) they are scrupulously clean. Be aware that shipping conditions in most climates and seasons can be dehydrating and are likely to result in less than ideal egg temperatures, affecting the eggs and embryos adversely. Ensure that eggs are packed carefully in insulated containers and are shipped by the fastest method. Monitoring strips and data loggers are available for recording temperature inside the containers (Cold Ice, Inc., Oakland, CA; Onset Computer Corporation, Pocasset, MA). Whatever their source, eggs must be collected as soon as possible after oviposition and handled correctly (Brake *et al.*, 1997). Before use or storage, sanitize the shell surface with a disinfectant designed specifically for hatching eggs (e.g., BioPhene, BioSentry Inc.). Once the eggs are clean and dry, they can be used immediately or stored, broad end up on clean, plastic egg racks. Eggs for chimera production will commonly be obtained in advance and temporarily stored. Store fertile eggs at 15 to 18° for no longer than 5 days before use to maintain optimum embryo viability. Storage conditions for surrogate shell eggs are less stringent, as only the quality of the shell and shell membranes is important. They can be exposed safely to 25° for up to several hours during transportation and/or preparation for use. Longer-term storage should be at 4° for a maximum of 4 weeks. Store all eggs in a moist but noncondensing environment, maintaining at least 75% relative humidity to preserve albumen quality and reduce evaporative weight loss through the porous shell. Condensation on the shells of cold eggs brought into a warmer or more humid environment can encourage microbial contamination and should be avoided.

Timeline of Chimera Production

A trained team with a well-equipped laboratory can process up to 250 recipient embryos twice weekly. Day 1 is the most labor-intensive, requiring two cES cell injectors and two technicians to seal the system II cultures. Transferring the embryos to the second shell (day 10) will require two people. One person can normally manage the hatching stage.

Day 0 or Day 1

    Match fertile recipient eggs with their corresponding surrogate shell
       eggs.
    Prepare first surrogate shells.

Day 1

    Irradiate recipient embryos.
    Transfer recipients to first shell.
    Prepare cES cells.
    Inject cES cells.
    Seal system II cultures and place in storage for 4 days.

Day 5

    Move system II cultures to incubator and incubate for 4 days.

Day 9

    Prepare second surrogate shells.
    Transfer recipients to second shell.
    Seal system III culture and incubate for 6 days.

Day 15

    Turn off rocking and incubate system III cultures with trays level
       for 9 days.

Day 24

Transfer system III cultures to hatcher.

Days 25 to 27

Assist hatching of chimeras.

Efficiency of Chimera Production Using Embryonic Stem Cells

The hatch rate of embryos derived from compromised recipients and genetically modified cES cells varies from 1 to 40% with an average of about 15%. The greatest loss is before day 3 of incubation. After day 10 the loss is minimal until hatch at which time approximately 50% of the embryos emerge successfully from the surrogate shell. The two main factors that affect the viability of manipulated embryos are the quality of the cES cells and the quality of the recipient embryos. Some cell lines or cell cultures yield fewer embryos surviving to day 14. As presented in Table I, cell line A exhibited a markedly lower survival and generated fewer high-grade chimeras compared to the other cell lines. In addition, 25% of the embryos generated with this cell line were abnormal.

The yield of chimeric embryos is optimized when recipient embryos are derived from hens older than 30 weeks and younger than 60 weeks. Eggs from younger and older hens generally result in lower survivability. The technical proficiency of the injector also influences the frequency and extent of chimerism; novice injectors tend to produce fewer high-grade chimeras.

TABLE I
FREQUENCY AND EXTENT OF CHIMERISM IN FOUR DIFFERENT CELL LINES[a]

Cell line	# embryos injected	Survival at D14 (%)	# chimeric embryos (%)[b]	# embryos with >65% feather chimerism (%)[c]
A	1835	516 (28)	423 (82)	106 (6)
B	602	258 (43)	232 (90)	171 (28)
C	814	297 (36)	224 (75)	149 (18)
D	457	232 (51)	177 (76)	111 (24)

[a] Chimerism rates were evaluated from all embryos that survived until D14 by feather pigmentation.
[b] Percentage of chimeras present at day 14.
[c] Percentage calculated as the number of high-grade (>65% feather chimerism) embryos per number of embryos injected.

Although there may be no apparent differences in morphology or obvious chromosomal differences between cell lines, their contribution to the recipient embryo can vary. When good cell lines are injected (Table I), one can expect approximately 20 to 25% of the injected embryos to be high-grade chimeras at day 14.

Using feather pigmentation as an index of chimerism, all feather follicles appear to be populated by donor-derived melanocytes in some cases (Fig. 5E). These data have been corroborated by quantitative estimates of the number of cells in brain, muscle, and liver that express green fluorescent protein (GFP) under the control of a ubiquitous promoter following injection of GFP-expressing cES cells (van de Lavoir *et al.*, 2006). To date, we have observed contributions to all somatic tissues that have been examined and have found chimeras to be very useful in the evaluation of transgene functionality. Transgenes can be evaluated rapidly, as the time from transfection of cES cells to a hatched transgenic chimeric chick is approximately 7 to 8 weeks. For example, we have determined that a transgene encoding a human monoclonal antibody under the control of a promoter derived from the ovalbumin locus was expressed in a tissue-specific and developmentally regulated fashion in 2-week-old estrogen-treated chimeric hens (Zhu *et al.*, 2005). The application of this technology in other experimental paradigms is limited only by the imagination and resources of the investigator.

Acknowledgments

The authors thank Jennifer Diamond for her contribution in developing the protocols. This work was funded by the Small Business Innovation Research Program of the National Institutes of Child Health and Human Development Grant 2R44 HD39583-02 and the Small Business Innovation Research Program of the U.S. Department of Agriculture, Grant 2003-33610-13933.

References

Borwornpinyo, S., Brake, J., Mozdziak, P., and Petitte, J. (2005). Culture of chicken embryos in surrogate eggshells. *Poult. Sci.* **84,** 1477–1482.

Brake, J., Walsh, T., Benton, C. E. J., Petitte, J., Meijerhof, R., and Penalva, G. (1997). Egg handling and storage. *Poult. Sci.* **76,** 144–151.

Carsience, R., Clark, M., Verrinder Gibbins, A., and Etches, R. (1993). Germline chimeric chickens from dispersed donor blastodermal cells and compromised recipient embryos. *Development* **117,** 669–675.

Deeming, D. (1989). Characteristics of unturned eggs: Critical period, retarded embryonic growth and poor albumen utilisation. *Br. Poult. Sci.* **30,** 239–249.

Etches, R. J., Clark, M. E., Toner, A., Liu, G., and Gibbins, A. M. (1996). Contributions to somatic and germline lineages of chicken blastodermal cells maintained in culture. *Mol. Reprod. Dev.* **45,** 291–298.

Eyal-Giladi, H., and Kochav, S. (1976). From cleavage to primitive streak formation: A complimentary normal table and a new look at the first stages of the development of the chick. *Dev. Biol.* **49**, 321–337.

Hadjantonakis, A. K., Gertsenstein, M., Ikawa, M., Okabe, M., and Nagy, A. (1998). Generating green fluorescent mice by germline transmission of green fluorescent ES cells. *Mech. Dev.* **76**, 79–90.

Hamburger, V., and Hamilton, H. (1951). A series of normal stages in the development of the chick embryo. *J. Morph.* **88**, 49–92.

Hooper, M. L. (1987). Isolation of genetic variants and fusion hybrids from embryonal carcinoma cell lines. *In* "Teratocarcinomas and Embryonic Stem Cells: A Practical Approach" (E. Robertson, ed.), pp. 51–70. IRL Press, Oxford.

Ikawa, M., Yamada, S., Nakanishi, T., and Okabe, M. (1998). "Green mice" and their potential usage in biological research. *FEBS Lett.* **430**, 83–87.

Kagami, H., Tagami, T., Matsubara, Y., Harumi, T., Hanada, H., Maruyama, K., Sakurai, M., Kuwana, T., and Naito, M. (1997). The developmental origin of primordial germ cells and the transmission of the donor-derived gametes in mixed-sex germline chimeras to the offspring in the chicken. *Mol. Reprod. Dev.* **48**, 501–510.

Lawitts, J., and Biggers, J. (1992). Joint effects of sodium chloride, glutamine, glucose in mouse preimplantation embryo culture media. *Mol. Reprod. Dev.* **31**, 189–194.

Love, J., Gribbin, C., Mather, C., and Sang, H. (1994). Transgenic birds by DNA microinjection. *Biotechnology* **12**, 60–63.

Mather, C., and Laughlin, K. (1979). Storage of hatching eggs: The interaction between parental age and early embryonic development. *Br. Poult. Sci.* **20**, 595–604.

McGrew, M. J., Sherman, A., Ellard, F. M., Lillico, S. G., Gilhooley, H. J., Kingsman, A. J., Mitrophanous, K. A., and Sang, H. M. (2004). Efficient production of germline transgenic chickens using lentiviral vectors. *EMBO Rep.* **5**, 728–733.

Miura, K., Sueyoshi, M., Jinbu, M., and Oka, M. (1991). Chick embryo culture using duck egg shell: First successful hatch. *Jikken Dobutsu* **40**, 251–254.

Niwa, H., Yamamura, K., and Miyazaki, J. (1991). Efficient selection for high-expression transfectants with a novel eukaryotic vector. *Gene* **108**, 193–199.

Pain, B., Chenevier, P., and Samarut, J. (1999). Chicken embryonic stem cells and transgenic strategies. *Cells Tissues Organs* **165**, 212–219.

Pain, B., Clark, M., Nakazawa, H., Sakurai, M., Samarut, J., and Etches, R. (1996). Long-term *in vitro* culture and characterization of avian embryonic stem cells with multiple morphogenetic potentialities. *Development* **122**, 2339–2348.

Perry, M. (1988). A complete culture system for the chick embryo. *Nature* **331**, 70–72.

Perry, M., and Mather, C. (1991). Satisfying the needs of the chick embryo in culture, with emphasis on the first week of development. *In* "Avian Incubation" (S. G. Tullet, ed.), Vol. 22, pp. 91–106. Butterworth-Heinemann, London.

Perry, M. M. (1991). *In vitro* embryo culture technique USA patent 5,011,780.

Petitte, J., and Yang, Z. (1993). Culture of ESC-like cells from the chicken blastoderm. *Poult. Sci.* **72**.

Petitte, J. N. (2004). Isolation and maintenance of avian ES cells. *In* "Handbook of Stem Cells" (R. Lanza, ed.), pp. 471–477. Academic Press, San Diego.

Petitte, J. N., Clark, M. E., Liu, G., Verrinder Gibbins, A. M., and Etches, R. J. (1990). Production of somatic and germline chimeras in the chicken by transfer of early blastodermal cells. *Development* **108**, 185–189.

Petitte, J. N., Liu, G., and Yang, Z. (2004). Avian pluripotent stem cells. *Mech. Dev.* **121**, 1159–1168.

Ratcliffe, M. J. H. (1989). Development of the avian B lymphocyte lineage. *Poult. Biol.* **2**, 207–234.

Robertson, E. (1987). Embryo-derived stem cell lines. *In* "Teratocarcinomas and Embryonic Stem Cells: A Practical Approach" (E. Robertson, ed.), pp. 71–112. IRL Press, Oxford.

Stern, C. D. (2005). The chick: A great model system becomes even greater. *Dev. Cell* **8,** 9–17.

van de Lavoir, M.-C., Mather-Love, C., Leighton, P. A., Diamond, J. H., Heyer, B. S., Roberts, R., Zhu, L., Winters-Digiacinto, P., Kerchner, A., Gessaro, T., Swanberg, S. E., Delany, M. E., and Etches, R. J. (2006). High grade transgenic somatic chimeras from chicken embryonic stem cells. *Mech. Dev.* **123,** 31–41.

Watt, J. M., Petitte, J. N., and Etches, R. J. (1993). Early development of the chick embryo. *J. Morphol.* **215,** 165–182.

Yang, Z., and Petitte, J. N. (1994). Use of avian cytokines in mammalian embryonic stem cell culture. *Poult. Sci.* **73,** 965–974.

Zhu, L., van de Lavoir, M.-C., Albanese, J., Beenhouwer, D. O., Cardarelli, P. M., Cuison, S., Deng, D. F., Deshpande, D., Diamond, J. H., Green, L., Halk, E. L., Heyer, B. S., Hudson, D., Kay, R. M., Kerchner, A., Leighton, P. A., Mather, C. M., Morrison, S. L., Nikolov, Z. L., Passmore, D. B., Pradas-Monné, A., Preston, B. T., Rangan, V. S., Sharkov, N., Shi, M., Srinivasan, M., White, S. G., Winters-Digiacinto, P., Wong, S., Zhou, W., and Etches, R. (2005). Production of human monoclonal antibody in eggs of chimeric chickens. *Nature Biotechnol.* **23,** 1159–1169.

[4]   Zebrafish Embryonic Stem Cells

By LIANCHUN FAN and PAUL COLLODI

Abstract

Methods are presented for the derivation of zebrafish embryonic stem (ES) cell cultures that are initiated from blastula and gastrula stage embryos. To maintain pluripotency, the ES cells are cocultured with rainbow trout spleen cells from the RTS34st cell line. ES cells maintained for multiple passages on a feeder layer of growth-arrested RTS34st exhibit *in vitro* characteristics of pluripotency and produce viable germ cells following transplantation into a host embryo. The ES cells are able to undergo targeted plasmid insertion by homologous recombination, and methods are described for the introduction of a targeting vector by electroporation. Two strategies are described for the efficient isolation of homologous recombinants using a visual marker screen and positive-negative selection.

Introduction

The zebrafish possesses characteristics that make it an ideal model for studies of embryo development and human disease, including a short generation time, external fertilization, and optically clear embryos that complete development in approximately 96 h (Fishman, 2001; Nusslein-Volhard, 1994; Stern and Zon, 2003). For genetic studies, methods are

METHODS IN ENZYMOLOGY, VOL. 418   0076-6879/06 $35.00
                                        DOI: 10.1016/S0076-6879(06)18004-0

established for conducting large-scale forward mutagenesis screens using chemical (Mullins *et al.*, 1994; Pelegri and Mullins, 2004), radiation (Walker, 1999), viral (Amsterdam *et al.*, 1999, 2004; Golling *et al.*, 2002), or transposon (Ivics *et al.*, 1999; Kawakami, 2005) induced mutations. The extensive linkage map that is available, together with the nearly complete zebrafish genomic sequence (Geisler *et al.*, 1999; Hukriede *et al.*, 2001; Woods *et al.*, 2000, 2005), makes it possible to isolate some of the altered genes by candidate and positional cloning methods (Coutinho *et al.*, 2004; Makino *et al.*, 2005; Miller *et al.*, 2004; Schulte-Merker *et al.*, 1994; Yuan and Joseph, 2004; Zhang *et al.*, 1998). Genes that are disrupted by retroviral or transposon insertion are tagged by the inserted sequence, thereby facilitating the cloning process (Golling *et al.*, 2002). Reverse genetic methods have also been used to study zebrafish gene function. Transient inhibition of gene expression is routinely accomplished in zebrafish using morpholino-modified oligonucleotides (Ekker and Larson, 2001; Nasevicius and Ekker, 2000), and a target-selected mutagenesis approach is used to screen for randomly introduced mutations that occur at a specific locus (Berghmans *et al.*, 2005; Wienholds and Plasterk, 2004; Wienholds *et al.*, 2002).

In order to complement these genetic approaches and further enhance the utility of the zebrafish model for studies of gene function, it would be valuable to establish gene-targeting methods for the production of knock-out and knock-in lines of fish. The ability to permanently disrupt zebrafish gene expression by targeted insertional mutagenesis would make it possible to functionally characterize genes expressed late in development and in adult fish, thereby complementing the more transient antisense-based gene knockdown strategies. Gene-targeting methods in mice have been available for nearly two decades and involve the introduction of vector DNA into a specific locus in mouse embryonic stem (ES) cell cultures by homologous recombination (Capecchi, 1989; Doetschman *et al.*, 1987). Colonies of ES cells that harbor the targeted insertion are selected and expanded in culture and the cells are introduced into a host embryo where they participate in development and contribute to the germ cell lineage. The resulting germ line chimera carrying the targeted insertion is used to establish the knockout/knock-in line. Although successful in mice, the ES cell-based gene-targeting strategy has not been successfully applied to other species. The lack of success with nonmurine species is due to the absence of ES cell lines that are capable of contributing to the germ cell lineage of a host embryo.

Our laboratory has been working to develop zebrafish ES cell lines that are suitable for use in gene-targeting experiments (Fan *et al.*, 2004a,b; Ma *et al.*, 2001; Sun *et al.*, 1995). Methods have been developed to derive zebrafish ES cell lines that remain germ line competent for multiple passages in culture, providing a sufficient length of time to introduce a plasmid and to select colonies of cells that have incorporated the vector in a targeted fashion by homologous recombination (Fan et al., 2004a, 2006). Germ line chimeras have been produced by introducing embryonic stem cells that were maintained in culture for up to six passages into host embryos (Fan et al., 2004a). A requirement for maintaining germ line competency is to culture the embryonic stem cells on a feeder layer of growth-arrested rainbow trout spleen cells from the RTS34st cell line (Ganassin and Bols, 1999). The spleen cells produce factors that maintain the zebrafish cultures in a germ line-competent condition (Fan et al., 2004a; Ma et al., 2001). Due to their pluripotent and germ line-competent characteristics, zebrafish ES cell cultures have the potential to form the basis of a gene-targeting strategy. This chapter describes methods for the derivation, maintenance, and genetic manipulation of zebrafish pluripotent ES cell cultures.

Materials

Reagents

1. The following cell culture media are available from GIBCO-BRL (Grand Island, NY): Leibowitz's L-15, Ham's F12, and Dulbecco's modified Eagle's media. One liter of each medium is prepared separately by dissolving the powder in ddH$_2$O and adding HEPES buffer (final concentration 15 mM, pH 7.2), penicillin G (120 μg/ml), ampicillin (25 μg/ml), and streptomycin sulfate (200 μg/ml). LDF medium is prepared by combining Leibowitz's L-15, Dulbecco's modified Eagle's, and Ham's F12 media (50:35:15) and supplementing with sodium bicarbonate (0.180 g/liter) and sodium selenite ($10^{-8}\,M$). The medium is filter sterilized before use.

2. Phosphate-buffered saline (PBS) (GIBCO-BRL)

3. TE buffer: 10 mM Tris-HCl, 1 mM EDTA, pH 8.0

4. Fetal bovine serum (FBS) (Harlan Laboratories, Indianapolis, IN)

5. Calf serum (GIBCO-BRL)

6. Trout plasma (SeaGrow, East Coast Biologics, Inc., North Berwick, ME) is sterile filtered, heat treated (56°, 25 min), and centrifuged (10,000g, 10 min) before use.

7. Trypsin/EDTA solution (2 mg/ml trypsin, 1 mM EDTA) is prepared in PBS. The solution is filter sterilized before use. Trypsin and EDTA are available from Sigma (St. Louis, MO).

8. Human epidermal growth factor (EGF, Invitrogen, Carlsbad, CA). The stock EGF solution is prepared at 10 μg/ml in ddH$_2$O.

9. Human basic fibroblast growth factor (bFGF, Invitrogen). The stock bFGF solution is prepared at 10 μg/ml in 10 mM Tris-HCl, pH 7.6.

10. Bovine insulin (Sigma). Stock insulin is prepared at 1 mg/ml in 20 mM HCl.

11. Bleach (Clorox) solution is prepared fresh at 0.5% in ddH$_2$O from a newly opened bottle.

12. Zebrafish embryo extract is prepared by homogenizing approximately 500 embryos in 0.5 ml of LDF medium and centrifuging (20,000g, 10 min) to remove debris. The supernatant is collected and filter sterilized, and the protein is measured. The extract is diluted to 10 mg protein/ml and stored frozen ($-20°$) in 0.2-ml aliquots.

13. Geneticin (G418 sulfate, GIBCO-BRL). The G418 stock solution is prepared at 100 mg/ml in ddH$_2$O and filter sterilized before use.

14. Pronase (Sigma) is prepared at 0.5 mg/ml in Hank's solution.

15. Egg water: 60 μg/ml aquarium salt.

16. Freezing medium: 80% FD medium (1:1 mixture of Ham's F12 and DMEM), 10% FBS, 10% dimethyl sulfoxide.

Feeder Cell Lines

Growth-Arrested Feeder Cells RTS34st cells (Ganassin and Bols, 1999) are cultured (18°) in Leibowitz's L-15 medium (Sigma) supplemented with 30% calf serum. To prepare growth-arrested cells, a confluent culture of RTS34st cells contained in a flask (25 cm^2) or dish (100 mm diameter) is irradiated (3000 rad), harvested by trypsinization, and frozen in liquid nitrogen within 24 h after irradiation. To recover frozen growth-arrested cells, the vial is thawed briefly in a water bath (37°) and the cells are collected by centrifugation and resuspended in L-15 medium. Cells from one frozen vial are distributed into two 25-cm^2 flasks or four wells of a six-well plate. After the cells have attached to the culture surface, the medium is supplemented with calf serum (30%). After 24 h the growth-arrested cells should be spread on the culture surface and used immediately as feeder layers. Before using the growth-arrested cells as a feeder layer, the L-15 medium is removed and the cells are rinsed one time.

Drug-Resistant Feeder Cells A G418-resistant feeder cell line is prepared by transfecting RTS34st with the pBKRSV plasmid, which contains the aminoglycoside phosphotransferase gene (*neo*) under the control of RSV promoter. Colonies of cells that stably express *neo* are selected in G418 (500 μg/ml). The *neo*-resistant cell line, RTS34st(*neo*), is cultured in L15 medium supplemented with 30% calf serum plus 200 μg/ml G418. Growth-arrested RTS34st(*neo*) cells are prepared using the same methods described for RTS34st.

RTS34st Cell-Conditioned Medium Conditioned medium is prepared by adding fresh L-15 plus 30% FBS to a confluent culture of RTS34st cells and incubating for 3 days (18°). The medium is removed, filter sterilized, and stored frozen (−20°).

Derivation of Zebrafish ES Cell Cultures

Overview

Zebrafish germ line chimeras have been generated by injecting host embryos with pluripotent ES cell cultures derived from either blastula- or gastrula-stage embryos (Fan *et al.*, 2004a; Ma *et al.*, 2001). Because the blastula consists entirely of nondifferentiated cells, it is the optimal stage to use for initiating the ES cell cultures (Fig. 1). To maintain pluripotency and germ line competency, the ES cells are cultured on a feeder layer of growth-arrested rainbow trout spleen cells derived from the established RTS34st cell line (Gannassin and Bols, 1999; Ma *et al.*, 2001). ES cells cocultured in the presence of the RTS34st feeder layer exhibit several *in vitro* characteristics associated with pluripotency, including alkaline phosphatase activity, recognition by the SSEA-1 antibody, and the capacity to form differentiated embryoid bodies in suspension culture (Fig. 2). The zebrafish ES cells also maintain the capacity to contribute to the germ cell lineage of a recipient embryo when they are grown on a feeder layer for at least six passages (6 weeks) in culture, a sufficient period of time *in vitro* for electroporation and selection of homologous recombinants (Fan *et al.*, 2004a, 2006). Polymerase chain reaction (PCR) analysis of DNA isolated from different tissues of adult germ line chimeric fish has demonstrated that, in addition to the germ cell lineage, cultured ES cells contribute to multiple tissues of the recipient embryo, including muscle, head, fin gut, and liver (Fan and Collodi, 2002; Fan *et al.*, 2004a).

Primary cell cultures that are initiated from zebrafish blastulas and maintained on a feeder layer of growth-arrested RTS34st consist of multiple dense aggregates of embryo cells, with each aggregate possessing a homogeneous appearance and lacking any morphological indication of differentiation. The cell aggregates form within 24 h after the culture is initiated. As the cells proliferate, the aggregates continue to increase in size without losing their homogeneous, dense appearance (Fig. 1A). The primary culture must be passaged on days 4 to 6 to prevent the cell aggregates from becoming too large and differentiating. During the first passage, the aggregates are partially dissociated and added to a fresh feeder layer of growth-arrested RTS34st cells. The aggregates become easier to dissociate

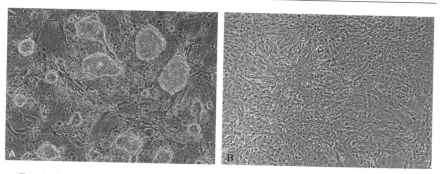

Fig. 1. (A) Primary culture of zebrafish embryo cells showing a dense homogeneous cell aggregate attached to a feeder layer of growth-arrested RTS34st rainbow trout spleen cells. (B) After approximately four passages, zebrafish ES cells begin to proliferate as a monolayer. A late passage monolayer culture growing on growth-arrested RTS34st feeder cells is shown.

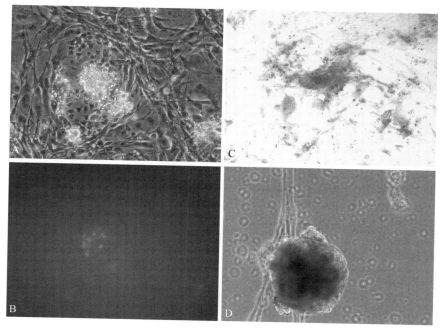

Fig. 2. Zebrafish ES cell cultures exhibit *in vitro* characteristics of pluripotency, including (A and B) expression of the SSEA-1 antigen, (C) alkaline phosphatase activity, and (D) formation of differentiated embryoid bodies in suspension culture. The same culture is shown in A and B before and after immunostaining using anti-SSEA-1 antiserum. (See color insert.)

with each passage and eventually grow to form a monolayer by the fourth passage (Fig. 1B).

Pluripotent ES cell cultures can also be initiated from zebrafish embryos that are at the germ ring stage of development (approximately 6 h postfertilization; Ma *et al.*, 2001). As with the blastula-derived cultures, cell aggregates will also form in primary cultures initiated from germ ring stage embryos; however, because cell differentiation has begun to occur during gastrulation, the majority of cell aggregates in the cultures from the later-stage embryos will not appear homogeneous and will contain recognizable differentiated cell types, including pigmented melanocytes, neural cells, and fibroblasts. Even though cell differentiation is pervasive in the gastrula-derived primary culture, pluripotent ES cells can be obtained by manually selecting the small number of cell aggregates that possess an ES-like morphology characterized by a compact and homogeneous appearance. The ES-like cell aggregates are dissociated and passaged to initiate the long-term culture.

Protocol for Initiation of ES Cell Culture from Zebrafish Blastulas

1. Collect zebrafish embryos, rinse several times with water to remove debris, and transfer the embryos to a Petri dish containing egg water (28°). Allow the embryos to develop to the blastula stage (approximately 4 h postfertilization) and then divide them into groups of approximately 50 individuals. Transfer each group of embryos into a 2.5-ml Eppendorf microfuge tube that has the bottom cut off and replaced with fine mesh netting attached with a rubber band. Submerge the bottom of the tube containing embryos sitting on the net into 70% ethanol for 10 s and then rinse immediately by submerging the embryos into a beaker of sterile egg water. Transfer the embryos from the tube into a 60-mm Petri dish containing egg water and remove any dead individuals. Remove the egg water and rinse the embryos three times with LDF medium (Collodi *et al.*, 1992). Remove the LDF, add approximately 2 ml of bleach solution to the dish, and let the embryos sit for 2 min. Remove the bleach and rinse immediately with LDF. Repeat the bleach treatment and rinse two additional times. It is important not to expose the embryos to the bleach solution for periods longer than 2 min without rinsing. Following the final bleach treatment, rinse the embryos three additional times with LDF medium.

2. Remove the chorions by incubating each group of embryos in 3 ml of pronase solution for approximately 15 min or until the chorions begin to break apart. Gently swirl the suspended embryos in the dish to release them from the digested chorion. Use a pipettor to remove the pronase solution along with the floating chorions and gently rinse the dechorionated embryos

with LDF medium, taking care to always keep the embryos suspended in medium. Transfer the embryos from one dish to a 15-ml polypropylene centrifuge tube (Corning), add 3 ml of trypsin/EDTA solution to the tube, and incubate for 1 to 2 min with occasional gentle pipetting to dissociate the cells. Add approximately 200 μl of FBS to the tube to stop the action of the trypsin and collect the cells by centrifugation ($500g$; 5 min). Resuspend the cell pellet obtained from each group of approximately 50 embryos in 1.8 ml LDF medium and transfer the cell suspension to a single well of a six-well tissue culture plate (Falcon) containing a confluent monolayer of growth-arrested RTS34st cells. Let the plate sit undisturbed for 30 min to allow the embryo cells to attach to the RTS34st monolayer. After the cells have attached, add the following factors to each well: 150 μl of FBS, 15 μl of zebrafish embryo extract, 30 μl of trout plasma, 30 μl of insulin stock solution, 15 μl of EGF stock solution, 15 μl of bFGF stock solution, and 945 μl of RTS34st-conditioned medium. If the cells are not attached after 30 min the plate can be incubated for 1 to 2 h before adding the factors. EGF can be added immediately after plating the cells to enhance cell attachment. Incubate the culture for 5 days (22°). During this time the cell aggregates should increase in size as the cells proliferate. As the cell aggregates become larger they should continue to possess a homogeneous appearance and be composed of tightly adherent cells. Although zebrafish cell cultures are normally propagated at 26° (Collodi et al., 1992), the embryo cell cultures are maintained at 22° to accommodate the feeder layer of trout spleen cells.

3. To passage the primary culture, harvest the cells from each well by adding 2 ml of trypsin/EDTA solution per well and incubating 30 s before transferring the cell suspension to a 15-ml polypropylene centrifuge tube (Corning). Pipette the cell suspension up and down several times to partially dissociate the cell aggregates and add 0.2 ml of FBS to stop the action of the trypsin. The cell aggregates cannot be completely dissociated during the first passage. Collect the cells by centrifugation ($500g$; 5 min) and resuspend the pellet in 3.6 ml of LDF medium. Add 1.8 ml of the cell suspension to each of two wells of a six-well plate (Falcon) containing a confluent monolayer of growth-arrested RTS34st cells and add the factors listed in step 2 to each well. Incubate the six-well plate for 5 days (22°) and harvest the cells with trypsin/EDTA as described earlier. Combine the cells harvested from two wells, collect the cells by centrifugation, and resuspend the cell pellet in 3.6 ml of LDF medium. The suspension will still contain a large number of cell aggregates.

4. Add the cell suspension to a 25-cm^2 tissue culture flask (Falcon) containing a confluent monolayer of growth-arrested RTS34st cells. Let the flask sit undisturbed for 1 to 3 h to allow the cells to attach to the feeder layer.

5. Add the following factors to the flask: 300 μl of FBS, 30 μl of zebrafish embryo extract, 60 μl of trout serum, 60 μl of bovine insulin stock solution, 30 μl of EGF stock solution, 30 μl of bFGF stock solution, and 1.890 ml of RTS34st-conditioned medium.

6. Incubate the flask for 7 days (22°), harvest the cells in trypsin/EDTA as described earlier, and seed them into two flasks that each contains a confluent monolayer of growth-arrested RTS34st cells. With each passage the cell aggregates become easier to dissociate and fewer aggregates are present in the culture. Continue to passage the culture approximately every 7 days.

7. The cultures can be cryopreserved as soon as they begin to grow as a monolayer (about passage 4) and a portion of the culture can be frozen at each passage. Harvest the cultures in trypsin/EDTA and resuspend the cell pellet obtained from one 25-cm^2 flask in 1 ml freezing medium. Transfer the cell suspension to a cryovial (Nalgene), place the vial in Styrofoam insulation, and incubate the vial at 4° for 10 min followed by −80° for at least 1 h and then submerge and store the vial in liquid nitrogen.

Protocol for Initiation of ES Cell Cultures from Zebrafish Gastrula Stage Embryos

1. Initiate primary cell cultures from embryos at the germ ring stage of development (approximately 6 h postfertilization) (Kimmel *et al.*, 1995) using the methods described earlier. After the primary culture has been growing for approximately 5 days, use a drawn-out Pasteur pipette or a micropipettor (Rainin) to remove aggregates of densely packed cells that appear homogeneous without morphological indications of differentiation. Combine 30 to 50 of the isolated cell aggregates in a sterile 2-ml centrifuge tube containing LDF medium.

2. Collect the cell aggregates by centrifugation (500g, 5 min), resuspend the pellet in 1.0 ml of trypsin/EDTA solution, and incubate 2 min while occasionally pipetting the cell suspension through a 5-ml pipette to partially dissociate the aggregates. Add 0.1 ml of FBS to stop the action of the trypsin and collect the cells by centrifugation (500g, 5 min).

3. Resuspend the cell pellet in 1.8 ml of LDF medium and add to a single well of a six-well plate containing a monolayer of growth-arrested RTS34st feeder cells.

4. Let the plate sit undisturbed for 5 h to allow the embryo cells to attach and add the factors listed in step 2 under the protocol for initiation of ES cell culture from zebrafish blastulas. The culture consists of small cell aggregates and some single embryo cells attached to the RTS34st cells.

5. Incubate the plate (22°) for 7 days. As the cells proliferate the cell aggregates should become larger while continuing to exhibit a homogeneous appearance and be composed of tightly adherent cells. The culture is passaged every 7 days as described earlier. The cell aggregates will become easier to dissociate and eventually grow as a monolayer after approximately four passages.

Homologous Recombination in ES cells and Selection of Colonies Carrying a Targeted Plasmid Insertion

Overview

A key requirement for the development of an ES cell-based gene-targeting approach in zebrafish is that the cells are able to incorporate vector DNA in a targeted fashion by homologous recombination and that methods are available to identify and select cell colonies that carry the targeted insertion (Capecchi, 1989). We have demonstrated that zebrafish ES cell cultures are able to incorporate a plasmid by homologous recombination by targeting the inactivation of two genes, *no tail* (*ntl*) and *myostatin 1* (*mstn 1*) (Fan *et al.*, 2006). Two different selection methods were used to isolate colonies of cells that had undergone the targeting event. Both of the selection methods are based on the positive-negative selection (PNS) strategy used commonly with mouse ES cells (Capecchi, 1989). The first screening method involves the use of a targeting vector that contains the bacterial selectable marker gene *neo* located within a region of the vector that is homologous to the gene being targeted and the red fluorescent protein gene (*RFP*) located outside of the homologous region (Fig. 3A). The targeting vector is introduced into the ES cell cultures by electroporation, and cells that have incorporated the plasmid are selected in G418. The G418-resistant colonies that incorporated the vector randomly are RFP^+ and those that had undergone homologous recombination and targeted insertion of the plasmid are RFP^- (Fig. 3B). The targeted colonies are identified by examining the cultures by fluorescence microscopy approximately 5 weeks after initiating G418 selection. Each colony that completely lacks *RFP* expression is selected manually from the dish using a micropipettor and is transferred to a single well of a 24-well plate. The second method to isolate homologous recombinants involves use of the diphtheria toxin A-chain gene (*dt*) to select against cells that carry random insertions (McCarrick *et al.*, 1993). The *dt* gene is located outside of the homologous region on the vector so cells that incorporate the plasmid in a targeted fashion by homologous recombination lose *dt* and survive whereas cells that undergo random insertion carry *dt*, making

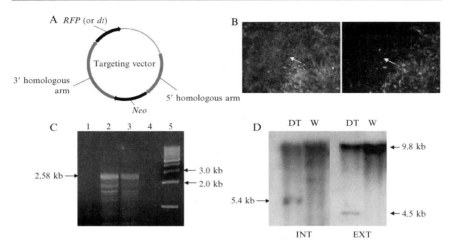

FIG. 3. Zebrafish ES cells incorporate plasmid DNA in a targeted fashion by homologous recombination. (A) A targeting vector contains neo flanked by arms that are homologous to the targeted gene and either RFP or dt located outside of the homologous region. (B) When RFP is used, following G418 selection, the potential homologous recombinants are identified by the loss of RFP expression (arrow). The targeting vector was introduced into ES cells that constitutively express the green fluorescent protein and the same two G418-resistant colonies are shown by fluorescence microscopy using a green (left) and red (right) filter. The RFP-negative colony (arrow) was removed from the dish, expanded, and confirmed to have undergone homologous recombination by PCR and Southern blot analysis. (C) Following electroporation of a targeting vector that contained dt located outside of the homologous arms, three surviving colonies were isolated and examined by PCR (lanes 1–3) for the presence of a 2.58-kb junction fragment created by targeted insertion of the plasmid. Two of the colonies were found to be homologous recombinants. (D) Southern blot analysis of one dt-resistant colony using probes that hybridized to sequences of the targeted gene that were either internal (INT) or external (EXT) to the homologous arms on the vector. Both probes hybridized to a 9.8-kb fragment corresponding to the nontargeted allele. Each probe also hybridized to a smaller restriction fragment (5.4-kb INT or 4.5-kb EXT) that corresponded to the targeted allele (Fan *et al.* [2006], B, C, and D reprinted with permission). (See color insert.)

them nonviable (Fig. 3A, C, and D). Both selection methods were used to successfully isolate zebrafish ES cell colonies carrying targeted insertions in *ntl* or *mstn 1* (Fan *et al.*, 2006).

Protocol for ES Cell Electroporation and Selection of Homologous Recombinants

1. Once the ES cells begin to grow as a monolayer (approximately passage 4) the cells can be efficiently transformed with plasmid DNA by electroporation and colonies of stable transformants isolated by drug

selection. To prepare the ES cells for electroporation, harvest the cells by trypsinization, wash two times with PBS, and suspend them at a density of 6×10^6 cells in 0.75 ml of PBS in a 0.4-cm electroporation cuvette.

2. Add 50 μg of sterile, linearized, plasmid DNA dissolved in 50 μl TE buffer. In addition to the gene of interest, the plasmid should contain a selectable marker gene such as *neo* under the control of a constitutively expressed promoter.

3. Electroporate the cells (950 μF, 300 V) and then suspend them in 20 ml of LDF medium. Add the cell suspension to two 100-mm diameter culture dishes that each contain a confluent layer of growth-arrested RTS34st(*neo*) cells. After allowing the cells to attach, add the medium and supplements described in step 2 under the protocol for initiation of ES cell culture from zebrafish blastulas. Two days later add 5 μl/ml of the G418 stock solution and change the medium every 5 days, adding fresh G418. Colonies will begin to appear 2 to 3 weeks after G418 selection is initiated.

4. To perform gene targeting by homologous recombination, the cells are electroporated with a targeting vector containing *neo* flanked by 5' and 3' arms that are homologous to the targeted gene along with *RFP* located outside of the homologous region (Fig. 3A). After electroporation the cells are selected in G418 as described in step 3 and the resulting colonies are examined by fluorescence microscopy. Potential homologous recombinants are identified by the absence of RFP expression (Fig. 3B). The RFP negative colonies are removed manually from the plate using a Pipetman micropipettor (Rainin) approximately 5 weeks after the start of G418 selection. The individual selected colonies are transferred to single wells of a 24-well plate containing growth-arrested RTS34st(*neo*) feeder cells. The individual colonies are cultured for 2 to 3 weeks before passaging into single wells of a 12-well plate. During passage, a portion of the cells from each colony are harvested and DNA isolated for PCR analysis to confirm that homologous recombination had occurred. A PNS strategy can be used by inserting *dt* into the targeting vector outside of the homologous arms (Fig. 3A). Following G418 selection, each of the surviving dt-resistant colonies represent potential homologous recombinants. Each colony is expanded and examined by PCR and Southern blot analysis for targeted plasmid insertion (Fig. 3C and D).

Acknowledgments

This work was supported by grants from NIH (R01 GM69384), USDA (2005-35206-15261), and Illinois-Indiana SeaGrant (02-340).

References

Amsterdam, A., Burgess, S., Golling, G., Chen, W., Sun, Z., Townsend, K., Farrington, S., Haldi, M., and Hopkins, N. (1999). A large-scale insertional mutagenesis screen in zebrafish. *Genes Dev.* **13,** 2713–2724.

Amsterdam, A., Sadler, K. C., Lai, K., Farrington, S., Bronson, R. T., Lees, J. A., and Hopkins, N. (2004). Many ribosomal protein genes are cancer genes in zebrafish. *PLoS Biol.* **2,** 690–698.

Berghmans, S., Murphey, R. D., Wienholds, E., Neuberg, D., Kutok, J. L., Fletcher, C. D., Morris, J. P., Liu, T. X., Schulte-Merker, S., Kanki, J. P., Plasterk, R., Zon, L. I., and Look, A. T. (2005). *Tp53* mutant zebrafish develop malignant peripheral nerve sheath tumors. *Proc. Natl. Acad. Sci. USA* **102,** 407–412.

Capecchi, M. (1989). Altering the genome by homologous recombination. *Science* **244,** 1288–1292.

Collodi, P., Kamei, Y., Ernst, T., Miranda, C., Buhler, D. R., and Barnes, D. W. (1992). Culture of cells from zebrafish (*Brachydanio rerio*) embryo and adult tissues. *Cell Biol. Toxicol.* **8,** 43–61.

Coutinho, P., Parsons, M. J., Thomas, K. A., Hirst, E. M., Saude, L., Campos, I., Williams, P. H., and Stemple, D. L. (2004). Differential requirements for COPI transport during vertebrate early development. *Dev. Cell* **7,** 547–558.

Doetschman, T., Gregg, R. G., Maeda, N., Hooper, M. L., Melton, D. W., Thompson, S., and Smithies, O. (1987). Targeted correction of a mutant HPRT gene in mouse embryonic stem cells. *Nature* **330,** 576–578.

Ekker, S., and Larson, J. (2001). Morphant technology in model developmental systems. *Genesis* **30,** 89–93.

Fan, L., and Collodi, P. (2002). Progress towards cell-mediated gene transfer in zebrafish. *Brief. Funct. Genom. Proteom.* **1,** 131–138.

Fan, L., Crodian, J., Alestrom, A., Alestrom, P., and Collodi, P. (2004a). Zebrafish embryo cells remain pluripotent and germ-line competent for multiple passages in culture. *Zebrafish* **1,** 21–26.

Fan, L., Alestrom, A., Alestrom, P., and Collodi, P. (2004b). Development of cell cultures with competency for contributing to the zebrafish germ line. *Crit. Rev. Euk. Gene Exp.* **14,** 43–51.

Fan, L., Moon, J., Crodian, J., and Collodi, P. (2006). Homologous recombination in zebrafish ES cells. *Transgen. Res.* **15,** 21–30.

Fishman, M. C. (2001). Zebrafish: The canonical vertebrate. *Science* **294,** 1290–1291.

Ganassin, R., and Bols, N. C. (1999). A stromal cell line from rainbow trout spleen, RTS34st, that supports the growth of rainbow trout macrophages and produces conditioned medium with mitogenic effects on leukocytes. *In Vitro Cell Dev. Biol. Anim.* **35,** 80–86.

Geisler, R., Rauch, G. J., Baier, H., van Bebber, F., Brobeta, L., Dekens, M. P., Finger, K., Fricke, C., Gates, M. A., Geiger, H., Geiger-Rudolph, S., Gilmour, D., Glaser, S., Gnugge, L., Habeck, H., Hingst, K., Holley, S., Keenan, J., Kirn, A., Knaut, H., Lashkari, D., Maderspacher, F., Martyn, U., Neuhauss, S., Haffter, P., *et al.* (1999). A radiation hybrid map of the zebrafish genome. *Nature Genet.* **23,** 86–89.

Golling, G., Amsterdam, A., Zhaoxia, S., Antonelli, M, Maldonado, E., Chen, W., Burgess, S., Haldi, M., Artzt, K., Farrington, S., Lin, S-Y., Nissen, R. M., and Hopkins, N. (2002). Insertional mutagenesis in zebrafish rapidly identifies genes essential for early vertebrate development. *Nature Genet.* **31,** 135–140.

Hukriede, N., Fisher, D., Epstein, J., Joly, L., Tellis, P., Zhou, Y., Barbazuk, B., Cox, K., Fenton-Noriega, L., Hersey, C., Miles, J., Sheng, X., Song, A., Waterman, R., Johnson, S. L., Dawid, I. B., Chevrette, M., Zon, L. I., McPherson, J., and Ekker, M. (2001). The LN54 radiation hybrid map of zebrafish expressed sequences. *Genome Res.* **11,** 2127–2132.

Ivics, Z., Izsvak, Z., and Hackett, P. B. (1999). Genetic applications of transposons and other repetitive elements in zebrafish. *Methods Cell Biol.* **60,** 99–131.

Kawakami, K. (2005). Transposon tools and methods in zebrafish. *Dev. Dyn.* **234,** 244–254.

Kimmel, C. B., Ballard, W. W., Kimmel, S. R., Ullmann, B., and Schilling, T. F. (1995). Stages of embryonic development of the zebrafish. *Dev. Dyn.* **203,** 253–310.

Ma, C., Fan, L., Ganassin, R., Bols, N., and Collodi, P. (2001). Production of zebrafish germ-line chimeras from embryo cell cultures. *Proc. Natl. Acad. Sci. USA* **98,** 2461–2466.

Makino, S., Whitehead, G. G., Lien, C. L., Kim, S., Jhawar, P., Kono, A., Kawata, Y., and Keating, M. T. (2005). Heat-shock protein 60 is required for blastema formation and maintenance during regeneration. *Proc. Natl. Acad. Sci. USA* **102,** 14599–14604.

McCarrick, J. W., Parnes, J. R., Seong, R. H., Solter, D., and Knowles, B. B. (1993). Positive-negative selection gene targeting with the diphtheria toxin A-chain gene in mouse embryonic stem cells. *Transgen. Res.* **2,** 183–190.

Miller, C. T., Maves, L., and Kimmel, C. B. (2004). *Moz* regulates *Hox* expression and pharyngeal segmental identity in zebrafish. *Development* **131,** 2443–2461.

Mullins, M. C., Hammerschmidt, M., Hafter, P., and Nusslein-Volhard, C. (1994). Large-scale mutagenesis in the zebrafish: In search of genes controlling development in a vertebrate. *Curr. Biol.* **4,** 189–202.

Nasevicius, A., and Ekker, S. C. (2000). Effective targeted gene "knockdown" in zebrafish. *Nature Genet.* **26,** 216–220.

Nusslein-Volhard, C. (1994). Of flies and fishes. *Science* **266,** 572–574.

Pelegri, F., and Mullins, M. C. (2004). Genetic screens for maternal-effect mutations. *Methods Cell Biol.* **77,** 21–51.

Schulte-Merker, S., van Eeden, F. J. M., Halpern, M. E., Kimmer, C. B., and Nusslein-Volhard, C. (1994). *no tail (ntl)* is the zebrafish homologue of the mouse T (*Brachyury*) gene. *Development* **120,** 1009–1015.

Stern, H. M., and Zon, L. I. (2003). Cancer genetics and drug discovery in the zebrafish. *Nature Rev. Cancer* **3,** 533–539.

Sun, L., Bradford, S., Ghosh, C., Collodi, P., and Barnes, D. W. (1995). ES-like cell cultures derived from early zebrafish embryos. *Mol. Mar. Biol. Biotech.* **4,** 193–199.

Walker, C. (1999). Haploid screens and gamma-ray mutagenesis. *Methods Cell Biol.* **60,** 43–70.

Wienholds, E., Schulte-Merker, S., Walderich, B., and Plasterk, R. H. A. (2002). Target-selected inactivation of the zebrafish *rag1* gene. *Science* **297,** 99–102.

Wienholds, E., and Plasterk, R. H. (2004). Target-selected gene inactivation in zebrafish. *Methods Cell Biol.* **77,** 69–90.

Woods, I. G., Kelly, P. D., Chu, F., Ngo-Hazelett, P., Yan, Y. L., Huang, H., Postlethwait, J. H., and Talbot, W. S. (2000). A comparative map of the zebrafish genome. *Genome Res.* **10,** 1903–1914.

Woods, I. G., Wilson, C., Friedlander, B., Chang, P., Reyes, D. K., Nix, R., Kelly, P. D., Chu, F., Postlethwait, J. H., and Talbot, W. S. (2005). The zebrafish gene map defines ancestral vertebrate chromosomes. *Genome Res.* **15,** 1307–1314.

Yuan, S., and Joseph, E. M. (2004). The small heart mutation reveals novel roles of Na+/K+-ATPase in maintaining ventricular cardiomyocyte morphology and viability in zebrafish. *Circ. Res.* **95,** 595–603.

Zhang, J., Talbot, W. S., and Schier, A. F. (1998). Positional cloning identifies zebrafish *one-eyed pinhead* as a permissive EGF-related ligand required during gastrulation. *Cell* **92,** 241–251.

[5]   Human Embryonic Stem Cells

By Hidenori Akutsu, Chad A. Cowan, and Douglas Melton

Abstract

Human embryonic stem cells hold great promise in furthering our treatment of disease and increasing our understanding of early development. This chapter describes protocols for the derivation and maintenance of human embryonic stem cells. In addition, it summarizes briefly several alternative methods for the culture of human embryonic stem cells. Thus, this chapter provides a good starting point for researchers interested in harnessing the potential of human embryonic stem cells.

Introduction

In 1981, two groups succeeded in cultivating pluripotent cell lines from mouse blastocysts (Evans and Kaufman, 1981; Martin, 1981). These cell lines, termed embryonic stem (ES) cells, originate from the inner cell mass (ICM) or epiblast and could be maintained *in vitro* without an apparent loss of developmental potential. The ability of these cells to contribute to all cell lineages has been demonstrated repeatedly both *in vitro* and *in vivo* (reviewed by Wobus and Boheler, 2005). Once established, ES cells display an almost unlimited proliferative capacity while retaining their developmental potential (reviewed by Smith, 2001). The first successful derivation of human ES (hES) cell lines was reported in 1998 (Thomson *et al.*, 1998). The establishment of hES cell lines provides a unique new research tool with widespread potential clinical applications.

Under specific *in vitro* culture conditions, hES cells also proliferate indefinitely without senescence and are able to differentiate into almost all tissue-specific cell lineages. These properties make hES cells an attractive candidate for cell replacement therapy and open exciting new opportunities to model human embryonic development *in vitro* (reviewed by Keller, 2005). In addition to developmental biology and cell-based therapy, the ES cell model has widespread applications in the areas of drug discovery and drug development (reviewed by Gorba and Allsopp, 2003).

Derivation of hES cell lines has not had a common uniform procedure among laboratories. Moreover, the culture and manipulation of hES cells differ considerably between laboratories and pose several unique challenges.

METHODS IN ENZYMOLOGY, VOL. 418   0076-6879/06 $35.00

To help facilitate research with hES cells we describe in detail the protocols used in our laboratory for the derivation and maintenance of hES cell lines (Cowan *et al.*, 2004; Klimanskaya and McMahon, 2004; http://mcb.harvard.edu/melton/hues). In addition, we briefly discuss alternative approaches to the maintenance of hES cells. Thus, this chapter provides a starting point for researchers interested in establishing and working with hES cell lines.

Derivation of hES Cell Lines

Since the initial derivation of human ES cell lines by Thomson *et al.* (1998), several additional hES cell lines have been established and characterized (Table I; www.stemcells.nih.gov/registry/index.asp). We reported previously the derivation and maintenance of 17 new hES cell lines that can be maintained in culture by enzymatic dissociation with trypsin (Cowan, 2004). Our complete protocol has been described previously in detail (Klimanskaya and McMahon, 2004). The general utility and success of our approach have been validated by the transfer of this technique to several researchers and their subsequent derivation of new hES cell lines (Melton and Eggan, unpublished data). This chapter presents our most concise and current protocol for the derivation of hES cell lines.

Planning and Considerations

In our experience, hES cell derivation can be rather time-consuming and demanding. Until the isolated cells are frozen and thawed, they must be continually passaged and maintained. On average, one can expect 3 to 6 weeks of uninterrupted culture from the point of initiating an attempt to isolate hES cells from blastocyst embryos. Before deriving any new hES cell lines, we recommend that all of the reagents necessary for culture and derivation of the cells be obtained and, if possible, tested by routine culture of preexisting hES cell lines. Our standard derivation protocol makes use of mouse embryonic fibroblast cells as a feeder layer, and we also recommend the isolation and testing of these cells before attempting to isolate new hES cell lines. Finally, our protocol is designed to derive hES cells from blastocyst stage embryos, and while we have derived several cell lines from embryos frozen at early cleavage stages, they are always first cultured until they mature into blastocysts. In the following sections we will attempt to walk the reader through a stepwise protocol for deriving hES cell lines and, when necessary, to provide specifics details as to the suppliers of certain essential reagents.

TABLE I
PUBLISHED HUMAN EMBRYONIC STEM CELL DERIVATIONS[a]

No. of established lines	No. of plated embryos	Karyotype		Feeder source	Isolation of ICM	Medium for isolation of ICM	References
		46,XX	46,XY				
5	14	3	2	irrad-MEF	IS	20% FBS	Thomson et al. (1998)
2	4	2	0	mitoC-MEF	IS	20% FBS+LIF	Reubinoff et al. (2000)
1	1	0	1	mitoC-HFM	IS	20% HS+ITS	Richards et al. (2002)
1	4	1	0	irrad-HFF	IS	20% FBS+LIF	Hovatta et al. (2003)
3	30	1	2	mitoC-STO	IS	20% FBS+LIF	Park et al. (2003)
2	19	0	2	inact-MEF	IS	20% FBS+LIF+bFGF	Mitalipova et al. (2003)
6	N/A[b]	1[c]	5[d]	mitoC-MEF	IS/WB	20% SR+bFGF+HA	Heins et al. (2004)
1	1	1	0	mitoC-MEF	WB	20% FBS+LIF	Baharvand et al. (2004)
17	97	8	9	irrad-MEF	IS	8% SR+8% plasmanate+ LIF+bFGF	Cowan et al. (2004)
1	9	1	0	inact-MEF	WB	20% FBS	Suss-Toby et al. (2004)
1	7	1	0	irrad-MEF	IS	10% FBS	Stojkovic et al. (2004)
9	20	4	5	mitoC-STO	IS	20% FBS+LIF+bFGF	Park et al. (2004)
9	19	6	3	mitoC-MEF	IS	20% SR	Kim et al. (2005)
2	10	1	1	irrad-HFF	IS	20% SR+bFGF	Inzunza et al. (2005)
2	16	2	0	HPF	IS	20% SR+bFGF	Simon et al. (2005)

2	111	1	1	irrad-HPF	WB	20% SR+bFGF	Genbacev et al. (2005)
3	7	3	0	mitoC-HUE	IS	20% SR+bFGF	Lee et al. (2005)
3	10	1	2	mitoC-STO	IS/WB	20% SR+bFGF	Oh et al. (2005)
1	5	1	0	free[e]	IS	8% SR+8% plasmanate+ LIF+bFGF	Klimanskaya et al. (2005)
2	19	1	1	mitoC-MEF	IS	20% FBS+bFGF	Chen et al. (2005)
2	55	0	2	irrad-MEF	IS	20% FBS+LIF+bFGF	Mateizel et al. (2005)
3	14	2	1	irrad-MEF	IS	20% SR+bFGF	Mateizel et al. (2005)
4	14	2	2	irrad-MEF	IS	20% SR+bFGF	Hong-mei and Gui-an (2006)
1	33	1	0	inact-MEF	WB	20% FBS+bFGF	Sun et al. (2006)
2	5	0	2[f]	free[g]	IS	TeSR1[h]	Ludwig et al. (2006)

[a] irrad, irradiated; mitoC, mitomycin C; MEF, mouse embryonic feeders; HFM, human fetal muscle; HFF, human foreskin fibroblasts; STO, STO cells; HPF, human placental fibroblasts; HUE, human uterine endometrial cells; IS, immunosurgery; WB, whole blastocyst; FBS, fetal bovine serum, HS, human serum; SR, Serum Replacement; ITS, insulin transferring selenium; HA, hyaluronic acid.

[b] Information not available from published sources.

[c] XX karyotype with a trisomy 13.

[d] One line was a triploid, 69,XXY.

[e] Mouse extracellular matrix coated.

[f] One line was a 47,XXY.

[g] Human extracellular matrix coated.

[h] TeSR1 was highly defined medium, which was composed of a DMEM/F12 base supplemented with human serum albumin, vitamins, antioxidants, trace minerals, specific lipids, and growth factors.

Preparation of Mouse Embryo Fibroblasts (MEFs)

We use primary MEF cells, which have been mitotically inactivated by γ-irradiation, for derivation and propagation of hES cells. MEFs are harvested from 12.5-day postcoitum (dpc) fetuses of ICR mice (Cowan et al., 2004). The following reagents are required to follow our protocol for preparing MEFs.

Sterile phosphate-buffered saline (PBS), pH 7.2

MEF medium (90% Dulbecco's modified Eagle's medium [DMEM], 10% fetal bovine serum [FBS], 50 units/ml penicillin, and 50 μg/ml streptomycin)

0.25% trypsin

0.1% gelatin (made by dissolving 1 g of gelatin in 1000 ml of Milli-Q quality water, followed by sterile filtering)

Freezing medium (90% FBS, 10% dimethyl sulfoxide [DMSO])

10- and 15-cm tissue culture dishes

Sterile single-edged razor blade

Dissection and Primary Culture of MEFs

Prior to dissecting the mouse embryos, several 15-cm tissue culture plates (seven to eight plates per pregnant ICR female) should be coated with 0.1% gelatin. We typically cover the plates with a minimal amount of the gelatin solution (5 to 7 ml) and incubate them for 20 min at 37° with 5% CO_2. Using a microscope placed in a laminar flow hood, 12.5-dpc embryos are dissected into a 10-cm Petri dish containing sterile PBS solution. The embryos are then stripped of any maternal or extraembryonic tissues and eviscerated. Eviscerated embryos are transferred to a 15-cm dish and, using a sterile blade, minced. Ten milliliters of warm 0.25% trypsin is added per 10 to 14 minced embryos and collected in a 50-ml conical tube. The embryos are homogenized further by trituration (pipetting up and down) until no large pieces remain. This partially dissociated mixture is then incubated at 37° for 1 min followed by further trituration (pipetting 5 to 10 more times). Forty milliliters of prewarmed MEF medium is added to the dissociated embryos and the mixture is centrifuged for 10 min at 500 to 600g at room temperature. Aspirate media and then resuspend the pelleted cells with 30 ml prewarmed MEF medium. Plating density is 1.5 to 2 embryos per 15-cm gelatin-coated plate. The final volume of medium on each plate should be 20 ml. The primary MEFs are incubated at 37° with 5% CO_2 until confluent (typically 5 to 6 days). MEFs are expanded once after the initial plating (1:3 to 1:5 split) and then frozen (passage 1). Freeze MEFs in freezing medium (90% FBS and 10% DMSO) at a rate of $-1°$/min and store at $-80°$ or in liquid nitrogen.

γ-Irradiation and Plating

Thawed MEFs are only passaged once (passage 2) for expansion purposes prior to γ-irradiation. MEFs are trypsinized and resuspended in a volume of MEF medium that will be accommodated by the γ-irradiator. Irradiate the MEFs for 25 min at 247.3 rad/min for a total exposure of 6182.5 rad. After irradiation, spin cells in MEF medium for 5 min at 500 to 600g. To ensure a confluent monolayer, plate MEFs at a concentration of approximately 50,000 cells/cm^2. If there is no immediate need for mitotically inactivated MEFs, they can be frozen at a concentration of 4×10^6 to 1.2×10^7 cells/vial. MEFs feeder layers should be prepared and used within 3 days.

Preparing hES Derivation Medium

During the isolation and early stages of ES cell cultivation, hES derivation medium is used, which consists of 75% knockout DMEM (Invitrogen GIBCO), 10% KO-Serum Replacement (Invitrogen GIBCO), 10% plasmanate (Bayer), 5% fetal bovine serum (Hyclone), 2 mM Glutamax-I (Invitrogen GIBCO), 1% nonessential amino acids (Invitrogen GIBCO), 50 units/ml penicillin, and 50 μg/ml streptomycin (Invitrogen GIBCO), 0.055 mM β-mercaptoethanol (Invitrogen GIBCO), 12 ng/ml recombinant hLIF (Chemicon International), and 5 ng/ml bFGF (Invitrogen GIBCO). The medium is sterilized by 0.22-μm filtration. Screening of FBS, plasmanate, and Serum Replacement should be done and is described elsewhere (Klimanskaya and McMahon, 2004).

Isolation of Inner Cell Mass

Fresh or frozen-thawed human embryos are cultured to the blastocyst stage in sequential media, G1.2 and G2.2 (Gardner et al., 1998). We have derived several new human ES cell lines at relatively higher efficiency from blastocysts cultured in modified KSOM media. Blastocysts are treated with acid tyrodes (Specialty Media) for 30 to 90 s to dissolve the zona pellucida. When the zona pellucida starts to dissolve, remove the embryo and wash it three times in fresh hES derivation medium. The zona-stripped embryos are then cultured in hES derivation medium at 37° with 5% CO$_2$ until immunosurgical isolation of the ICM. The process of immunosurgery includes several stages and is performed essentially as described by Solter and Knowles (1975). Initially, the embryo is incubated for approximately 30 min in rabbit antihuman RBC antibodies (resuspended as per manufacturer's instructions, aliquoted, and stored at −80°, freshly diluted 1:10 in derivation medium, Inter Cell Technologies). Penetration of the antibodies into the blastocyst is prevented because of cell–cell connections within the outer layer of the

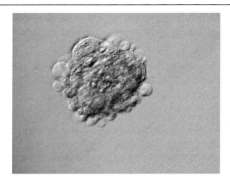

FIG. 1. "Bubbling" of trophoblast cells. Blastocyst after exposure to guinea pig sera complement is lysed and stop the incubation followed by removing the lyses trophoblast cells.

trophoblasts, leaving the ICM intact. After rinsing off any antibody residue (at least three washes with hES derivation media), the blastocyst is transferred into a guinea pig sera complement (resuspended as per manufacturer's instructions, aliquoted, and stored at −80°, freshly diluted 1:10 in derivation medium, Sigma), diluted in hES derivation medium, and incubated until cell lysis is notable, indicated by an apparent "bubbling" of the trophoblast cells (Fig. 1). Following selective removal of the trophectoderm cells by gentle mouth pipetting of the embryo in and out of a glass capillary, the intact ICM is cultured on MEF feeders plated on gelatin (Sigma)-coated tissue culture plates at a density of approximately 50,000 cells/cm². After 2 days add a few fresh drops of hES derivation medium and then every other day change one-half the total medium (e.g., for 500 µl total medium and then remove 250 µl of medium and add 250 µl of fresh medium to a final volume of 500 µl).

Dispersion of Inner Cell Mass

Six to 10 days after the initial plating, ICM outgrowths require mechanical dissociation. Two to three pieces are cut from the initial outgrowth using a narrow glass capillary and are left in the same well or moved to a new well (Fig. 2). When doing the initial dispersion, a part of the original colony should be left untouched as a backup, especially if the picked pieces are transferred into a new well. At this stage, it is better to concentrate on expanding the number of hES cell colonies versus freezing or proceeding to any downstream experiments. When the colonies are growing steadily, FBS is omitted from the culture media. Usually, mechanical passaging needs to be done every 5 to 6 days, but several larger colonies may need to be dispersed daily.

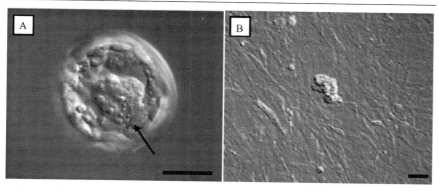

FIG. 2. Blastocyst and ICM outgrowth. (A) Cultured blastocyst is grade 4AA. Arrow indicates ICM. (B) Isolated ICM from the blastocyst (A) is just grown ICM at day 4 after plating on mitotically inactivated MEFs. Black bar: 50 μm.

Maintenance of hES Cells

Variability among human ES cell lines has been reported by several groups, including differences in growth characteristics, differentiation potential, karyotype, and gene expression pattern. In part, these differences might reflect the genetic heterogeneity of hES cell lines derived, as they are from a genetically diverse, outbred population (Abeyta *et al.*, 2004; Bhattacharya *et al.*, 2004). Further confounding researchers is the fact that human ES cell cultures are often heterogeneous because they contain both undifferentiated stem cells and spontaneously arising differentiated derivatives. While no single uniform protocol exists for the maintenance of hES cells in culture that adequately addresses all researchers' concerns, we will attempt to present an overview of the techniques currently used by a number of laboratories around the world. Again, we describe in detail our method for maintaining undifferentiated hES cell growth in culture and briefly review several alternative protocols.

Enzymatic Dissociation with Trypsin

Human ES cell colonies are passaged by mechanical dissociation until there are sufficient colonies (50 to 100 average-sized colonies) or cells (usually 1×10^5 cells) to passage enzymatically. Thereafter, hES cells are propagated by enzymatic dissociation with 0.05% trypsin/EDTA (Invitrogen GIBCO). During the first three passages with trypsin, it is a good idea to keep a backup well of mechanically passaged cells. A mechanical backup should always be maintained until the cells are frozen. Subconfluent cultures are generally split at a 1:3 ratio (i.e., one culture well is split into three new culture wells). It is important to split colonies prior to excessive differentiation.

Materials

For the routine culture of hES cell by enzymatic dissociation with trypsin we recommend the following media and reagents.

hES medium (80% knockout DMEM, 10% KO-Serum Replacement, 10% plasmanate, 2 mM Glutamax-I, 1% nonessential amino acids, 50 units/ml penicillin, 50 μg/ml streptomycin, 0.055 mM β-mercaptoethanol, and 5 ng/ml bFGF

Trypsin 0.05%

Sterile PBS, pH 7.2.

Trypsinization

1. Warm hES medium and trypsin in a 37° water bath and keep them warm until ready for use.

2. Place MEF plate from incubator in the hood and aspirate off the medium from the well followed by 1 ml prewarmed hES medium. Set the plate aside in the hood.

3. Carefully aspirate the hES medium from the culture to be split. Gently rinse the cells with a sufficient volume of PBS to completely cover the bottom of the culture dish (e.g., 5 ml for a 10-cm dish).

4. Aspirate the PBS and add a small volume of trypsin (usually 0.3 ml for a 35-mm well or 2 ml for a 10-cm dish) to the cells. Incubate in the hood at room temperature, frequently checking the cells under the microscope. MEFs surrounding the colonies should begin to retract (Fig. 3). When the MEFs are sufficiently shrunk and the borders of the colonies are roughly rounded up, add 10 volumes of prewarmed hES medium to the trypsinized colonies. Gently pipette up and down five to seven times until the MEF monolayer has completely detached. Extensive pipetting should be avoided.

5. Aliquot the hES cell solution dropwise, making sure to distribute the drops evenly about the well. Without shaking the plate, carefully return to the cells to a 37° incubator overnight to let the colonies seed.

The time in trypsin required for the cells to detach varies depending on the hES cell density, age of MEF monolayer, etc. We recommend checking the appearance of the hES culture under a stereomicroscope and determining the best incubation time for each well empirically.

Freezing hES Cells

1. Trypsinize the cells; see trypsinization section. Centrifuge the cells at 600g in 10 volumes of hES culture media.

2. Resuspend the pellet in cold freezing medium, which consists of 90% FBS and 10% DMSO.

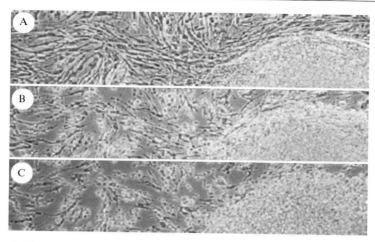

Fig. 3. Time-lapse series of photographs showing dissociation of hES cells and MEF feeder layer with trypsin. (A) Prior to addition of trypsin. (B) Approximately 30 s after addition of trypsin. (C) Approximately 60 s after addition of trypsin. Trypsinization should be stopped when cells appear as in C.

3. Aliquot the cell suspension into prechilled freezing vials and sandwich the vials between two Styrofoam racks, taping to prevent them from separating. Transfer to a −80° freezer overnight. Cryovials should be placed in liquid nitrogen for long-term storage.

Thawing hES Cells

Ensure that the MEF plate prepared is confluent and in good condition before thawing hES cells. Prewarm hES medium to 37°. Aliquot 10 ml hES medium into a sterile and labeled 15-ml conical tube for each cell line.

All procedures should be done quickly.

1. Thaw the vial in a 37° water bath. (Do not overthaw; the vial should be removed from the water bath with a small ice crystal still remaining.) It should take about 45 to 60 s before the cells are 80% thawed.

2. Bring the tube to a laminar flow hood; spray down with 70% isopropanol. Gently transfer cells to the 10 ml of prewarmed medium.

3. Centrifuge the 15-ml conical tube at 500 to 600g for 5 min.

4. Remove preplated MEFs from incubator to the hood. Aspirate off the MEF medium and aliquot prewarmed hES medium into each well of the plate, being careful not to disturb the attached MEFs.

5. After the spin is complete, carefully remove the medium without disturbing the pellet.

6. Gently resuspend the pellet in a small volume of prewarmed hES medium.

7. Transfer the hES cell solution, in a dropwise manner, to a prepared MEF plate well that already contains hES medium. Carefully return the plate to avoid swirling to a 37° incubator overnight to allow the hES cells to seed the MEFs.

8. The colonies usually begin to appear in 3 to 4 days and can be ready for splitting in 5 to 10 days.

Alternative Methods

Several alternative methods exist for the culture of hES cells, but few have been examined rigorously over a long period of time. We will attempt to summarize some of the more common alternative methods for maintaining hES cells in culture. For detailed protocols, we advise referring to the primary literature. In addition, several alternatives, such as feeder and serum-free culturing of hES cells, are described elsewhere in this volume.

Dissociation with Collagenase or Dispase

Quite possibly the most widespread method for maintaining hES cells in culture depends on their dissociation with either collagenase or dispase. For a detailed protocol, please see http://www.geron.com/PDF/scprotocols. pdf. The reported advantages of culture with these enzymes are reduced cell death and perhaps greater karyotypic stability. The disadvantages of enzymatic dissociation with collagenase or dispase include the inability to accurately assess cell number and the failure to generate definitive single cell clones.

Culture with Human Feeder Cells

Mouse embryonic fibroblast cells have generally been used as feeder layers to support the unlimited growth of hES cells, but the use of animal feeder cells is associated with risks such as pathogen transmission and viral infection (Amit et al., 2003, 2004; Richards et al., 2002; Rosler et al., 2004). Martin et al (2005) reported that hES cells could incorporate foreign sugars into the glycoproteins on the cell surface. They also showed that an immune reaction could occur following exposure of the cells to serum from adults with high level of the antibody. These reports and other concerns have prompted many researchers to seek alternatives to mouse feeder layers.

Several groups have reported that feeder layers composed of cells originating from human fetal and adult tissues support unlimited proliferation of hES cells without differentiation. The cell types used include human fetal skin fibroblasts, human muscle cells, adult fallopian tubal epithelial cells (Richards *et al.*, 2002), adult marrow cells (Cheng *et al.*, 2003), foreskin fibroblasts (Amit *et al.*, 2003; Hovatta *et al.*, 2003), human uterine endo-metrium cells, and breast parenchyma cell abortus fetus fibroblasts (Lee *et al.*, 2004). In perhaps the most comprehensive study, Richards *et al.* (2003) reported on the evaluation of various human adult, fetal, and neonatal tissues as feeder cells for supporting the growth of hES cells. In addition, feeder cells derived from hES cells can be used as an autogenic feeder system that efficiently supports the growth and maintenance of pluripotency of hES cells (Stojkovic *et al.*, 2005; Yoo *et al.*, 2005).

Conclusion

Human ES cells are viewed by many as a novel and unlimited source of cells and tissues for transplantation for the treatment of a broad spectrum of diseases (reviewed by Keller, 2005). Moreover, human ES cells represent an unprecedented system suitable for the identification of new molecular tar-gets and the development of novel drugs, which can be tested *in vitro* or used to predict or anticipate potential toxicity in humans. Finally, human ES cells can yield insight into the developmental events that occur during human embryogenesis, which are, for ethical reasons, nearly impossible to study in the intact embryo (reviewed by Dvash and Benvenisty, 2004).

Acknowledgments

We thank Jacob Zucker for providing the images used to illustrate the dissociation of hES cells by trypsin and Stephen Sullivan for suggestions and advice.

References

Abeyta, M. J., Clark, A. T., Rodriguez, R. T., Bodnar, M. S., Pera, R. A., and Firpo, M. T. (2004). Unique gene expression signatures of independently-derived human embryonic stem cell lines. *Hum. Mol. Genet.* **13,** 601–608.

Amit, M., Margulets, V., Segev, H., Shariki, K., Laevsky, I., Coleman, R., and Itskovitz-Eldor, J. (2003). Human feeder layers for human embryonic stem cells. *Biol. Reprod.* **68,** 2150–2156.

Amit, M., Shariki, C., Margulets, V., and Itskovitz-Eldor, J. (2004). Feeder layer- and serum-free culture of human embryonic stem cells. *Biol. Reprod.* **70,** 837–845.

Baharvand, H., Ashtiani, S. K., Valojerdi, M. R., Shahverdi, A., Taee, A., and Sabour, D. (2004). Establishment and *in vitro* differentiation of a new embryonic stem cell line from human blastocyst. *Differentiation* **72,** 224–229.

Bhattacharya, B., Miura, T., Brandenberger, R., Mejido, J., Luo, Y., Yang, A. X., Joshi, B. H., Ginis, I., Thies, R. S., Amit, M., Lyons, I., Condie, B. G., Itskovitz-Eldor, J., Rao, M. S., and Puri, R. K. (2004). Gene expression in human embryonic stem cell lines: Unique molecular signature. *Blood* **103**, 2956–2964.

Chen, H., Qian, K., Hu, J., Liu, D., Lu, W., Yang, Y., Wang, D., Yan, H., Zhang, S., and Zhu, Q. (2005). The derivation of two additional human embryonic stem cell lines from day 3 embryos with low morphological scores. *Hum. Reprod.* **20**, 2201–2206.

Cheng, L., Hammond, H., Ye, Z., Zhan, X., and Dravid, G. (2003). Human adult marrow cells support prolonged expansion of human embryonic stem cells in culture. *Stem Cells* **21**, 131–142.

Cowan, C. A., Klimanskaya, I., McMahon, J., Atienza, J., Witmyer, J., Zucker, J. P., Wang, S., Morton, C. C., McMahon, A. P., Powers, D., and Melton, D. A. (2004). Derivation of embryonic stem cell lines from human blastocysts. *N. Engl. J. Med.* **350**, 1353–1356.

Dvash, T., and Benvenisty, N. (2004). Human embryonic stem cells as a model for early human development. *Best Pract. Res. Clin. Obstet. Gynaecol.* **18**, 929–940.

Evans, M. J., and Kaufman, M. H. (1981). Establishment in culture of pluripotential cells from mouse embryos. *Nature* **292**, 154–156.

Gardner, D. K., Vella, P., Lane, M., Wagley, L., Schlenker, T., and Schookraft, W. B. (1998). Culture and transfer of human blastocysts increases implantation rates and reduces the need for multiple embryo transfers. *Fertil. Steril.* **69**, 84–88.

Genbacev, O., Krtolica, A., Zdravkovic, T., Brunette, E., Powell, S., Nath, A., Caceres, E., McMaster, M., McDonagh, S., Li, Y., Mandalam, R., Lebkowski, J., and Fisher, S. J. (2005). Serum-free derivation of human embryonic stem cell lines on human placental fibroblast feeders. *Fertil. Steril.* **83**, 1517–1529.

Gorba, T., and Allsopp, T. E. (2003). Pharmacological potential of embryonic stem cells. *Pharm. Res.* **47**, 269–278.

Heins, N., Englund, M. C. O., Sjöblom, C., Dahl, U., Tonning, A., Bergh, C., Lindahl, A., Hanson, C., and Semb, H. (2004). Derivation, characterization, and differentiation of human embryonic stem cells. *Stem Cells* **22**, 367–376.

Hong-mei, P., and Gui-an, C. (2006). Serum-free medium cultivation to improve efficacy in establishment of human embryonic stem cell lines. *Hum. Reprod.* **21**, 217–222.

Hovatta, O., Mikkola, M., Gertow, K., Strömberg, A., Inzunza, J., Hreinsson, J., Rozell, B., Andäng, M., and Ährlund-Richter, L. (2003). A culture system using human foreskin fibroblasts as feeder cells allows production of human embryonic stem cells. *Hum. Reprod.* **18**, 1404–1409.

Inzunza, J., Gertow, K., Strömberg, A., Matilainen, E., Blennow, E., Skottman, H, Wolbank, S., Ährlund-Richter, L., and Hovatta, O. (2005). Derivation of human embryonic stem cell lines in serum replacement medium using postnatal human fibroblasts as feeder cells. *Stem Cells* **23**, 544–549.

Keller, G. (2005). Embryonic stem cell differentiation: Emergence of a new era in biology and medicine. *Genes Dev.* **19**, 1129–1155.

Kim, S. J., Lee, J. E., Park, J. H., Lee, J. B., Kim, J. M., Yoon, B. S., Song, J. M., Roh, S. I., Kim, C. G., and Yoon, H. S. (2005). Efficient derivation of new human embryonic stem cell lines. *Mol. Cells* **19**, 46–53.

Klimanskaya, I., and McMahon, J. (2004). Approaches for derivation and maintenance of human ES cells: Detailed procedures and alternatives. *In* "Handbook of Stem Cells" (R. Lanza, J. Gearhart, B. Hogan *et al.*, eds.), Vol. 1, pp. 437–449. Elsevier/Academic Press, San Diego.

Klimanskaya, I., Chung, Y., Meisner, L., Johnson, J., West, M. D., and Lanza, R. (2005). Human embryonic stem cell derived without feeder cells. *Lancet* **365**, 1636–1641.

Lee, J. B., Lee, J. E., Park, J. H., Kim, S. J., Kim, M. K., Roh, S. I., and Yoon, H. S. (2005). Establishment and maintenance of human embryonic stem cell lines on human feeder cells derived from uterine endometrium under serum-free condition. *Biol. Reprod.* **72,** 42–49.

Lee, J. B., Song, J. M., Lee, J. E., Park, J. H., Kim, S. J., Kang, S. M., Kwon, J. N., Kim, M. K., Roh, S. I., and Yoon, H. S. (2004). Available human feeder cells for the maintenance of human embryonic stem cells. *Reproduction* **128,** 727–735.

Ludwig, T. E., Levenstein, M. E., Jones, J. M., Berggren, W. T., Mitchen, E. R., Frane, J. L., Grandall, L. J., Daigh, C. A., Conard, K. R., Piekarczyk, M. S., Llanas, R. A., and Thomson, J. A. (2006). Derivation of human embryonic stem cells in defined conditions. *Nature Biotechnol.* Advanced published Jan. 1.

Martin, G. R. (1981). Isolation of a pluripotent cell line from early mouse embryos cultured in medium conditioned by teratocarcinoma stem cells. *Proc. Natl. Acad. Sci. USA* **78,** 7634–7638.

Martin, M. J., Muotri, A., Gage, F., and Varki, A. (2005). Human embryonic stem cells express an immunogenic nonhuman sialic acid. *Nature Med.* **11,** 228–232.

Mateizel, I., De Temmerman, N., Ullmann, U., Cauffman, G., Sermon, K., Van de Velde, H., De Rycke, M., Degreef, E., Devroey, P., Liebaers, I., and Van Steirteghem, A. (2006). Derivation of human embryonic stem cell lines from embryos obtained after IVF and after PGD for monogenic disorders. *Hum. Reprod.* **21,** 503–511.

Mitalipova, M., Calhoun, J., Shin, S., Wininger, D., Schulz, T., Noggle, S., Venable, A., Lyons, I., Robins, A., and Stice, S. (2003). Human embryonic stem cell lines derived from discarded embryos. *Stem Cells* **21,** 521–526.

Oh, S. K., Kim, H. S., Ahn, H. J., Seol, H. W., Kim, Y. Y., Park, Y. B., Yoon, C. J., Kim, D. W., Kim, S. H., and Moon, S. Y. (2005). Derivation and characterization of new human embryonic stem cell lines: SNUhES1, SNUhES2, and SNUhES3. *Stem Cells* **23,** 211–219.

Park, J. H., Kim, S. J., Oh, E. J., Moon, S. Y., Roh, S. I., Kim, C. G., and Yoon, H. S. (2003). Establishment and maintenance of human embryonic stem cells on STO, a permanently growing cell line. *Biol. Reprod.* **69,** 2007–2014.

Park, S., Lee, Y. J., Lee, K. S., Shin, H. A., Cho, H. Y., Chung, K. S., Kim, E. Y., and Lim, J. H. (2004). Establishment of human embryonic stem cell lines from frozen-thawed blastocysts using STO cell feeder layers. *Hum. Reprod.* **19,** 676–684.

Reubinoff, B. E., Pera, M. F., Fong, C. Y., Trounson, A., and Bongso, A. (2000). Embryonic stem cell lines from human blastocysts: Somatic differentiation *in vitro*. *Nature Biotechnol.* **18,** 399–404.

Richards, M., Fong, C. Y., Chan, W. K., Wong, P. C., and Bongso, A. (2002). Human feeders support prolonged undifferentiated growth of human inner cell masses and embryonic stem cells. *Nature Biotechnol.* **20,** 933–936.

Rosler, E. S., Fisk, G. J., Ares, X., Irving, J., Miura, T., Rao, M. S., and Carpenter, M. K. (2004). Long-term culture of human embryonic stem cells in feeder-free conditions. *Dev. Dyn.* **229,** 259–274.

Simon, C., Escobedo, C., Valbuena, D., Genbacev, O., Galan, A., Krtolica, A., Asensi, A., Sanchez, E., Esplugues, J., Fisher, S., and Pellicer, A. (2005). First derivation in Spain of human embryonic stem cell lines: Use of long-term cryopreserved embryos and animal-free conditions. *Fertil. Steril.* **83,** 246–249.

Smith, A. G. (2001). Embryo-derived stem cells: Of mice and men. *Annu. Rev. Cell. Dev. Biol.* **17,** 435–462.

Solter, D., and Knowles, B. B. (1975). Immunosurgery of mouse blastocyst. *Proc. Natl. Acad. Sci. USA* **72,** 5099–5102.

Stojkovic, M., Lako, M., Stojkovic, P., Stewart, R., Przyborski, S., Armstrong, L., Evans, J., Herbert, M., Hyslop, L., Ahmad, S., Murdoch, A., and Strachan, T. (2004). Derivation of human embryonic stem cells from day-8 blastocysts recovered after three-step *in vitro* culture. *Stem Cells* **22,** 790–797.

Stojkovic, P., Lako, M., Stewart, R., Przyborski, S., Armstrong, L., Evans, J., Murdoch, A., Strachan, T., and Stojkovic, M. (2005). An autogeneic feeder cell system that efficiently supports growth of undifferentiated human embryonic stem cell. *Stem Cells* **23,** 306–314.

Sun, B. W., Yang, A. C., Feng, Y., Sun, Y. J., Zhu, Y. F., Zhang, Y., Jiang, H., Li, C. L., Gao, F. R., Zhang, Z. H., Wang, W. C., Kong, X. Y., Jin, G., Fu, S. J., and Jin, Y. (2006). Temporal and parental-specific expression of imprinted genes in a newly derived Chinese human embryonic stem cell line and embryoid bodies. *Hum. Mol. Genet.* **15,** 65–75.

Suss-Toby, E., Gerecht-Nir, S., Amit, M., Manor, D., and Itskovitz-Eldor, J. (2004). Derivation of a diploid human embryonic stem cell line from a mononuclear zygote. *Hum. Reprod.* **19,** 670–675.

Thomson, J. A., Itskovitz-Eldor, J., Shapiro, S. S., Waknitz, M. A., Swiergiel, J. J., Marshall, V. S., and Jones, J. M. (1998). Embryonic stem cell lines derived from human blastocysts. *Science* **282,** 1145–1147.

Wobus, A. M., and Boheler, K. R. (2005). Embryonic stem cells: Prospects for development biology and cell therapy. *Physiol. Rev.* **85,** 635–678.

Yoo, S. J., Yoon, B. S., Kim, J. M., Song, J. M., Roh, S. I., You, S., and Yoon, H. S. (2005). Efficinet culture system for human embryonic stem cells using autologous human embryonic stem cell-derived feeder cells. *Exp. Mol. Med.* **37,** 399–407.

[6] Embryonic Stem Cells from Morula

By NICK STRELCHENKO and YURY VERLINSKY

Abstract

It has been shown that it is possible to establish human embryonic stem cell (hESC) lines from morula. Details of the aforementioned injection method of morula under blastocyst are described in this chapter. This chapter also discloses the application of simultaneous staining for two markers, TRA-2-39 and Oct-4, for characteristics of nondifferentiated hESC derived from morula and gives a method. Technical approaches of freezing morula-derived hESC are discussed.

Introduction

Pluripotent cells have attracted the attention of researchers as a powerful tool for cell therapy. A report on isolation of stem cells directly from rabbit embryos was published in 1965 (Cole *et al.*, 1965), while most of the researchers worked with murine teratocarcinomas or embryonic cell lines (Stevens, 1970) after *in vivo* initiation, which helped determine renewing part population of stem cells possessing specific group of glycoproteins that share expression with the early stage so-called specific stage embryo antigen (SSEA) (Knowles *et al.*, 1977).

Pluripotent cells have acquired their modern name, *embryonic stem (ES) cells*, and have been described by Evans and Kaufman (1981) and Martin (1981). Both groups independently established murine embryonic stem cell lines from outgrowth inner cell mass (ICM) of delayed murine blastocysts (strain 129). That type of cell was positive for alkaline phosphatase and expressed SSEA markers. These cells have contributed germ line chimera animals because of the euploid karyotype (Bradley *et al.*, 1984), which is different compared to EC cells that contribute derivatives of three germ layers, except germ cells.

The ICM appears as a result of first embryo differentiation, but morula has already committed blastomeres for ES cells (Tesar, 2005). Also, murine ES cell lines have been established from the morula stage embryo by Eistetter (1989). These ES cells were similar to ES cells isolated from blastocysts. Several ES cell lines have been established from morula for several mammalian species (Sukoyan *et al.*, 1993; Stice *et al.*, 1996).

METHODS IN ENZYMOLOGY, VOL. 418
Copyright 2006, Elsevier Inc. All rights reserved.
0076-6879/06 $35.00
DOI: 10.1016/S0076-6879(06)18006-4

The essential differences between establishing teratocarcinoma and embryonic stem cells were conditions of initial embryo culture (*in vivo* for EC cells versus *in vitro* for ES cells). It emphasizes an important role of initial conditions for embryo culture.

Before publication (Thomson *et al.*, 1998), attempts have been made to produce human ES cells (hESC) from entire blastocysts (Bongso, 1994). The purpose of this chapter is to bring the scientific community details of methods and experience used at the RGI laboratory for establishing hESC lines from morula.

Blastomeres or cells taken from morula are very different compared to cells from ICM of blastocyst, not only in the size of the adjacent cytoplast but in the gene pattern expression. For instance, interferon-τ is an exclusive product of trophectoderm and serves for fetal–maternal recognition (Larson *et al.*, 2001). Serious changes in the expression pattern of the gene have been seen as a distinctive but unstable maternal methylation pattern that persists until the morula stage and disappears in the blastocyst stage, where low levels of methylation are present on most DNA strands independently from parental origin (Hanel and Wevrick, 2001). Bovine embryos display high sensitivity to ouabain (potent inhibitor of the Na/K-ATPase), and enzyme activity undergoes a ninefold increase from the morula to the blastocyst stage (Watson and Barcroft, 2001). Comparison of mRNA expression patterns has shown differences in mouse embryos at the two-, four-, and eight-cell/morula and blastocyst stages by differential display (Lee *et al.*, 2001).

After removal of the zona pellucida morula, including the compact morula stage, all cells are equal in terms of ability of differentiation, similar to ICM cells, because differentiation has not occurred yet. Morula-derived embryonic stem cells are supposed to be more pluripotent in terms of an ability to produce a variety of differentiated cells and to be more stable in terms of spontaneous differentiation because they were isolated before first embryo differentiation versus embryonic stem cells established from ICM of blastocysts.

Protocol for Isolation Human ES Cell Line from Morula

Protocols for establishing human ES cells from morula and blastocyst stages have been approved by the IRB at RGI. Patients who have given their consent have donated all embryos used for experiments to establish ES cell lines.

Materials and Methods

Pronase (3 mg/ml) in HTF-buffered HEPES supplemented with 5% plasmanate Bayer (Code 613-25)

Confluent mouse embryonic feeder layer on a 35-mm dish (Nunclon) inactivated by a mitomycin C confluent mouse embryonic feeder layer
Growth medium: αMEM (GIBCO) supplemented with 15% fetal bovine serum, mercaptoethanol (1 mM, GIBCO), and fibroblast growth factor-basic human F0291 (5 ng/ml)

Procedure

1. Dissolve morula zona pellucida in pronase and wash out pronase in HTF-plasmanate mixture.
2. Make four light cuts of feeder layer by a sharp sterile blade 4 to 6 mm in length.
3. Using a glass needle from the left side, lift up feeder layer and insert pipette with naked morula stage embryo under feeder layer.
4. Remove lifted needle carefully and feeder layer will stay in injection pipette.
5. By moving pipette up or down, carefully spill out naked morula under feeder layer.
6. As soon as the feeder layer engages with morula, remove injection pipette.
7. Morula cells proliferate and produce a plate of cells within 10 to 15 days.
8. Passage 0 has to be performed when plate reaches a size of around 500 μm.
9. For isolation of hES cells, wash out dish with 2 ml 0.2% EDTA on Ca, Mg-free phosphate-buffered saline (PBS) and add 1 ml of fresh 0.2% EDTA and leave at 37° in an incubator for 5 min.
10. Inactivate cell disaggregation by adding an equal volume of HTF-plasmanate mixture.
11. Using a pulled glass pipette with a tiny tip (60 to 70 μm), scratch cells in clamps under a microdissecting scope. It is important to remember that hES cells are very sensitive to mechanical influence.
12. Harvest clumps of hESC by a plastic pipette with a tiny plastic tip (20 μl) and transfer onto a fresh feeder layer equally spread at bottom well.
13. After 3 to 5 days, check Petri dish under an inverted microscope, preferably a phase contrast, and mark colonies with hESC morphology.
14. As cells will form colony, repeat passage as described earlier. The next several passages are devoted to selecting colonies with typical hES morphology.
15. Medium has to be replaced in full volume 3 ml for 35-mm dish every second day.
16. Between the fifth and seventh passages, cell lines can be frozen for storage and cells may be characterized.

Commentary to Protocol

The procedure of isolation hESC lines described earlier was designed for experienced cell biologists. For that reason, a description of isolating the murine embryonic feeder layer will be dropped. The efficiency of establishing cell lines from morula and blastocyst or ICM correlates with the quality of feeder cells. It is a widely held opinion about the influence of conditioned media for cells, as it contains growth factors. Because of that, usually, media for cells should be refreshed on half of volume or added from a different source of cells, such as BRL (secretion of LIF) and Sl/Sl^4-m220 (secretion of *c-kit* ligand). Qualities of conditioned media depend on numerous factors, such as the dominated type of cells in culture, age of culture since last passage, and mitotic inactivation procedure. It is hard to speculate, but no stem cell line has been established in suspension. If human morula is cultured in conditioned media without cell-to-cell contact with feeder in a couple of days, it just simply turns to blastocyst. It has been shown that ICM produces hESC in the absence of a feeder layer cell, but that a cellular matrix is still needed (Klimanskaya *et al.*, 2005). Placing morula under a feeder layer technically complicates the procedure and requires well-trained personnel. A chart of microtool disposition is shown in Fig. 1A (front view) and visual injection of morula is shown in Fig. 1B (bottom view). A very original alternative was offered by Tesar (2005) using sedimentation of inactivated cells together with morula. It is a great idea, especially when the murine feeder layer can be replaced with human cells, such as foreskin- or placenta-derived cells, and potentially allowed to get away from animal derivatives, but it has not been tested in the RGI laboratory yet.

Proliferation of morula blastomeres continued under the feeder layer, and after 18 to 24 h cell spread included large and small cells, which can be clearly observed with a phase-contrast microscope (Fig. 2A). Expansion of hESC produced a cell plate 400 to 600 μm in size like the one shown in Fig. 2B under phase contrast and in Fig. 2C under DIC optic. Further stabilization of the morula-derived hES cell line still had certain difficulties. After cell outgrowth (approximately 8 to 14 days), passage 0 is performed by 2 mM EDTA in Hank's or phosphate-buffered saline (PBS) Ca^{2+}, Mg^{2+}-free solutions. Only soft loose cell clumps are transferred into a new dish with the feeder layer. Within the next two to five passages, selection of uniform proliferating cells has been made and colonies of established ES cell lines are passaged with collagenase V at 1.5 μg/ml in HTF-HEPES or 2 mM EDTA and harvested with a cell lifter (Costar). The undifferentiated part of the ES cell population has been isolated with EDTA solution (the patent application for the isolation of ES cell lines from morula is pending). Careful selection of cells with hESC morphology and a reliable technique

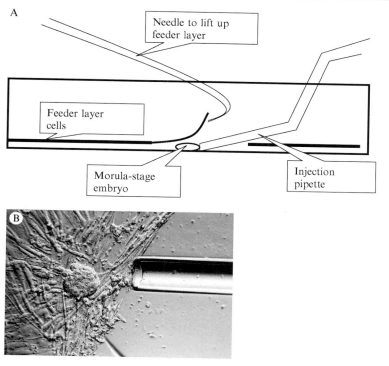

FIG. 1. (A) Chart of microtool disposition shows how to place naked morula under feeder layer. On left side is needle used to lift up cell layer. On right side is injection pipette. Position tool in front of view. (B) Microtool position after injection of naked human morula stage embryo under feeder layer. The bottom shows end of needle to hold feeder layer. Right side shows injection pipette. (See color insert.)

of pickup cell clamps are the tools to success. Isolated cells are extremely sensitive to mechanical exposure. Experience in our laboratory has shown that a good outgrowth rate does not always correlate with further proliferation. For example, 11 and 8 cell lines of hESC have been established from 46 plated morula outgrowth, while ICM produced five cell lines out of five outgrowth (Strelchenko et al., 2004).

Characteristics of Morula-Derived hESC Lines

A major marker for hESC is alkaline phosphatase. It is can be detected easily with a commercially available kit from Vector. The kit is based on a "know-how" immunocytochemical reaction. In the first 15 to 20 min it

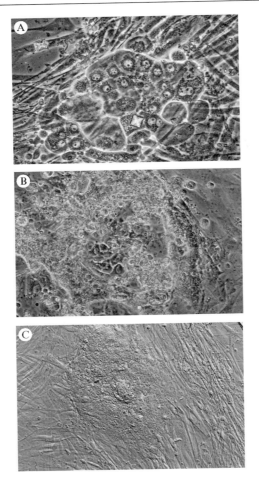

FIG. 2. (A) Human morula the next day after placing under feeder layer. Some cells probably have committed. There are cells with a small amount of cytoplasts, probably ancestors of ES cells, and with a large amount of cytoplasts targeted to trophoblast cells. (B) Cluster of cells derived from human morula on day 5 of culture. Different types of cells can be observed. The center shows small clusters of hESC. Phase contrast. (C) Cluster of cells derived from human morula on day 7 of culture. Differential interference contrast (Hoffman modulation contrast). (See color insert.)

appears in hESC colonies, but after a long stay in the mixture, a nonspecific light blue stain shows up even in a few feeder cells. It shows that the kit is specific to enzymatic activities of a big group enzyme called alkaline phosphatase but does not show what type of AP is active. For hESC, two commercially available antibodies have been developed: TRA-2-39 (Santa Cruz Biotechnologies, Inc.) and TRA-2-54 (Santa Cruz Biotechnologies, Inc.), which are specific to the epitope of a specific enzyme liver or L-alkaline phosphatase of 2102Ep human embryonic carcinoma cells. Simple comparisons of enzymatic staining (SK-5300) and subsequent staining with indirect immunofluorescence have shown almost a perfect match (see Fig. 3A and B). It has been shown (Strelchenko *et al.*, 2004) that 100% of hES cells in colonies possess both types of staining. Antibodies TRA-2-39 or TRA-2-54 target human origin L-AP, which is a very specific marker for cytoplasts of hESC.

The rest of the cytoplast markers, such as SSEA-3, SSEA-4 and TRA-1-60, TRA-1-80, are of embryonic derivative and human origin, respectively, and less constant in different hESC lines with varied expression around 70 to 95%.

Nuclei of pluripotent cells are supposed to have marker Oct-4 related to POU family transcription factors (Schöler *et al.*, 1991). Presence of the

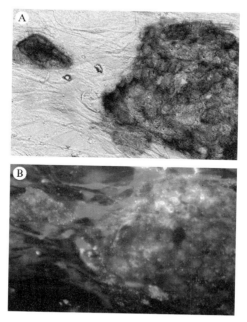

FIG. 3. (A) Human morula-derived stem cells stained for alkaline phosphatase (AP) general enzyme reaction (SK-5300). (B) The same cluster of morula-derived hESC stained with specific monoclonal antibody TRA-2-39. Both approaches in detection of L-AP match each other. (See color insert.)

marker can be found by two different approaches. One is the analysis of mRNA, which will identify the unique sequence of that gene; the second is the analysis of protein products in the nuclei. Expression of Oct-4 was assessed by use of the Gene Choice One Tube reverse transcriptase-polymerase chain reaction (RT-PCR) kit (Vector Laboratories, Inc.) using the primer sequences CACGAGATGCAAAGCAGAAACCCTCGG and TTGCCTCTCACTCGGTTCTCG, which generate a product of 73 bp in the presence mRNA of Oct-4. Figure 4A shows RT-PCR products from a cluster of cells from culture of hESC analyzed on a ABI3100 fluorescent sequencer and a 8% polyacrylamide gel using a size control of 100 bp line A and compares electrophoresis bands of RT-PCR products from hESC lines established from morula, blastocyst, and ICM of blastocysts (see Fig. 4B). However, the presence of mRNA does not always correlate with gene expression. Analysis of protein products of gene Oct-4 *in situ* has been performed by indirect immunofluorescence with polyclonal antibodies. Most of the hES cells in colonies of morula-derived cell lines possess the protein marker of gene Oct-4 (see Fig. 5A). Large-scale observation of over 100 hESC lines has shown that differentiation in the first line will disappear in Oct-4 in nuclei and later in L-AP. When a test is done on large colonies of hESC, the middle part of colonies may not be permeable for antibodies of Oct-4 because of tight packing cells, confusing researchers (see Fig. 5B). For a better description of hESC lines derived from morula stage embryos, the RGI hESC laboratory has developed internal standard or criteria for characteristic not differentiated hESC. The simultaneous presence of L-AP in cytoplasts and Oct-4 in nuclei represents the nondifferentiated part of the hESC population. For that purpose, to characterize a nondifferentiated hESC population, the original protocol from Santa Cruz Biotechnology was modified at the RGI laboratory and used to test new hESC on a routine basis. This test probably will have more performance for flow cytometry, but it has to be performed on single cell suspension and is more labor and time consuming; it has not been tested at the RGI laboratory yet.

Cell Cycle Analysis of Morula-Derived hESC Cell Lines

There are not many clearly published data on distribution hESC within cell cycle phases to compare with. Based on RGI laboratory established cell lines it was found that the distribution of morula-derived hESC is different compared to human differentiated cells. Most of the population of differentiated cells, such as human fibroblasts, are located in phase G0/G1 of the cell cycle ($\sim$60 to 80%), whereas morula-derived hESC lines at this stage

A

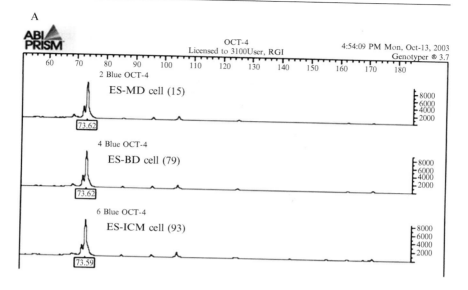

B

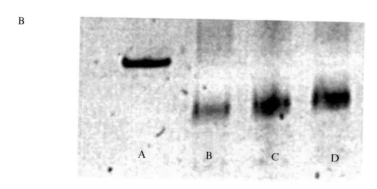

FIG. 4. (A) Demonstrates 73-bp RT-PCR products from a cluster of cells from a culture of hESC different derivation cell lines analyzed on a ABI3100 fluorescent sequencer; from top to bottom: track of morula-derived hESC line, 15; blastocyst-derived hESC line, 79: and ICM-derived hESC line, 93. (B) An 8% polyacrylamide gel using a size control of 100-bp (line A) and electrophoresis bands of RT-PCR products from a cluster of cells from culture hESC of different derivation cell lines: line B, morula-derived hESC line; line C, blastocyst-derived hESC line; and line D, ICM-derived hESC line.

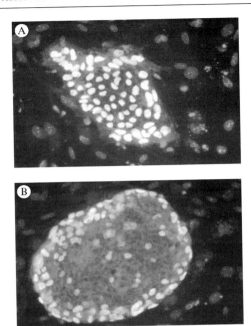

F𝐈G. 5. (A) Simultaneous presence of two markers, TRA-2-39 and Oct-4, in a colony of morula-derived hESC with green fluorescence (FITC) TRA-2-39, red fluorescence (TRITC) Oct-4, and blue fluorescence nonspecific nuclei (DAPI). (B) Large colony of morula-derived hESC where nuclei of hESC are nonpermeable for antibody Oct-4. (See color insert.)

have almost two times less the relative amount of hESC. It depends on the cell line and varies from 25 to 35%. Most of the hESC are located in phase G2/M and vary from 35 to 45%, while the S phase varies from 30 to 35%. Duration of the cell cycle was performed by hESC observation and was computed between 11 and 14 h.

Karyotyping of Morula-Derived hESC Lines

Monitoring of karyotype of ES cell lines is highly important for maintenance *in vitro*. For each cell line, 20 suitable metaphase spreads have been analyzed. Karyotyping has been performed *in situ* by plating hESC on a serum-treated surface of a coverslip glass (Mattek); after several days of proliferation, mitotic hESC are harvested by demecolcine, treated by hypotonic solution, and fixed *in situ* by methanol glacial acetic acid mixed 2:1. Standard procedure for G banding was applied. An example of a karyotype is shown in Fig. 6.

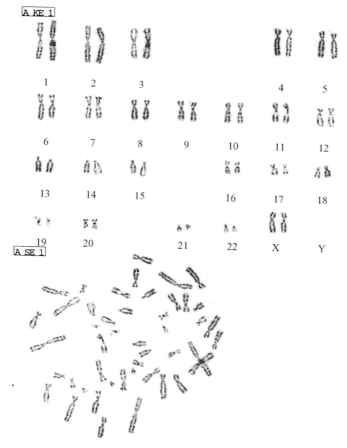

Fig. 6. ll xKaryotype of metaphasic spread hESC line 15.

Protocol of Immunofluorescence Staining for TRA-2-39 or TRA-2-54 and Oct-4

Equipment

Nunc plate 2 × 2 with assigned test ES cell line plated on feeder layer or Matrigel
Timer
Incubator at 37°

Materials

Blocking solution: 10% fetal bovine serum in PBS
Fixative solution (4% paraformaldehyde/PBS)
Rinse buffer: 0.75 ml fetal bovine serum and 50 ml PBS
Permeabilization buffer: 0.1% Triton X-100 in PBS
Primary antibodies available commercially from Santa Cruz Biotechnology, Inc. for TRA-2-39 or TRA-2-54 (isotype IgG_1) and polyclonal antibodies against Oct-4 (isotype IgG) and secondary antibodies from Santa Cruz Biotechnology, Inc. for detection of primary antibodies against L-AP used FITC antibodies and Oct-4, $F(ab')_2$ TRITC.

Procedure

1. For test hESC line, plate into Nunc 2×2 wells plate 2 to 3 days before analysis. Two wells are used for two lines.

2. Remove medium and wash with no serum containing PBS.

3. Fix cultured ES cells in fixative solution for 15 to 20 min at room temperature.

4. Permeabilize cells with 0.1% Triton X-100/PBS for 10 min at room temperature.

5. Wash three times with $1\times$ rinse buffer.

6. Apply blocking solution for 30 min at $37°$.

7. In an Eppendorf tube, dilute mixture of primary antibodies (1:20 for Oct-4 and 1:50 for TRA-2-39) to working concentrations in $1\times$ rinse buffer and centrifuge briefly 3 to 4 min to remove protein pellets. Add 150 μl per well. Incubate primary antibodies for 40 to 45 min at $37°$.

8. Wash three times with $1\times$ rinse buffer.

9. In an Eppendorf tube, dilute mixture of secondary antibodies (1:50 for FITC and 1:50 for TRITC) in $1\times$ rinse buffer just before use and centrifuge briefly 3 to 4 min to remove protein pellets. Add 150 μl per well. Incubate secondary antibodies for 40 to 45 min at $37°$.

10. Wash three times with $1\times$ rinse buffer and remove rinsing solution, but do not let dry.

11. After the staining procedure, cells should be covered with 50 μl (1 drop) of a mixture of antifade mounting solution (Vector) and DAPI standard solution, preliminary mixed 1:1, and then covered by a suitable round coverslip for better visualization.

12. Fluorescence hESC can be visualized with a fluorescence microscope, such as Nikon, with D/F/T set filters (Chroma). Usually fluorescence will be visible for the next 24 h.

Freezing Cell Lines

Storage of hESC is a very important part of establishing hESC lines. Low survival rates of ES cells after freezing could produce a "population bottle neck" and create conditions for selection and survival of differentiated types of cells. Dimethyl sulfoxide (DMSO) itself is a powerful tool for cell membrane permeabilization and differentiation agent that could induce nonspecific hemoglobin synthesis even in fibroblasts (Deisseroth et al., 1975). Glycerol is a more traditional cryoprotector and is less chemically affective for hESC, but its slow penetration via the cell membrane can be the cause of low survival rates (Bettler et al., 1977). Different ratios of glycerol and DMSO have been tested (data not shown). With light deviation from line to line, an optimal combination of DMSO and glycerol has been found to be 1:1 (w/w). It has produced stable results; on average, the number of recovered colonies is significantly higher compared to only DMSO or glycerol freezing medium.

Every batch of established morula-derived hESC lines has been tested for mycoplasma absence by PCR. For freezing ES cells has been used medium αMEM contained 5% DMSO, 5% glycerol and 10% FBS or 15% SR1. A freezing device, Planer, has lowered the temperature for freezing ES cell lines at $1°$ per minute from $4°$ to $-70°$, and cryovials were transferred and stored in liquid nitrogen ($-176°$). All morula-derived ES cell lines were frozen and thawed out successfully.

At a high volume of operation it is hard to keep appropriate timing; after adding the DMSO agent to the Petri dish, cells should not exposed at room temperature for longer than 30 min, leaving not much time to count and dilute cells. The following is an approximate cell ratio for freezing and thawing. Density of hESC: the distance between colonies approximately equal to colony diameter Nunc Petri dish 60 mm with hESC will be frozen four to five cryovials (2 ml each). One vial will be plated for one 35-mm Nunc Petri dish or 1 Nunc plate 2×2 wells.

Protocol for Freezing Morula-Derived hESC Lines

Required Equipment

  Aspirator
  Timer
  Incubator
  $-70°$ freezer Planer
  Vapor liquid nitrogen storage tank

Required Materials

PBS Ca^{2+}, Mg^{2+} free or Hank's Ca^{2+}, Mg^{2+} free

Sterile pipette transfer

Media with cryoprotectors: freshly made 5% DMSO solution (Sigma) and 5% glycerol (Sigma) in a 0.2-μm HTF-HEPES filter and add 15% fetal bovine serum or Serum Replacement (SR-1). *Note:* reagent DMSO is highly sensitive to oxidation and has to be stored in a freezer at $-20°$; we recommend the HybriMax grade of DMSO sold in glass vials sealed under argon gas. Until the vial is broken, it can be kept at room temperature, but once it has been opened, the vial must be kept in a freezer. The first sign of oxidation and nonusable DMSO is a light stinky smell; normal DMSO has no smell.

Cryovials, labeled

Freezing containers

Procedure for Freezing Human Embryonic Stem Cells

1. Add 5 ml of PBS Ca^{2+}, Mg^{2+}-free or Hank's Ca^{2+}, Mg^{2+}-free solution to 60 mm or 2 ml per 35-mm Petri dishes with ES cells that have reached high density.

2. Wash out cells and add fresh 2.5 ml PBS Ca^{2+}, Mg^{2+}-free or Hank's Ca^{2+}, Mg^{2+}-free solution to 60 mm or 1 ml per 35-mm Petri dishes

3. Incubate cells with PBS Ca^{2+}, Mg^{2+}-free or Hanks Ca^{2+}, Mg^{2+}-free solution for 5 min at $37°$.

4. Carefully, using the wall of the dish, add an equal volume of 2.5 ml freezing media by gentle rotation of dish, mix it up without distorting the cells, and turn on timer.

5. After 5 to 10 min remove mixture of media and PBS Ca^{2+}, Mg^{2+}-free or Hank's Ca^{2+}, Mg^{2+}-free mixture by aspirator and add 5 to 5.5 ml media with cryoprotectors for 60 mm and 3.2 ml per 35-mm Petri dishes.

6. Scrape cells with blade and extremely gently resuspend cells (on time with no bubbles) by pipetting them into sterile 1.5-ml cryovials (1 ml per unit).

7. On the label of every vial include cell line identification, passage number, and date of freezing.

8. A programmable freezer, Planer, should be preset at $4°$. Timer should show 25 to 30 min before vials with cells are moved in. Start ESC program after vials with cells are transferred into the freezer.

9. Remains of cell suspension left in the dish should be centrifuged and lysis buffer must be added (aliquot for mycoplasma analysis).

10. As soon as the temperature in the freezer has reached $-70°$, transfer vials into liquid vapor storage and record event into freezing cell form and attached temperature chart.

11. Make a record in chart and log book.

Thawing hESC Lines

Place vial into warm tap water ($35°$). As soon as frozen medium in vial thaws out around vial wall, transfer into warm media 10 to 12 ml (usually in a 15-ml conical tube). Gently move until the piece of ice melts out. No big air bubbles should be seen. Place tube in centrifuge for 5 min at 400 to 500 rpm, depending on radius rotor ($\sim50g$). Replace supernatant for growth media and transfer onto feeder layer carefully. From 50 to 100 colonies may be expected to appear on the third day of culture.

Acknowledgments

We are grateful to RGI professionals Koukharenko, Kalmanovich, Seckin, Verlinsky, and Zlatopolsky.

References

Bettler, M., Odavic, R., Deubelbeiss, K. A., and Bucher, U. (1977). The effect of various cryoprotective agents on the human bone marrow. *Schweiz. Med. Wochenschr.* **107**(41), 1459.

Bongso, A., Fong, C. Y., Ng, S. C., and Ratnam, S. (1994). Isolation and culture of inner cell mass cells from human blastocysts. *Hum. Reprod.* **9**(11), 2110–2117.

Bradley, A., Evans, M., Kaufman, M. H., and Robertson, E. (1984). Formation of germ-line chimaeras from embryo-derived teratocarcinoma cell lines. *Nature* **309**(5965), 255–256.

Cole, R. J., Edwards, R. G., and Paul, J. (1965). Cytodifferentiation in cell colonies and cell strains derived from cleaving ova and blastocysts of the rabbit. *Exp. Cell Res.* **37**, 501–504.

Deisseroth, A., Burk, R., Picciano, D., Anderson, W. F., Nienhuis, A., and Minna, J. (1975). Hemoglobin synthesis in somatic cell hybrids: Globin gene expression in hybrids between mouse erythroleukemia and human marrow cells or fibroblasts. *Proc. Natl. Acad. Sci. USA* **72**(3), 1102–1106.

Eistetter, H. R. (1989). Pluripotent embryonal stem cells can be established from disaggregated mouse morulae. *Dev. Growth Differ.* **31**, 275–282.

Evans, M. J., and Kaufman, M. H. (1981). Establishment in culture of pluripotential cells from mouse embryos. *Nature* **292**(5819), 154–156.

Hanel, M. L., and Wevrick, R. (2001). Establishment and maintenance of DNA methylation patterns in mouse Ndn: Implications for maintenance of imprinting in target genes of the imprinting center. *Mol. Cell. Biol.* **21**(7), 2384–2392.

Klimanskaya, I., Chung, Y., Meisner, L., Johnson, J., West, M. D., and Lanza, R. (2005). Human embryonic stem cells derived without feeder cells. *Lancet* **365**(9471), 1636–1641.

Knowles, B. B., Aden, D. P., and Solter, D. (1977). Monoclonal antibody detecting a stage-specific embryonic antigen (SSEA-1) on preimplantation mouse embryos and teratocarcinoma cells. *Curr. Top. Microbiol. Immunol.* **81,** 51–53.

Larson, M. A., Kimura, K., Kubisch, H. M., and Roberts, R. M. (2001). Sexual dimorphism among bovine embryos in their ability to make the transition to expanded blastocyst and in the expression of the signaling molecule IFN-tau. *Proc. Natl. Acad. Sci. USA* **98**(17), 9677–9682.

Lee, K. F., Chow, J. F., Xu, J. S., Chan, S. T., Ip, S. M., and Yeung, W. S. (2001). A comparative study of gene expression in murine embryos developed *in vivo*, cultured *in vitro*, and cocultured with human oviductal cells using messenger ribonucleic acid differential display. *Biol. Reprod.* **64**(3), 910–917.

Martin, G. R. (1981). Isolation of a pluripotent cell line from early mouse embryos cultured in medium conditioned by teratocarcinoma stem cells. *Proc. Natl. Acad. Sci. USA* **78,** 7634–7638.

Schöler, H. R., Ciesiolka, T., and Gruss, P. (1991). A nexus between Oct-4 and E1A: Implications for gene regulation in embryonic stem cells. *Cell* **66**(2), 291–304.

Stevens, L. C. (1970). The development of transplantable teratocarcinomas from intratesticular grafts of pre- and postimplantation mouse embryos. *Dev. Biol.* **21,** 364–382.

Stice, S. L., Strelchenko, N. S., Keefer, C. L., and Matthews, L. (1996). Pluripotent bovine embryonic cell lines direct embryonic development following nuclear transfer. *Biol. Reprod.* **54**(1), 100–110.

Strelchenko, N., Verlinsky, O., Kukharenko, V., and Verlinsky, Y. (2004). Morular-derived human embryonic stem cells. *Reprod. Biomed. Online* **9**(6), 623–629.

Sukoyan, M. A., Vatolin, S. Y., Golubitsa, A. N., Zhelezova, A. I., Semenova, L. A., and Serov, O. L. (1993). Embryonic stem cells derived from morulae, inner cell mass, and blastocysts of mink: Comparisons of their pluripotencies. *Mol. Reprod. Dev.* **36**(2), 148–158.

Tesar, P. J. (2005). Derivation of germ-line-competent embryonic stem cell lines from preblastocyst mouse embryos. *Proc. Natl. Acad. Sci. USA* **102**(23), 8239–8244.

Thomson, J. A., Itskovitz-Eldor, J., Shapiro, S. S., Waknitz, M. A., Swiergiel, J. J., Marshall, V. S., and Jones, J. M. (1998). Embryonic stem cell lines derived from human blastocysts. *Science* **282,** 1145–1147.

Watson, A. J., and Barcroft, L. C. (2001). Regulation of blastocyst formation. *Front. Biosci.* **6,** D708–D730.

[7] Embryonic Stem Cells from Single Blastomeres

By SANDY BECKER and YOUNG CHUNG

Abstract

The fact that deriving embryonic stem (ES) cells from a blastocyst prevents its further development as an embryo is a major issue in human ES cell research. Using eight-cell mouse embryos, we have developed a

METHODS IN ENZYMOLOGY, VOL. 418
Copyright 2006, Elsevier Inc. All rights reserved.

0076-6879/06 $35.00
DOI: 10.1016/S0076-6879(06)18007-6

method of deriving ES cells from a single blastomere, allowing the other seven to continue normal embryonic development. We remove one blastomere and coculture it with green fluorescent protein-labeled ES cells so that it is possible later to separate the clump of blastomere-derived ES cells to a feeder layer for further culturing. The removal of one blastomere is already performed on some human embryos during *in vitro* fertilization as part of prenatal genetic diagnosis.

Introduction

Embryonic stem (ES) cells are routinely derived from the inner cell mass (ICM) of both human and mouse blastocysts (Cowan *et al.*, 2004; Evans and Kaufman, 1981; Thomson *et al.*, 1998). Various attempts have been made to derive them from earlier-stage embryos, primarily to investigate the developmental potency of the blastomeres or to facilitate the process of prenatal genetic diagnosis (PGD) in mice, mink, cows, and humans (Delhaise *et al.*, 1996; Eisetter, 1989; Mitalipova *et al.*, 2001; Strelchenko *et al.*, 2004; Sukoyan *et al.*, 1993; Tesar, 2005). Delhaise *et al.* (1996) disaggregated eight-cell mouse embryos and then cultured the eight cells together on a feeder layer of mouse embryonic fibroblasts (MEFs), eventually deriving an ES cell line. Eisetter (1989) disaggregated 16-cell mouse embryos and cultured the cells from each embryo together on a MEF feeder layer, also deriving several ES cell lines. More recently, Tesar (2005) derived germ line-competent ES cells from preblastocyst mouse embryos, although the embryos were not disaggregated. A previous attempt by Wilton and Trounson (1989) to culture single mouse blastomeres on a variety of substrates produced trophoblast-like cell clumps but no ES cell lines, although the remaining $\frac{3}{4}$ embryos were able to develop normally to the blastocyst stage *in vitro*. However, none of these studies succeeded in deriving an ES line from one blastomere while allowing the others to continue normal embryonic development.

The process described here is far less efficient than deriving ES cells from blastocysts. Often the failure to derive an ES cell line occurs because a trophoblast stem (TS) cell line has emerged instead. This may be at least partly due to the fact that some of the eight-cell blastomeres, although indistinguishable at the light microscope level, are already somewhat disposed toward a trophoblast fate. These would, of course, be the cells at the outer surface of the ball of eight cells and thus the most accessible to the piezo pulse drill, or single blastomeres may be predisposed to become trophoblast cells if they are not in close contact with other blastomeres.

Research suggests that all cells of four- and eight-cell embryos can contribute to both embryonic and trophectoderm lineages when combined with cells from another embryo in chimeras (Tarkowski *et al.*, 2001).

The developmental capacity of blastomeres isolated from mammalian embryos has been studied extensively, and it is clear that they retain their pluripotency and, indeed, are capable of regular *in vivo* development upon transfer into mice (Tarkowski *et al.*, 2001), sheep (Willadsen, 1981), swine (Niemann and Reichelt, 1993), and primates (Chan *et al.*, 2000). However, other studies suggest that although the blastomeres are not irrevocably determined, they may be specified as early as the two- to four-cell stage (Fujimori *et al.*, 2003; Pintrowska-Nitsche *et al.*, 2005). Thus in a neutral *in vitro* environment some of the eight blastomeres may proceed to TS cells, although they could contribute to the embryo proper if they received the proper signals, as, for example, in a chimera.

Derivation of ES Cells from a Single Blastomere

Protocol

1. Biopsy eight-cell stage 129/Sv-*ROSA26:LacZ* mouse embryos through a hole in the zona pellucida using Piezo-pulse drilling. The use of LacZ$^+$ embryos as the source for the blastomeres facilitates confirmation that the resulting cell lines are indeed derived from the single blastomere rather than from any contaminating material. (Of course if this procedure is attempted with human or other mammalian embryos, lacZ labeling will not be available.)

2. Transfer the biopsied embryos to the oviducts of 1.5-dpc synchronized surrogates to establish that they can continue their development. In our hands the blastomere-biopsied embryos develop to term without a reduction in their developmental capacity (49% [23/47] live young versus 51% [38/75] for control nonbiopsied embryos [χ^2 test, $p = 0.85$]). These results are consistent with human data, which indicate that normal and PGD-biopsied embryos develop into blastocysts with comparable efficiency.

3. Aggregate each separated blastomere with a small clump ($\leq$100 cells) of green fluorescent protein (GFP)-positive ES cells in a 300-μm depression created by pressing a needle into the bottom of a plastic tissue culture plate, as described in Nagy *et al.* (2003). The use of GFP-labeled ES cells for coculture facilitates separating the blastomere-derived nascent ES cell clumps from the cocultured cells. GFP-labeled ES cells are available from ATCC or can be made either by transfecting an ES cell line or by deriving a new line from a GFP$^+$ mouse strain. It is also possible to label the coculture cells with a fluorescent cell surface marker such as PKH26 (Sigma) to identify them during the initial steps, but this is less satisfactory, as it may leak into the blastomeres and confuse both the initial steps and the later confirmation.

4. Incubate for 48 to 72 h in ES cell growth medium supplemented with 2000 U/ml mouse leukemia inhibitory factor (LIF) (Chemicon, Temecula,

CA) and 50 μM MEK1 inhibitor (Cell Signaling Technology, Beverly, MA). ES cell growth medium is described in detail in Nagy *et al.* (2003). In brief, it is Dulbecco's modified Eagle's medium (DMEM), supplemented with glutamine, MEM nonessential amino acids, β-mercaptoethanol (BME), antibiotics, 15% fetal calf serum (FCS),[1] and LIF. After 24 to 48 h a growing "bud" of GFP-negative cells should be observed on the sides of the majority of GFP-labeled ES clusters, as shown in Fig. 1A and B. Note that at this stage it is possible to "lose" some buds of blastomere-derived nascent ES cells, as they may be hidden behind the larger floating aggregate of GFP-labeled cocultured ES cells.

5. Aggregates containing visible buds of GFP-negative cells should be plated onto mitomycin C-treated MEFs and cultured in ES cell growth medium until GFP-negative clumps become large enough for dispersion, as shown in Fig. 1C and D. At this point it is again possible to lose some of the single blastomere-derived ES cell clumps, as some may be hidden beneath the larger (and still growing) clump of GFP-labeled cocultured cells. Of course care should be taken to expose the culture to only the minimum ultraviolet light required to visualize the GFP-labeled cells.

6. When the single blastomere-derived, GFP-negative clumps are large enough—about 20 cells or more—separate them from GFP$^+$ ES cells by hand with a microcapillary under a fluorescence microscope. Figure 1E and F show a recently picked GFP-negative colony, along with a few contaminating GFP$^+$ cells. Continue to expand the cells using mechanical and enzymatic methods, while further selecting by eye colonies morphologically resembling ES cells and excluding any GFP$^+$ cells.

7. At this point many of the GFP-negative blastomere outgrowths will morphologically resemble trophoblast or extraembryonic endoderm rather than ES cells. Be patient and observe each well carefully every day. Sometimes colonies of ES cells will appear later in dishes that initially seem to contain cells of trophoblast or endoderm morphology, and these ES-like colonies can be picked mechanically to another dish of feeders for further culture.

8. If TS cells are of interest, they can be cultured further in the ES cell medium with 50 ng/ml fibroblast growth factor (FGF-4) and will produce TS-like cells that can be maintained under these conditions and passaged with trypsin. Detailed protocols for culturing TS cells are provided in Nagy *et al.* (2003). TS cells, like ES cells, are positive for alkaline phosphatase

---

[1] Note that it is important to use a good batch of fetal calf serum. We have found FCS from Hyclone, ES qualified, gives good results, but some batches are better than others. Unfortunately, the only way to test a batch of serum for this application is to use it for this application. Batches tested and found suitable for growing established lines are not necessarily optimal for this derivation protocol.

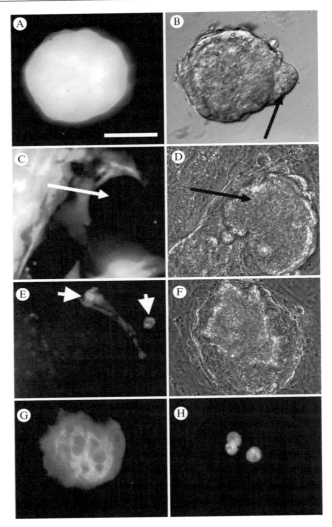

FIG. 1. Stages of single blastomere growth in the presence (A-F) or absence (G and H) of mES cells. (A [green fluorescence] and B [Hoffman modulation optics]) Clump of GFP mES cells 48 h after aggregation with single blastomeres; arrow in B shows a protruding cluster of GFP-negative cells not visible in A. (C [green fluorescence] and D [phase contrast]) Outgrowth of GFP-negative cells aggregated with GFP$^+$ mES cells, after being plated on MEF; arrows point to GFP-negative cells. (E [green fluorescence] and F [phase contrast]) Growth of GFP$^+$ mES cells and cells arising from a single blastomere after mechanical dissociation of initial outgrowth (P1; see text); arrows show remaining GFP$^+$ mES cells. (G) Cells derived from a single blastomere grown on MEF alone for 4 days without ES cells, stained with Troma1, which labels trophoblast cells. (H) Same cells as G, stained with DAPI to show the three nuclei. Scale bar: 100 μm. From Chung et al. (2006), reproduced with permission of Nature Publishing Group.

(Fig. 2E and F). Their morphology as stem cells is similar to that of ES cells, but the colonies are flatter and the colony margins are less smooth. (Fig. 2A and B). They can be confidently distinguished from ES cells because they are negative for Oct4 and positive for Troma1, while ES cells are the reverse (Fig. 2G, H, and J).

Analysis of Cell Lines

1. To establish that the cell lines are indeed derived from the lacZ-positive single blastomere and not from GFP$^+$ helper ES cells, extract DNA and perform polymerase chain reaction (PCR) analysis for both LacZ and GFP. This analysis avoids false negatives associated with possible loss of lacZ or GFP expression. Absence of GFP also confirms that the blastomere-derived ES cells have not fused with the GFP-labeled helper ES cells. We used a QIAamp DNA minikit (Qiagen, Valencia, CA) and 100 ng per reaction for both GFP and LacZ gene amplification. For the GFP gene we used primers 5′-TTGAATTCGCCACCATGGTGAGC-3′ (forward) and 5′-TTGAATTCTTACTTGTACAGCTCGTCC-3′ (reverse), with reaction parameters of 95° for 9 min (1 cycle) and 94° for 45 s, 59° for 1 min, and 72° for 1.5 min for 37 cycles. Separate PCR products on a 1.5% agarose gel and visualize by ethidium bromide staining. We performed LacZ gene genotype analysis with primers and PCR parameters recommended by The Jackson Laboratory (Bar Harbor, ME).

2. LacZ staining on fixed putative single blastomere-derived lines should also be done. We use the Gal-S staining kit from Sigma, according to the manufacturer's instructions. We also use an antibody against β-Gal so that double staining can be done.

3. To establish that the cells are pluripotent ES cells (or TS cells), both RT-PCR and antibody staining can be performed. For RT-PCR, we isolated total RNA from ES and TS cells using an RNAeasy minikit (Qiagen) and subjected 1 μg RNA to first-strand cDNA synthesis with an oligo(dT) primer, using AMV reverse transcriptase (Promega). Use one-tenth of the RT reaction for PCR amplification. PCR conditions for all genes are 95° for 9 min (1 cycle), 94° for 45 s, 62° for 1 min, and 72° for 1.5 min with 2 mM Mg^{2+} concentration. Primers are Oct-4 (expressed by ES cells), forward 5′-CTGAGGGCCAGGCAGGAGCACGAG-3′ reverse 5′-CTGTAGG-GAGGGCTTCGGGCACTT-3′ (484 bp); Nanog (expressed by ES cells), forward 5′-AGGGTCTGCTACTGAGATGCTCTG-3′, reverse 5′-CAAC-CACTGGTTTTTCTGCCACCG-3′ (363 bp); Cdx2 (expressed by TS cells), forward 5′-GGCGAAACCTGTGCGAGTGGATGCGGAA-3′, reverse 5′-GATTGCTGTGCCGCCGCCGCTTCAGACC-3′ (492 bp), and Rex-1 (expressed by ES cells), forward 5′-AGCAAGACGAGGCAAGGCCAG-TCCAGAATA-3′, reverse 5′-GAGGACACTCCAGCATCGATAAGA-

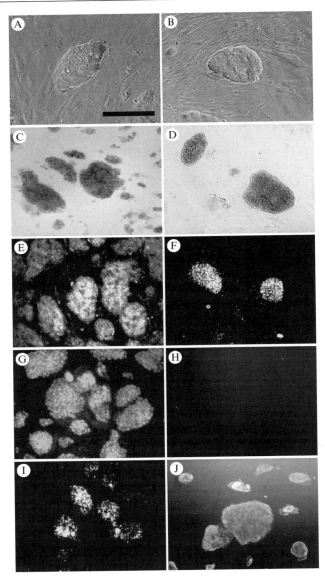

FIG. 2. Comparison of putative ES (left column) and TS (right column) cell lines derived from single blastomeres. (A and B) Phase-contrast photograph of typical colonies. (C and D) Lac-Z-stained colonies showing their single blastomere origin. (E and F) Alkaline phosphatase staining. (G and H) Indirect immunofluorescence with antibodies to Oct-4. (I) SSEA-1 staining of putative ES cells. (J) TROMA-1 staining of putative TS cells (same field as H). Scale bar: 200 μm. From Chung *et al.* (2006), reproduced with permission of Nature Publishing Group. (See color insert.)

CACCAC-3′ (423 bp). Separate PCR products on a 1.5% agarose gel and visualize by ethidium bromide staining.

4. For antibody staining, grow the cells on four-well tissue culture plates according to normal ES cell culture protocols. These are described in detail in Nagy *et al.* (2003). Fix in freshly made 2% paraformaldehyde for 10 min, permeabilize with 0.1 % Nonidet P-40 for 10 min, block with 10% goat serum + 10% donkey serum (Jackson Immunoresearch) in phosphate-buffered saline (PBS) for an hour, and incubate with primary antibodies overnight. (Cells can be stored for several weeks immediately after fixation at 4° in PBS.) In the morning rinse 3× 15 min in PBS containing 0.1% Tween 20 (Sigma), followed by secondary antibodies for 1 h at room temperature, then 3× 15-min washes with PBS/Tween. Mount specimens in Vectashield with 4′,6-diamidino-2-phenylindole (DAPI) (Vector Laboratories, Burlingame, CA). We have used the following primary antibodies: Oct-4 (Santa Cruz Biotechnology, Santa Cruz, CA) identifies ES cells, SSEA-1 (developed by Solter and Knowles and obtained through the DSHB of the University of Iowa, Iowa City, IA) stains stem cells, Troma-1 (raised by Brulet and Kemler and obtained through DSHB) stains trophoblast stem cells and extraembryonic endoderm, α-feto protein (DAKO) stains extraembryonic visceral endoderm, and β III tubulin (Covance, Berkeley, CA) stains neural cells and muscle actin (Abcam, Cambridge, MA). For alkaline phosphatase staining we use the Vector Red kit from Vector Laboratories. This kit yields both a visible (reddish) signal and a red fluorescent signal so cells can be double stained with a green-labeled primary antibody.

5. Like existing ES cell lines, these can be induced to differentiate *in vitro* into cell types representing all three germ layers. They can also form teratomas and contribute to chimeras. To make teratomas inject approximately 1 million ES cells into the rear thigh of a NOD-SCID mouse in 100 μl of DMEM. After approximately 2 months sacrifice the mice and excise the teratomas, fix them in 4% paraformaldehyde, embed in paraffin, section, and stain. Our lines produced teratomas showing examples of tissue from all three embryonic germ layers.

6. To make chimeras, either inject cells into CD-1 mouse blastocysts or aggregate with eight-cell stage morulae and transfer to recipient females. Nagy *et al.* (2003) provides detailed protocols for making chimeras. X-Gal staining of the harvested fetuses will reveal the contribution of the blastomere-derived ES cell lines. The ultimate proof of the totipotency of your ES lines will be their contribution to the germ line of a chimeric male and subsequent birth of his live offspring, which are lacZ positive.

References

Chan, A. W., Dominko, T., Luetjens, C. M., Neuber, E., Martinovich, C., Hewitson, L., Simerly, C. R., and Schatten, G. P. (2000). Clonal propagation of primate offspring by embryo splitting. *Science* **287**, 317–319.

Chung, Y., Klimanskaya, I., Becker, S., Marh, J., Lu, S. J., Johnson, J., Meisner, L., and Lanza, R. (2006). Embryonic and extraembryonic stem cell lines derived from single mouse blastomeres. *Nature* **439**, 216–219.

Cowan, C. A., Klimanskaya, I., McMahon, J., Atienza, J., Witmyer, J., Zucker, J. P., Wang, S., Morton, C. C., McMahon, A. P., Powers, D., and Melton, D. A. (2004). Derivation of embryonic stem-cell lines from human blastocysts. *N. Engl. J. Med.* **350**, 1353.

Delhaise, F., Bralion, V., Schuurbiers, N., and Dessy, F. (1996). Establishment of an embryonic stem cell line from 8-cell stage mouse embryos. *Eur. J. Morph.* **34**, 237–243.

Eistetter, H. R. (1989). Pluripotent embryonal stem cell lines can be established from disaggregated mouse morulae. *Dev. Growth Differ.* **31**, 275–282.

Evans, M. J., and Kaufman, M. H. (1981). Establishment in culture of pluripotential cells from mouse embryos. *Nature* **292**, 154.

Fujimori, T., Kurotaki, Y., Miyazaki, J., and Nabeshima, Y. (2003). Analysis of cell lineage in two- and four-cell mouse embryos. *Development* **130**, 5113–5122.

Mitalipova, M., Beyhan, Z., and First, N. L. (2001). Pluripotency of bovine embryonic cell line derived from precompacting embryos. *Cloning* **3**, 59–67.

Nagy, A., Gertsenstein, M., Vintersten, K., and Behringer, R. (eds.) (2003). "Manipulating the Mouse Embryo: A Laboratory Manual," 3rd Ed. Cold Spring Harbor Laboratory Press, Cold Spring Harbor, NY.

Niemann, H., and Reichelt, B. (1993). Manipulating early pig embryos. *J. Reprod. Fertil. Suppl.* **48**, 75–94.

Piotrowska-Nitsche, K., Perea-Gomez, A., Haraguchi, S., and Zernicka-Goetz, M. (2005). Four-cell stage mouse blastomeres have different developmental properties. *Development* **132**, 479–490.

Strelchenko, N., Verlinsky, O., Kukharenko, V., and Verlinsky, Y. (2004). Morula-derived human embryonic stem cells. *Reprod. Biomed.* **9**, 623–629.

Sukoyan, M. A., Vatolin, S. Y., Golubitsa, A. N., Zhelezova, A. I., Semenova, L. A., and Serov, O. L. (1993). Embryonic stem cells derived from morulae, inner cell mass, and blastocysts of mink: Comparisons of their pluripotencies. *Mol. Reprod. Dev.* **36**, 148–158.

Tarkowski, A. K., Ozdzenski, W., and Czolowska, R. (2001). Mouse singletons and twins developed from isolated diploid blastomeres supported with tetraploid blastomeres. *Int. J. Dev. Biol.* **45**, 591–596.

Tesar, P. J. (2005). Derivation of germ-line-competent embryonic stem cell lines from preblastocyst mouse embryos. *Proc. Natl. Acad. Sci. USA* **102**, 8239.

Thomson, J. A., Itskovitz-Eldor, J., Shapiro, S. S., Waknitz, M. A., Swiergiel, J. J., Marshall, V. S., and Jones, J. M. (1998). Embryonic stem cell lines derived from human blastocysts. *Science* **282**, 1145.

Willadsen, S. M. (1981). The development capacity of blastomeres from 4- and 8-cell sheep embryos. *J. Embryol. Exp. Morphol.* **15**, 165–172.

Wilton, L., and Trounson, A. (1989). Biopsy of preimplantation mouse embryos: Development of micromanipulated embryos and proliferation of single blastomeres *in vitro*. *Biol. Reprod.* **40**, 145–152.

[8] Embryonic Stem Cells from Parthenotes

By Jose B. Cibelli, Kerrianne Cunniff, and Kent E. Vrana

Abstract

While human embryonic stem cells (hESCs) hold tremendous therapeutic potential, they also create societal and ethical dilemmas. Adult and placental stem cells represent two alternatives to the hESC, but may have technical limitations. An additional alternative is the stem cell derived from parthenogenesis. Parthenogenesis is a reproductive mechanism that is common in lower organisms and produces a live birth from an oocyte activated in the absence of sperm. However, parthenogenetic embryos will develop to the blastocyst stage and so can serve as a source of embryonic stem cells. Parthenogenetic ESCs (pESCs) have been shown to have the properties of self-renewal and the capacity to generate cell derivatives from the three germ layers, confirmed by contributions to chimeric animals and/or teratoma formation when injected into SCID mice. Therefore, this mechanism for generating stem cells has the ethical advantage of not involving the destruction of viable embryos. Moreover, the cells do not involve the union of male and female and so genetic material will be derived exclusively from the female oocyte donor (with the attendant potential immunological advantages). This chapter describes the biology underlying parthenogenesis, as well as provides detailed technical considerations for the production of pESCs.

Introduction

There are no reports of mammalian reproduction by parthenogenesis. However, this form of reproduction—through activation of the unfertilized oocyte—is common in insects and other lower organisms. Within the laboratory, we can generate mammalian preimplantation embryos and fetuses that, although incapable of developing to term, can go through gastrulation and early stages of organogenesis when transferred to a surrogate uterus. We and others have taken advantage of this phenomenon and generated a number of preimplantation embryos in a variety of mammalian species, such as mouse, rat, rabbit, pig, goat, cow, monkey, and human (Table I). This chapter focuses on ways to activate mammalian oocytes parthenogenetically and on how to derive embryonic stem cells (ESCs) from them. These parthenogenetic ESCs (pESCs) have been shown to have the properties of self-renewal and the capacity to generate cell derivatives from the three

METHODS IN ENZYMOLOGY, VOL. 418
0076-6879/06 $35.00
DOI: 10.1016/S0076-6879(06)18008-8

TABLE I

SELECTED GROUP OF OOCYTE ACTIVATION PROTOCOLS EFFECTIVE IN PRODUCING BLASTOCYSTS FOR DIFFERENT SPECIES

Author	Year	Species	Agent	% Blastocyst
Lin Liu and Xiangzhong Yang (1998)	1998	Bovine	Ethanol + DMAP	36
			Ionomycin + DMAP	40
			Ionomycin + cycloheximide	7
			Ionomycin + CHX + Cyto D	36
Meo et al. (2005)	2005	Bovine	Strontium with or without ionomycin	8–13
Loi et al. (1998)	1998	Sheep	Ionomycin + DMAP	58.4
			EtOH + DMAP	19.1
Ongeri et al. (2001)	2000	Goat	Ionomycin + DMAP	50.1
			EtOH + DMAP	49.8
Lan et al. (2005)	2005	Goat	Ionomycin + DMAP	41.9
Zhu et al. (2002)	2002	Pig	Various electrical pulses	15–41
Grupen et al. (2002)	2002	Pig	Electrical pulses + DMAP (different times and concentrations)	20–34
Yi and Park (2005)	2005	Pig	EtOH + CHX + Cyto B + DMAP	25
Ozil (1990)	1990	Rabbit	22 double pulses gradually decreasing duration	89
Mitalipov et al. (1999)	1999	Rabbit	Electroporation of 25 mM IP3 + DMAP	50
			Ionomycin + DMAP	5.7
			Multiple pulses	30
Liu et al. (2002)	2002	Rabbit	A23187 + DMAP	36
			EtOH + DMAP	47
Chesne et al. (2002)	2002	Rabbit	Electrical pulses then CHX + DMAP (1 h)	90
Liu et al. (2004)	2004	Rabbit	Pulses then CHX + DMAP then pulses then CHX + DMAP	64

Reference	Year	Species	Method	%
Liu et al. (2005)	2005	Rabbit	Pulses then pulses + IP3 then DMAP	72.7
Roh et al. (2003)	2003	Rat	Strontium + Cyto B + CHX	28.2
Krivokharchenko et al. (2003)	2003	Rat	Electrical pulses + Cyto B	29.6
			EtOH + Cyto B	25
Mizutani et al. (2004)	2004	Rat	Strontium (2 h) + Cyto B	19.6
			Electrical pulses + DMAP	42
			Ionomycin + DMAP	42
Cuthbertson (1983)	1983	Mouse	8.6% EtOH 8'	62.2
			0.32% benzyl alcohol 8'	85.1
Bos-Mikich et al. (1997)	1997	Mouse	Strontium (different times)	73–87
Uranga and Arechaga (1997)	1996	Mouse	A23187 + OAG + Cyto D	55
Ozil et al. (2005)	2005	Mouse	Various electrical pulses	75–79
Toth et al. (2006)	2006	Mouse	Various electrical pulses	62–75
Mitalipov et al. (2001)	2001	Monkey	Ionomycin + DMAP	58
			Electroporation + CHX + Cyto B	48
			Ionomycin + roscovitine	25
Cibelli et al. (2002)	2002	Monkey	Ionomycin + DMAP	22
Cibelli et al. (2001)	2001	Human	Ionomycin + DMAP	27
Rogers et al. (2004)	2004	Human	0.1 μg/ml PLCzeta cRNA injection	16.67
Lin et al. (2003)	2003	Human	Ionophore + DMAP	28.57
			Ionophore + puromycin	1 morula (20%)
			Sham ICSI + 50 μM ionophore 15'	1 blast (16.67%)

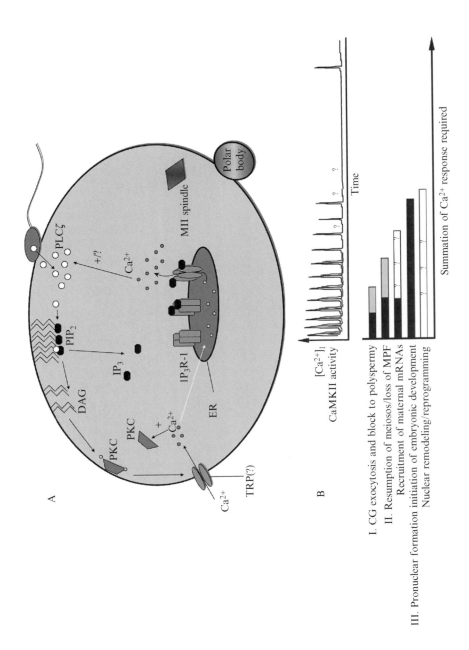

germ layers, confirmed by contributions to chimeric animals and/or teratoma formation when injected into SCID mice (Cibelli *et al.*, 2002; Kaufman *et al.*, 1983; Vrana *et al.*, 2003).

In the absence of sperm, mammalian oocytes will remain arrested at the metaphase II stage of meiosis II until they are fertilized. This blockage in development is possibly an evolutionary safety mechanism to avoid parthenogenetic development inside the female. During fertilization, the sperm enters the egg and triggers a series of events that will ultimately lead to the first cell division. We refer to this process throughout the manuscript as oocyte "activation." Over the years, laboratory protocols have been developed and refined to release the oocyte from its arrest in the absence of sperm. These protocols are highly effective, but differ among species. Our primary objective here is to describe the available methods to activate mammalian oocytes. Emphasis will be placed on describing in detail those protocols that are effective whether or not they resemble the activation triggered by the sperm.

We know now a great deal about the mechanisms by which the mammalian oocyte is arrested at metaphase II. This phenomenon is mediated by the activity of maturation promoting factor (MPF) and cytostatic factor (CSF). Upon sperm entry, phospholipase C (PLC-ζ) is released into the oocyte cytosol and a series of signaling pathways are coordinately activated (Fig. 1A). This results in release of calcium into the cytosol at an amplitude and frequency that is dependent on the species. These periodic calcium oscillations can be maintained for a few hours and up to 20 h. PLC-ζ will act upon phosphatidylinositol 4,5-bisphosphate and release inositol 1,4,5-trisphosphate (IP$_3$) from the cell membrane; IP$_3$, in turn, will open the calcium channels in the endoplasmic reticulum and calcium is released. Moreover, there is an influx of calcium from the extracellular medium as well. IP$_3$ also activates 1, 2-diacylglycerol, which then activates protein kinase C (PKC). The opening of the cell membrane calcium channels is thought to be mediated by PKC and transient receptor potential ultimately to be inactivated by calcium. The cycle repeats itself for as long as there is PLC-ζ available in the cytosol. The increase in the concentration of intracellular calcium will translate into an increase in the activity of Ca^{2+}/calmodulin-dependent protein kinase II (CaMKII), which also mediates inactivation of MPF and CSF. Once these two complexes are inactivated—

FIG. 1. (A) Proposed mechanisms of [Ca2+]i oscillations during mammalian fertilization. (B) Egg activation events exhibit different [Ca2+]i requirements. Adapted from Malcuit *et al. Journal of Cellular Physiology*, 2006.

a process that takes several hours—a series of events will have occurred that indicate the oocyte is activated. The hallmarks for oocyte activation are cortical granular exocytosis, resumption of meiosis (loss of MPF), and finally pronuclear formation. The recruitment of maternal RNA present in the cytosol has been added to the list of events linked to oocyte activation (Malcuit et al., 2006) (Fig. 1B).

In the context of this chapter, we describe different protocols that are effective at bypassing some of these well-characterized mechanisms. Some of the compounds that are commonly used to activate the oocytes will inactivate MPF and CSF directly, whereas others will reduce the synthesis of these proteins, ultimately releasing the oocyte from the MII arrest. A simplistic approach for classifying the methods of activation of oocytes is by dividing them based on their nature: (1) mechanical, (2) physical, and (3) chemical. Mechanical methods are not currently used in mammals, although they are extremely effective in amphibians. The sole act of touching oocytes with a needle can trigger activation in lower organisms. Physical methods include temperature fluctuations, electrical pulses, and altering osmolarity. Chemical activation, the most sophisticated method available today, includes incubation with protein kinase inhibitors (specific or broad action), protein synthesis inhibitors, and microfilament inhibitors, as well as calcium ionophore and/or strontium.

The literature on oocyte activation for different species is vast. We have selected a few examples that illustrate protocols that have been successful at generating blastocysts (Table I); however, it is worth mentioning that only two species (mouse and monkey) have generated putative parthenogenetic ESCs (Cibelli et al., 2002; Robertson et al., 1983; Vrana et al., 2003).

Mouse Parthenogenetic ESCs

Mouse pESCs were first described by Robertson and colleagues in 1983. These cells have been studied extensively and have shed light onto issues of parental contribution to the overall phenotype of the animal. The early activation protocols for mouse oocytes were based on the use of a solution of 8% ethanol for a short period of time (approximately 6 min). This treatment triggers calcium oscillations and activates CaMKII, ultimately inducing the oocytes to cleave. Although reliable, this protocol produces parthenogenetic embryos that are haploid, as the second polar body is extruded (Borsuk et al.,1996; Winston and Maro, 1995). There are two well-described methods to generate diploid embryos in the mouse. One is to let the oocyte extrude the second polar body and then later fuse the two blastomeres following the first cleavage (Rougier and Werb, 2001).

The second option is to use a reversible, nontoxic microfilament inhibitor such as cytochalasin B (Allen *et al.*, 1994). The effectiveness of the different methods for the generation of diploid parthenogenetic blastocysts is depicted in Table I. Briefly, the most effective method that seems to be applicable for all mouse strains is a combination of strontium and cytochalasin B (see details later). Strontium triggers intracellular calcium oscillations that resemble those produced in response to sperm entry, whereas cytochalasin B prevents the extrusion of the second polar body. This protocol, described by Kono *et al.* (1996), is still the best protocol to date for the generation of morphologically normal diploid mouse parthenogenetic blastocysts and ESCs (Fig. 2).

Parthenogenetic mouse ESCs have been instrumental in the understanding of imprinted gene function, but, more importantly, they offer an opportunity to test the role of the female genome during development. Perhaps one of the most interesting manuscripts published in this area is by Allen *et al.* (1994). In their work, a comparison between parthenogenetic ESCs and parthenogenetic blastomeres was made in order to determine their potential to contribute to the three different germ layers of a developing embryo/fetus (Fig. 3). Chimeric mice produced using parthenogenetic blastomeres manifested significant growth retardation that was directly correlated with the degree of parthenogenetic contribution. Based on these results, it is tempting to conclude that parthenogenesis will never be a viable form of mammalian reproduction. However, a more intriguing finding from the same group was also reported. When parthenogenetic ESCs were used instead of parthenogenetic blastomeres, the results were dramatically different. Parthenogenetic ESCs that were cultured *in vitro* for several passages and later transferred to the host embryo produced chimeric offspring with a more normal phenotype. There was no growth retardation even though the level of chimerism was very high and, in some instances, germ line transmission was obtained from these ESCs. Their results strongly suggest that the female genome can

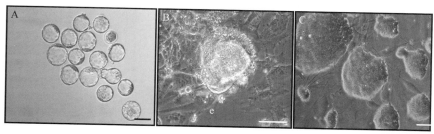

FIG. 2. Derivation of mouse parthenogenetic ES cells. (A) parthenogenetic mouse embryos 5 days after activation; (B) ICM outgrowth 5 days after embryo plating (200×), (C) ESCs colonies after 20 passages (200×). Scale bar = 100 μm.

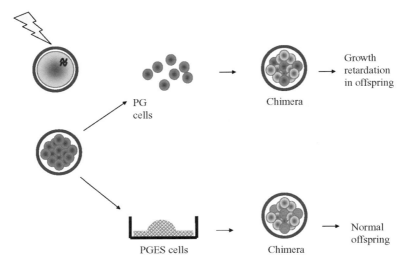

FIG. 3. Parthenogenetic blastomeres (PG) when used to make chimeras, generate offspring with growth retardation; however, when a parthenogenetic embryo is used to make ESCs and then those ESCs are used for chimeras, the offspring is normal. *Adapted from Allen et al. Development 1994.* (See color insert.)

indeed be remodeled or reprogrammed to sustain development. Professor Surani and colleagues did not elaborate on the mechanism by which these cells were capable of improving their developmental potential. Answers arose from later work conducted by Kono and colleagues (2002). They found that H19 was the most deregulated gene in parthenogenetic embryos. Their finding proved to be of central importance, as a subsequent report showed that by using a hemizygous knockout line for H19, parthenogenetic development was improved up until day 17.5, whereas the control produced live fetuses until day 10.5 only. This work culminated with the report of viable adult offspring following recombinant manipulation of H19 and generation of chimera from different stages of culture (Kono *et al.,* 2004). These results highlight the significance of one gene in the overall developmental program of mammalian embryos. Furthermore, it stimulates this field of research in terms of the potential use of parthenogenetic ESCs in primates for regenerative medicine.

Nonhuman Primate Parthenogenetic ESCs

While exploring the possibility of conducting somatic cell nuclear transfer in cynomolgus macaque, we developed an activation protocol that has proven to be effective for the generation of blastocysts (see later). Our experiences, as well as others, have shown that a combination of ionomycin and 6-dimethylaminopurine (DMAP) treatments is quite effective. Briefly, the protocol calls for the use of 10 μM of ionomycin for 8 min followed by incubation in 2 mM DMAP for 4 h. Ionomycin elicits a large intracellular calcium spike, while DMAP inhibits protein phosphorylation. Four hours of incubation in DMAP has previously proven effective in bypassing normal cytokinesis of the metaphase II oocyte and inhibiting the extrusion of the second polar body. Upon its removal, the oocyte resumes cytokinesis, but this time going directly into the first mitotic division and therefore maintaining a diploid state. Using this protocol, Mitalipov and colleagues (2001) activated rhesus monkey oocytes, while our group worked with the cynomolgus macaque (Cibelli *et al.*, 2002; Vrana *et al.*, 2003). Parthenogenetic activation produced development to blastocyst stage in 58 and 22% of the activated oocytes, respectively. Our study included 18 MII oocytes from which we obtained four blastocysts, one of which developed into a parthenogenetically ESC line—the Cyno 1 line (Fig. 4 [Cibelli *et al.*, 2002; Vrana *et al.*, 2003]).

In our case, cynomolgus macaque oocytes are removed from ovaries and matured *in vitro* in maturation medium. After a short period of equilibration, all the oocytes that have extruded the first polar body are placed in hamster embryo culture media with the addition of HEPES (HECM-HEPES) plus 10 μM ionomycin for 4 min. Subsequently, the oocytes are moved to a solution of Cooks cleavage medium with 2 mM DMAP for 4 h and finally placed in Cooks cleavage medium for 72 h. Cook blastocyst culture medium with 10% fetal bovine serum cocultured with mouse embryonic fibroblasts is used for the last 3 days of culture. All incubations are performed at 37° in 5% CO_2 in air.

Using a mild incubation in pronase in HECM-HEPES, we remove the zona pellucida and plate the blastocysts in a layer of mitotically inactivated mouse fetal fibroblast (MEF). After 2 weeks in culture, a colony appears in

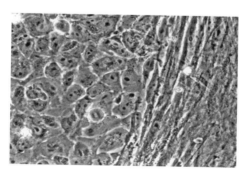

FIG. 4. Cyno1 cells (left) growing on top of MEF (right).

one of the plates (Cyno-1). Morphologically, these cells are similar to human ESCs. They have large nuclei with prominent nucleoli and very small cytoplasm (Fig. 4). The first passages are performed mechanically as described elsewhere (Thomson et al., 1998). Briefly, using a glass pipette, pieces of the colony are cut and moved to fresh MEF. Cyno-1 cells are characterized extensively for the presence of pluripotency markers. They are positive for SSEA1, SEEA3, TRA1-60, TRA1-81, telomerase, alkaline phosphatase, Oct4, and Nanog. During the first few passages, Cyno-1 cells spontaneously differentiate into contractile beating cells with cardiomyo-cyte characteristics and ciliated epithelia. In vivo differentiation is tested by injecting 1×10^6 cells into SCID mice. Apparently, the location of injection has no effect on the outcome, as we found that these cells are capable of forming teratomas inside the peritoneal cavity, intramuscularly, or subcu-taneously. Upon histological examination, these teratomas were shown to have several highly differentiated tissues, such as gut, respiratory epitheli-um, bone, cartilage, skeletal and smooth muscle, ganglion, and hair (Fig. 5). Direct differentiation studies into particular lineages were performed by our group and others. We found that the neuronal derivatives are the easiest to obtain (Vrana et al., 2003). Cyno-1 cells can differentiate into tyrosine hydroxylase (TH)-positive neurons, glial cells, and serotonergic neurons (Fig. 6). TH-positive cells were also transplanted into the brain of Parkinsonian rats and shown to reduce their symptoms significantly when compared to sham injection (Sanchez-Pernaute et al., 2005). Whether parthenogenetically derived ESCs will have a role in regenerative medicine remains to be determined. One of the most important questions that remains unanswered is to what extent the cells can be considered self when transplanted back into the female that donated the oocytes. Considering the fact that these cells were derived after the first meiotic division and that chromosome recombination had already taken place, some alleles that were expressed in the mother will be expressed in the ESCs. If we have to predict the outcome of the transplantation experiment into the oocyte donor, parthenogenetically derived cells will be less antigenic than allogeneic cells but more than isogenic cells.

The Importance of Choosing the Right Activation Protocol

As described earlier, the majority of the activation protocols rely on an increase of intracellular calcium triggered either by calcium itself or with the aid of exogenous IP_3 or adenophostin (Jellerette et al., 2004), plus the addition of microfilament inhibitors or kinase inhibitors. Several combina-tions of the aforementioned protocols do work for most of the mammalian species and can be optimized easily by changing the concentration of the

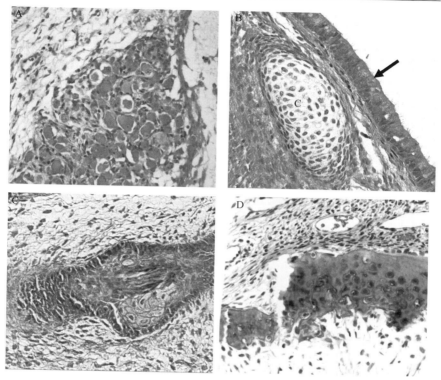

Fig. 5. Teratoma produced from Cyno1 cells 15 weeks after injection into SCID mice. (A) Ganglion; (B) Cartilage (c) and Respiratory epithelium (arrow); (C) Hair follicle; and (D) Bone. (See color insert.)

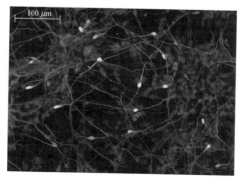

Fig. 6. Double ICC for MAP-2 and Serotonin after differentiation of Cyno1. Nuclei are stained with Hoechest dye. (See color insert.)

drug or their incubation length. The end point generally used in almost all of the *in vitro* experiments is development to the blastocyst stage; in rare cases the total cell number in the parthenogenetic blastocysts is reported. Ozil and Huneau (2001) published a seminal paper describing a method of activation that marked a turning point on the way we look at oocyte activation in relation to gene expression. They used rabbit oocytes, subjected them to electrical pulses of different amplitude and length, and later transferred them to the uterus of recipient females. Although most of the different treatments were similar for activating oocytes (measured by cleavage rate), the outcome was very different when these parthenogenetic embryos were transferred to the uterus. Only a particular set of electrical pulses of a precise amplitude and frequency was capable of giving rise to parthenogenetic fetuses morphologically similar to biparental ones. No parthenogenetic offspring were obtained, however.

Further studies in mice, using the same approach, have shown that not only is the *in vivo* developmental potential altered depending on the protocol used, but the early markers for oocyte activation, such as cortical granule exocytosis, were also affected. There was also a remarkable difference in the gene expression profile of these activated oocytes between treatments (Ozil *et al.*, 2005). Taken together, these results strongly suggest that more work needs to be conducted to understand the best activation conditions to mimic the activation produced by sperm. Further, it is clear that development to blastocyst stage and/or total cell number is not a reliable marker indicator for proper activation.

Imprinting and Parthenogenesis

Several properties of parthenogenetic ESCs make them appealing for *in vitro* studies of imprinting. These include the following: (1) pESCs lack any paternal contribution; (2) when cultured properly, they can divide indefinitely; and (3) when induced to differentiate, they can make all the tissues of the adult body. Caution must be exerted, however. Our laboratories have extensively characterized Cyno1 cells for a large battery of imprinted genes (Table II). Focusing on H19 and IGF2 (maternally and paternally expressed, respectively), we found that H19 is always upregulated when compared to biparental control cells (either adult skin fibroblasts or biparental macaque ESCs). Surprisingly, we also found that IGF2 was upregulated when the cells are pluripotent, but that it is later downregulated when differentiation takes place. H19 is always expressed at higher levels when compared to controls irrespective of the differentiation status of the cells. More studies are required to understand the mechanism of gene regulation of these cells.

TABLE II

GENE EXPRESSION COMPARISON OF IMPRINTED GENES IN BIPARENTAL (BP) CYNOMOLGUS
ES CELLS AND CYNO-1 PESCs USING QUANTITATIVE REAL-TIME PCR ANALYSIS

Gene[a]	BP	CYNO-1	Expression	Folds upregulated in Cyno1	Folds downregulated in Cyno1
p57	4.56E-05	0.000461319	Maternal	10.1	—
Peg10	0.001094	1.40292E-05	Paternal	0	10
Ndn	0.000282	1.00E-10	Paternal	0	—
Snrpn	0.013069	1.00E-10	Paternal	0	—
Igf2	0.004152	0.009617525	Paternal	2.3	—
H19	0.000411	0.00536016	Maternal	13.0	—

[a] Most notably, Igf2 was upregulated 2.3 times the normal expression level found in BP cells even though it is expressed paternally. Additionally, Peg10 was downregulated 10-fold.

Protocols for Derivation of Parthenogenetic ESCs

Mouse

Oocytes are collected from superovulated CD1 females 20 h after hCG injection using HECM-HEPES-buffered medium (Table III). A brief exposure to hyaluronidase (1 mg/ml in HECM-HEPES) is used to obtain denuded oocytes.

Oocytes with proper morphology are activated using 10 mM strontium chloride (Sigma) and 5 μg/ml cytochalasin B (Sigma) in calcium-free potassium simplex-optimized medium (KSOM) (Lawitts and Biggers, 1993) for 5 h at 37° and 5% CO_2 in air.

Oocytes are rinsed thoroughly in HECM-HEPES three times.

Oocytes are cultured in KSOM (Chemicon) at 37° and 5% CO_2 in air for 3 to 4 days until they reach the expanded/hatching blastocyst stage (Fig. 3A).

Parthenogenetic blastocysts that have not hatched are subjected to acidic Tyrode's solution (Sigma) to remove the zona pellucida.

Whole embryos (trophoblast and inner cell mass) are plated on mitotically inactivated MEF in mouse ESC culture medium at 37° and 5% CO_2 in air and allowed to attach until the inner cell mass (ICM) starts to grow.

Once a visible mound has grown from the ICM, it is picked from the plate and dissociated with trypsin EDTA (GIBCO) and replated on a fresh feeder layer.

Medium is replaced daily and colonies are passaged using trypsin EDTA approximately every 3 days.

TABLE III
REAGENTS AND SUPPLIERS

(a) HECM-HEPES (HH) adapted from McKiernan et al. (1991)[a]

Component	mM	Company	Unit	Unit
NaCl	114.0	Sigma	6.662 g	3.331 g
KCl	3.2	Sigma	0.239 g	0.1195 g
$CaCl_2 \cdot 2H_2O$	2.0	Sigma	0.294 g	0.147 g
$MgCl_2 \cdot 6H_2O$	0.5	Sigma	0.102 g	0.051 g
MEM nonessential amino acids	—	Sigma	10 ml	5 ml
Lactic acid		Sigma	1.44 ml	0.72 ml
Sodium pyruvate	0.1	Sigma	0.011 g	0.0055 g
$NaHCO_3$	2.0	Sigma	0.168 g	0.084 g
HEPES	10.0	Sigma	2.38 g	1.19 g
Phenol red	—	Sigma	5 mg	2.5 mg
Penicillin/streptomycin	—	GIBCO	5 ml	2.5 ml
Bovine serum albumin	3 mg/ml	Sigma	3 g	1.5 g
Double-distilled water	—	—	To 1 liter	To 0.5 liter

(b) Maturation medium[b]

Component	Company	Unit
CMRL-1066	GIBCO	
FBS	Hyclone	20%
PMSG	Sigma	10 IU/ml
hCG	Sigma	10 IU/ml
Penicillin/streptomycin	GIBCO	50 μl

(c) Ionomycin (5 mM) 500× stock solution adapted from Susko-Parrish et al. (1994)[c]

Component	Company	Unit
Ionomycin	Cal-Biochem	1 mg
Dimethyl sulfoxide (DMSO)	Sigma	267.6 μl

(d) DMAP (200 mM) 100× stock solution adapted from Susko-Parrish et al. (1994)[d]

Component	Company	Unit
DMAP (200 mM)	Sigma	1 g
D-PBS	GIBCO	30.64 ml

(e) Strontium 100 × stock solution[e]

Component	Company	Unit
$SrCl_2 \ 6H_2O$	Sigma	0.2666 g
dH2O		1 ml

(continued)

TABLE III (*continued*)

(f) Cytochalasin B (5 mg/ml) 1000× stock solution adapted from Presicce and Yang (1994)[f]

Component	Company	Unit
Cytochalasin B	Sigma	1 mg
DMSO	Sigma	200 μl

(g) mESC medium[g]

Component	Concentration	Company	Unit
Knockout D-MEM		GIBCO	400 ml
Knockout Serum Replacement	20%	GIBCO	100 ml
L-Glutamine	2 mM	GIBCO	5 ml
MEM nonessential amino acids	—	Sigma	5 ml
β-Mercaptoethanol	0.1 mM	Sigma	4 μl
b-FGF	4 ng/ml	Invitrogen	1 ml

(h) mESC freezing medium[h]

Component	Concentration	Company
mESC medium	80%	
FBS	10%	GIBCO
DMSO	10%	Sigma

[a] Use plastic containers; pH 7.3–7. Osmolarity: 275 $\pm$ 10 mOsm/kg 0.22-μm Millipore filter.
[b] Filter. Store at 4°.
[c] Five-microliter aliquots. Store at −20° up to 12 months.
[d] Dissolve DMAP in 90° water bath; ~12-μl aliquots. Store at −20° up to 6 months.
[e] Filter with 0.2-μm filter. Aliquot at 15 μl. Store at −20°.
[f] Five-microliter aliquots. Store at −20° <36 months.
[g] Store at 4°.
[h] Filter and store at 4°.

Nonhuman Primate

Cynomolgus macaques used in our study are not superovulated. Ovaries are removed and oocytes aspirated from the follicles. Oocyte maturation is performed using the maturation medium described in Table III at 37°, 5% CO_2 in air for 40 h.

Subsequently, the oocytes were treated as described in the mouse protocol.

For activation, oocytes are placed in 10 μM ionomycin (Calbiochem) in HECM-HEPES for 4 min.

Oocytes are transferred to a solution of 2 mM DMAP in Cooks cleavage medium (www.cookobgyn.com) for 4 h at 37°, 5% CO_2 in air.

Activated oocytes are rinsed thoroughly in HECM-HEPEs three times.

Activated oocytes are placed in Cooks cleavage medium (www. cookobgyn.com) at 37° and 5% CO_2 in air for 6 to 7 days until they reach blastocyst stage.

Zona pellucida is removed using a mild pronase treatment.

Parthenogenetically derived blastocysts are plated in mitotically inactivated MEF in Cyno culture medium at 37° and 5% CO_2 in air.

Once attachment is observed, cells with the morphology depicted in Fig. 4 will start to grow after 2 weeks. Culture medium must be changed three times per week.

Passage of these cells is done mechanically by breaking the colony into small pieces using a glass pipette or a 21-gauge needle. Passages are usually performed every 2 weeks.

Summary

Several methods of mammalian oocyte activation are already available in the published literature. Although not described in this chapter, efforts are currently under way to obtain human parthenogenetic ESCs with moderate success (Lin *et al.*, 2003). Parthenogenetic ESCs are an excellent tool for understanding the differentiation process in monoparental cells. To what extent these cells will have an impact on regenerative medicine remains to be determined.

Acknowledgments

The authors thank Neli Ragina, Ramon Rodriquez, and Pablo Ross for their scientific insights and technical assistance and Dr. Professor Norio Nakatsuji for providing RNA from biparental cynomolgus macaque ESCs.

References

Allen, N. D., Barton, S. C., Hilton, K., Norris, M. L., and Surani, M. A. (1994). A functional analysis of imprinting in parthenogenetic embryonic stem cells. *Development* **120**, 1473–1482.

Borsuk, E., Szollosi, M. S., Besombes, D., and Debey, P. (1996). Fusion with activated mouse oocytes modulates the transcriptional activity of introduced somatic cell nuclei. *Exp. Cell Res.* **225**, 93–101.

Bos-Mikich, A., Whittingham, D. G., and Jones, K. T. (1997). Meiotic and mitotic Ca^{2+} oscillations affect cell composition in resulting blastocysts. *Dev. Biol.* **182**, 172–179.

Chesne, P., Adenot, P. G., Viglietta, C., Baratte, M., Boulanger, L., and Renard, J. P. (2002). Cloned rabbits produced by nuclear transfer from adult somatic cells. *Nature Biotechnol.* **20,** 366–369.

Cibelli, J., Grant, K., Chapman, K., Cunniff, K., Worst, T., Green, H., Walker, S., Gutin, P., Vilner, L., Tabar, V., Dominko, T., Kane, J., Wettstein, P., Lanza, R., Studer, L., Vrana, K., and West, M. (2002). Parthenogenetic stem cells in nonhuman primates. *Science* **295,** 819.

Cibelli, J. B., Kiessling, A. A., Cuniff, K., Richards, C., Lanza, R. P., and West, M. D. (2001). Somatic cell nuclear transfer in humans: Pronuclear and early embryonic development. *J. Regen. Med.* **2,** 25–31.

Cuthbertson, K. S. (1983). Parthenogenetic activation of mouse oocytes *in vitro* with ethanol and benzyl alcohol. *J. Exp. Zool.* **226,** 311–314.

Grupen, C. G., Mau, J. C., McIlfatrick, S. M., Maddocks, S., and Nottle, M. B. (2002). Effect of 6-dimethylaminopurine on electrically activated *in vitro* matured porcine oocytes. *Mol. Reprod. Dev.* **62,** 387–396.

Jellerette, T., Kurokawa, M., Lee, B., Malcuit, C., Yoon, S. Y., Smyth, J., Vermassen, E., De Smedt, H., Parys, J. B., and Fissore, R. A. (2004). Cell cycle-coupled [Ca(2+)](i) oscillations in mouse zygotes and function of the inositol 1,4,5-trisphosphate receptor-1. *Dev. Biol.* **274,** 94–109.

Kaufman, M. H., Robertson, E. J., Handyside, A. H., and Evans, M. J. (1983). Establishment of pluripotential cell lines from haploid mouse embryos. *J. Embryol. Exp. Morphol.* **73,** 249–261.

Kono, T., Jones, K. T., Bos-Mikich, A., Whittingham, D. G., and Carroll, J. (1996). A cell cycle-associated change in Ca^{2+} releasing activity leads to the generation of Ca^{2+} transients in mouse embryos during the first mitotic division. *J. Cell Biol.* **132,** 915–923.

Kono, T., Obata, Y., Wu, Q., Niwa, K., Ono, Y., Yamamoto, Y., Park, E. S., Seo, J. S., and Ogawa, H. (2004). Birth of parthenogenetic mice that can develop to adulthood. *Nature* **428,** 860–864.

Kono, T., Sotomaru, Y., Katsuzawa, Y., and Dandolo, L. (2002). Mouse parthenogenetic embryos with monoallelic H19 expression can develop to day 17.5 of gestation. *Dev. Biol.* **243,** 294–300.

Krivokharchenko, A., Popova, E., Zaitseva, I., Vil'ianovich, L., Ganten, D., and Bader, M. (2003). Development of parthenogenetic rat embryos. *Biol. Reprod.* **68,** 829–836.

Lan, G. C., Han, D., Wu, Y. G., Han, Z. B., Ma, S. F., Liu, X. Y., Chang, C. L., and Tan, J. H. (2005). Effects of duration, concentration, and timing of ionomycin and 6-dimethylaminopurine (6-DMAP) treatment on activation of goat oocytes. *Mol. Reprod. Dev.* **71,** 380–388.

Lawitts, J. A., and Biggers, J. D. (1993). Culture of preimplantation embryos. *Methods Enzymol.* **225,** 153–164.

Lin, H., Lei, J., Wininger, D., Nguyen, M. T., Khanna, R., Hartmann, C., Yan, W. L., and Huang, S. C. (2003). Multilineage potential of homozygous stem cells derived from metaphase II oocytes. *Stem Cells* **21,** 152–161.

Lin Liu, J.-C., and Xiangzhong Yang, J. (1998). Parthenogenetic development and protein patterns of newly matured bovine oocytes after chemical activation. *Mol. Reprod. Dev.* **49,** 298–307.

Liu, C. T., Chen, C. H., Cheng, S. P., and Ju, J. C. (2002). Parthenogenesis of rabbit oocytes activated by different stimuli. *Anim. Reprod. Sci.* **70,** 267–276.

Liu, J. L., Sung, L. Y., Du, F., Julian, M., Jiang, S., Barber, M., Xu, J., Tian, X. C., and Yang, X. (2004). Differential development of rabbit embryos derived from parthenogenesis and nuclear transfer. *Mol. Reprod. Dev.* **68,** 58–64.

Liu, S. Z., Jiang, M. X., Yan, L. Y., Jiang, Y., Ouyang, Y. C., Sun, Q. Y., and Chen, D. Y. (2005). Parthenogenetic and nuclear transfer rabbit embryo development and apoptosis after activation treatments. *Mol. Reprod. Dev.* **72,** 48–53.

Loi, P., Ledda, S., Fulka, J., Jr, Cappai, P., and Moor, R. (1998). Development of parthenogenetic and cloned ovine embryos: Effect of activation protocols. *Biol. Reprod.* **58,** 1177–1187.

Malcuit, C., Kurokawa, M., and Fissore, R. A. (2006). Calcium oscillations and mammalian egg activation. *J. Cell. Physiol.* **206,** 565–573.

McKiernan, S. H., Bavister, B. D., and Tasca, R. J. (1991). Energy substrate requirements for in-vitro development of hamster 1- and 2-cell embryos to the blastocyst stage. *Hum. Reprod.* **6,** 64–75.

Meo, S. C., Yamazaki, W., Leal, C. L. V., Oliveira, J. A. D., and Garcia, J. M. (2005). Use of strontium for bovine oocyte activation. *Theriogenology* **63,** 2089–2102.

Mitalipov, S. M., Nusser, K. D., and Wolf, D. P. (2001). Parthenogenetic activation of rhesus monkey oocytes and reconstructed embryos. *Biol. Reprod.* **65,** 253–259.

Mitalipov, S. M., White, K. L., Farrar, V. R., Morrey, J., and Reed, W. A. (1999). Development of nuclear transfer and parthenogenetic rabbit embryos activated with inositol 1,4,5-trisphosphate. *Biol. Reprod.* **60,** 821–827.

Mizutani, E., Jiang, J. Y., Mizuno, S., Tomioka, I., Shinozawa, T., Kobayashi, J., Sasada, H., and Sato, E. (2004). Determination of optimal conditions for parthenogenetic activation and subsequent development of rat oocytes *in vitro. J. Reprod. Dev.* **50,** 139–146.

Ongeri, E. M., Bormann, C. L., Butler, R. E., Melican, D., Gavin, W. G., Echelard, Y., Krisher, R. L., and Behboodi, E. (2001). Development of goat embryos after *in vitro* fertilization and parthenogenetic activation by different methods. *Theriogenology* **55,** 1933–1945.

Ozil, J. P. (1990). The parthenogenetic development of rabbit oocytes after repetitive pulsatile electrical stimulation. *Development* **109,** 117–127.

Ozil, J. P., and Huneau, D. (2001). Activation of rabbit oocytes: The impact of the Ca^2+ signal regime on development. *Development* **128,** 917–928.

Ozil, J. P., Markoulaki, S., Toth, S., Matson, S., Banrezes, B., Knott, J. G., Schultz, R. M., Huneau, D., and Ducibella, T. (2005). Egg activation events are regulated by the duration of a sustained $[Ca2+]cyt$ signal in the mouse. *Dev. Biol.* **282,** 39–54.

Presicce, G. A., and Yang, X. (1994). Parthenogenetic development of bovine oocytes matured *in vitro* for 24 hr and activated by ethanol and cycloheximide. *Mol. Reprod. Dev.* **38,** 380–385.

Robertson, E. J., Evans, M. J., and Kaufman, M. H. (1983). X-chromosome instability in pluripotential stem cell lines derived from parthenogenetic embryos. *J. Embryol. Exp. Morphol.* **74,** 297–309.

Rogers, N. T., Hobson, E., Pickering, S., Lai, F. A., Braude, P., and Swann, K. (2004). Phospholipase Czeta causes Ca^{2+} oscillations and parthenogenetic activation of human oocytes. *Reproduction* **128,** 697–702.

Roh, S., Malakooti, N., Morrison, J. R., Trounson, A. O., and Du, Z. T. (2003). Parthenogenetic activation of rat oocytes and their development (*in vitro*). *Reprod. Fertil. Dev.* **15,** 135–140.

Rougier, N., and Werb, Z. (2001). Minireview: Parthenogenesis in mammals. *Mol. Reprod. Dev.* **59,** 468–474.

Sanchez-Pernaute, R., Studer, L., Ferrari, D., Perrier, A., Lee, H., Vinuela, A., and Isacson, O. (2005). Long-term survival of dopamine neurons derived from parthenogenetic primate embryonic stem cells (cyno-1) after transplantation. *Stem Cells* **23,** 914–922.

Susko-Parrish, J. L., Leibfried-Rutledge, M. L., Northey, D. L., Schutzkus, V., and First, N. L. (1994). Inhibition of protein kinases after an induced calcium transient causes transition of bovine oocytes to embryonic cycles without meiotic completion. *Dev. Biol.* **166,** 729–739.

Thomson, J., Itskovitz-Eldor, J., Shapiro, S., Waknitz, M., Swiergiel, J., Marshall, V., and Jones, J. (1998). Embryonic stem cell lines derived from human blastocysts. *Science* **282,** 1145–1147.

Toth, S., Huneau, D., Banrezes, B., and Ozil, J. P. (2006). Egg activation is the result of calcium signal summation in the mouse. *Reproduction* **131,** 27–34.

Uranga, J. A., and Arechaga, J. (1997). Cell proliferation is reduced in parthenogenetic mouse embryos at the blastocyst stage: A quantitative study. *Anat. Rec.* **247,** 243–247.

Vrana, K., Hipp, J., Goss, A., McCool, B., Riddle, D., Walker, S., Wettstein, P., Studer, L., Tabar, V., Cunniff, K., Chapman, K., Vilner, L., West, M., Grant, K., and Cibelli, J. (2003). Nonhuman primate parthenogenetic stem cells. *Proc. Natl. Acad. Sci. USA* **100,** 11911–11916.

Winston, N. J., and Maro, B. (1995). Calmodulin-dependent protein kinase II is activated transiently in ethanol-stimulated mouse oocytes. *Dev. Biol.* **170,** 350–352.

Yi, Y. J., and Park, C. S. (2005). Parthenogenetic development of porcine oocytes treated by ethanol, cycloheximide, cytochalasin B and 6-dimethylaminopurine. *Anim. Reprod. Sci.* **86,** 297–304.

Zhu, J., Telfer, E. E., Fletcher, J., Springbett, A., Dobrinsky, J. R., De Sousa, P. A., and Wilmut, I. (2002). Improvement of an electrical activation protocol for porcine oocytes. *Biol. Reprod.* **66,** 635–641.

[9]   Embryonic Stem Cells Using Nuclear Transfer

By YOUNG CHUNG and SANDY BECKER

Abstract

Despite the fact that embryonic stem (ES) cells are able to differentiate into multiple therapeutically useful cell types, a prominent obstacle to their use is immune rejection by the recipient. One possible solution could be the derivation of ES cells from embryos generated by cloning using somatic cell nuclei from individual patients. The first section of this chapter describes progress in optimizing procedures for cloning, using the mouse as a model. The second section describes procedures for establishing ES cell lines from cloned mouse embryos.

Introduction

An important clinical use for cloned human embryos would be to estab-lish embryonic stem (ES) cell lines that are an immunological match for the donor nucleus. Despite the much-publicized retraction of a recent paper claiming to have established patient-specific human ES cell lines via somatic cell nuclear transfer (SCNT), this technology still holds great promise for cell-based treatments.

METHODS IN ENZYMOLOGY, VOL. 418
0076-6879/06 $35.00
DOI: 10.1016/S0076-6879(06)18009-X

The inefficiency of cloning, however, has inhibited its implementation and also raises concerns about the validity of conclusions drawn regarding basic biological questions. It is clear that procedural differences likely exist between laboratories and that the effects of these procedural differences have not been documented. In truth it may be difficult to judge the impact of any given aspect of a cloning protocol. The number of studies and the number of laboratories to conduct such studies remain quite small, and the number of variables and experimental parameters remaining to be tested is quite large. This chapter describes a method that has worked well both for deriving ES cells and for generating live cloned pups. Parts of the cloning procedure have been published elsewhere and are reproduced here with permission (Chung *et al.*, 2006b).

Cloning by somatic cell nuclear transfer is a complex procedure that is dependent on correct interactions between oocyte and donor cell genome. These interactions require minimal insult to either the oocyte or the transplanted nucleus. Available data also indicate that reprogramming the donor cell genome may be slow so that the cloned embryo expresses genes typical of the donor cell, and thus has different characteristics from normal embryos (Gao *et al.*, 2003).

There is considerable heterogeneity among cloned embryo constructs and between cells in a single embryo (mosaicism). Boiani *et al.* (2002) documented heterogeneity in *Oct4* expression, and we have observed mosaic expression of Dnmt1 proteins in 100% of eight-cell stage cloned embryos examined (Chung *et al.*, 2003). Blelloch *et al.* (2006) demonstrated that both the differentiation state and the methylation state of the donor nucleus strongly influence the efficiency of deriving ES cells from cloned embryos. Thus it appears that most cloned embryos do not recapitulate a truly embryonic mode of development and gene expression so that only a minority would be expected to initiate postimplantation development and only a very small fraction of these would be expected to develop to term, as is observed (Wakayama *et al.*, 1998, Wakayama and Yanagimachi, 1999). Success in cloning and in deriving ES cells from cloned embryos could likely be improved by (1) reducing the heterogeneity between embryos and between blastomeres and (2) shifting the balance in favor of successful reprogramming. A schematic diagram of the cloning procedure in mice is given in Fig. 1.

Somatic Cell Nuclear Transfer

Chemicals, Media, and Equipment

Polyvinyl pyrrolidone (PVP, ICN)
Strontium chloride (Sigma)

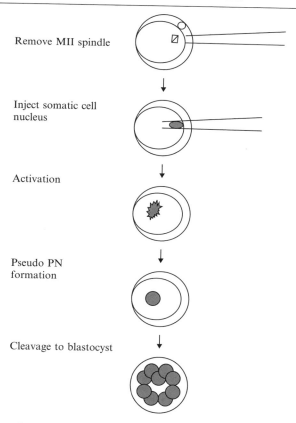

Remove MII spindle

Inject somatic cell
nucleus

Activation

Pseudo PN
formation

Cleavage to blastocyst

Fig. 1. Schematic outline of cloning procedure in mice. The spindle-chromosome complex is removed by pinching off without penetrating the plasma membrane. Donor nuclei are injected. After injection the donor nuclear membrane breaks down and chromosomes condense. Activation induces pseudopronucleus (PN) formation, and the embryo begins to cleave. Blastocyst stage embryos are used to derive ES cells.

Cytochalasin B (5.0 mg/ml for 1000× solution in dimethyl sulfoxide, stored at −70°; Sigma)
Mannitol (Sigma)
HEPES (Sigma)
Human chorionic gonadotropin (hCG; Sigma)
Pregnant mare gonadotropin (PMSG; Calbiochem)
Hyaluronidase (ICN)
Bovine serum albumin, fraction V (BSA; ICN)
Modular plastic incubator (Billups-Rothenberg, Del Mar, CA)

ECM 2001 embryo manipulation system fusion machine (BTX/Gene-tronics, San Diego, CA)

Flaming-Brown pipette puller (Model P-87 or equivalent, Sutter Instruments, Novato, CA)

Olympus stereomicroscope

Micromanipulation equipment: Nikon Diaphot-inverted microscope equipped with Hoffman optics and objectives. Narishige three-axis hanging joystick micromanipulators with coarse manipulators and Narishige IM-6 and IM-9B microinjectors

Piezo pipette driver (PMM Controller, Prime Tech, Ibaraki, Japan)

Dulbecco's modified Eagles medium (DMEM) (GIBCO/Invitrogen)

KO-DMEM (GIBCO/Invitrogen)

CZB medium (Chatot et al., 1989) augmented with glucose to 5.5 mM (CZB-G) (Kuretake et al., 1996)

ES cell-qualified fetal bovine serum (FBS) (Hyclone)

HEPES-CZB-G medium: HEPES is added to 20 mM and sodium bicarbonate is reduced to 5 mM

Calcium-free CZB-G medium

Dulbecco's phosphate-buffered saline (PBS) (GIBCO/Invitrogen)

KSOM medium (Calbiochem)

L-Glutamine (100×)

MEM nonessential amino acids (NEAA) (100×)(GIBCO/Invitrogen)

Mouse leukemia inhibitory factor (LIF) (ESGRO) (Chemicon)

Mitomycin C (Sigma)

β-Mercaptoethanol (BME) (Sigma)

M2 medium (Hogan et al., 1986)

0.05% trypsin EDTA (GIBCO/Invitrogen)

Hyaluronidase stock: hyaluronidase of 600 U/mg or greater activity at a concentration of approximately 600 U/ml prepared in M2 medium

Electrofusion medium: 270 mM mannitol, 0.1 mM $MgSO_4$, 0.05 mM $CaCl_2$, and 0.3% BSA, pH 7.2–7.4.

Media Preparation

As with any embryological procedure, the culture medium employed is of fundamental importance. In our opinion, one important prerequisite for maximizing yield in cloning procedures is to use only culture medium that is 2 weeks or less in age.

Exclusive use of either disposable tissue cultureware or dedicated glassware that is rinsed and dried thoroughly between uses is recommended to avoid any contamination with detergents or other chemicals. Dry chemicals should only be weighed with disposable utensils. The water should be of the highest possible purity, such as that from GIBCO. Any reasonable effort should be made to minimize the degree to which water-borne trace contaminants enter culture media.

Prepare CZB-G, HEPES-CZB, and Ca^{2+}-free CZB media from dry powders, store at $4°$, and use for up to 2 weeks. Gas the air spaces in bottles containing bicarbonate-buffered media with a 5% CO_2 mixture to avoid equilibration to an alkaline pH. The correct osmolarity of the culture medium is also important, as differences in osmolarity can affect embryo development.

The source and manufacturing lot of BSA can also affect medium quality and should be batch tested. We have had success with BSA from ICN Chemicals.

Culture Dishes and Incubator

We prefer the use of nontissue culture plastic Petri dishes for all of our embryo cultures. Embryos can be cultured using a system of droplets of medium under light mineral oil in 35- or 60-mm Petri dishes. The source and manufacturing lot of mineral oil must be tested. We have had good reproducibility with the light mineral oil from Fisher Scientific. Maintain dishes in a CO_2 cell culture incubator while the experiment is in progress and then transfer to a plastic modular incubator, which is then gassed and maintained at $37°$. The gas mixture that we generally use for culturing cloned embryos is 5% CO_2, 21% O_2, and a balance of N_2. Although embryo culture conditions often involve a reduced amount (5%) of O_2, we have not found that reduced O_2 is beneficial for cloned embryos and may, in some cases, be detrimental.

Protocol

Oocyte Isolation

The mouse oocyte is a very fragile cell and can easily have its competence to support long-term development compromised. As cloning is obviously dependent on the oocyte for success, care must be taken not to fail in the experiment during its first step. One consideration is the age of oocytes. We obtained the best cloning results when oocytes collected 13 to 15 h after hCG injection were used as recipients.

1. To obtain the recipient oocytes, inject 5 IU of PMSG and 5 IU of hCG into 8- to 10-week old B6D2F1 female mice (Charles River Laboratories), 48 h apart.

2. On the day of cloning, dilute the hyaluronidase to 100 to 150 U/ml from 600 U/ml stock in M2 medium (Hogan *et al.*, 1986) and prepare a large droplet in a dish with an oil overlay.
3. Isolate the oviducts in M2 medium.
4. Place the oviducts directly into the oil and drag them into the enzyme solution one at a time until all cumulus–oocyte complexes are released. It is also acceptable to work without the oil overlay; however, desiccation must be avoided.
5. After a brief enzymatic treatment, aspirate the cumulus–oocyte complexes with a pipette of the appropriate size (200 to 300 μM) so as to the cumulus cells without damaging the oocytes.
6. Wash oocytes extensively in M2 medium to remove both cumulus cells and enzyme, wash them through several changes of CZB-G medium (Kuretake *et al.*, 1996), and culture in CZB-G until use.

Removing the Spindle–Chromosome Complex (SCC)

The SCC can be difficult to visualize (see Fig. 2). However, excellent visualization of the SCC can be achieved with Hoffman modulation contrast optics, preferable to that achieved with DIC optics.

1. Manipulate oocytes in HEPES-buffered CZB-G medium (add HEPES to 20 mM and reduce sodium bicarbonate to 5 mM) supplemented with 2.5 to 3.0 μg/ml cytochalasin B. It is important that the oocytes be exposed to room temperature for only about 10 min during the removal of the spindle. Thus, groups of oocytes must be small when an operator is first learning, but can eventually be increased to about 20 to 25 oocytes.

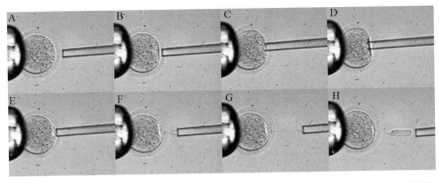

FIG. 2. Removal of the SCC. (A–C) Penetrating the zona adjacent to the SCC (visible as a pale dot). (D–F) Removing the SCC without penetrating the plasma membrane. (G and H) Breaking the cytoplasmic bridge to release the SCC.

2. Use a piezo pipette driver to penetrate the zona pellucida rapidly without disturbing the ooplasm. Insert a blunt pipette of about 8 to 10 μm inside diameter through the zona pellucida using multiple piezo pulses. Speed and intensity settings are generally in the range of 2 to 4 for penetrating the zona and a setting of 1 for penetrating the plasma membrane. The amount and position of mercury in the pipette will affect actual settings. We find that a 2- to 3-mm-long bead of mercury inside the main bore of the pipette is optimal. Insert the mercury into the pipette near the beginning of the taper using a 36-gauge spinal tap needle. Then insert the pipette into the injection holder. Using the injector, move the mercury to the tip of the pipette after mounting on the micromanipulator and microscope.

3. Take care not to penetrate the plasma membrane during removal of the SCC. Typically, it is easiest if the SCC is positioned in the hemisphere opposite from that where the oocyte is held by the holding pipette. The tip of the pipette is positioned near the SCC and mild suction is applied in order to aspirate a portion of the plasma membrane, the SCC, and a minimal volume of cytoplasm (see Fig. 2). A microinjector with very fine control is preferable, with no more than 5 μl per turn.

4. Gently withdraw the pipette after the SCC is aspirated and expel the SCC. The SCC need not be entirely within the pipette in order to be withdrawn.

Introduction of Nuclei

The introduction of nuclei can be achieved by either of two methods: injection or electrofusion. The size of the donor nucleus is the determining factor. Nuclei larger than about 7 to 8 μm in diameter may be damaged by the injection pipette. In such cases, electrofusion is preferable.

Injection. Introducing nuclei by injection is performed most easily using a piezo-driven pipette. The size and the shape of the pipette are critical. We find that an elongated, gradually tapering pipette with an inside diameter of about 5 μm works well.

1. Prepare dishes for injection using the lids of standard 10-cm plastic disposable Petri dishes.

2. Injections can be performed using either HEPES-buffered CZB-G or bicarbonate-buffered CZB-G. Drops of different media are prepared on the dish and covered with oil. These include drops of 10% PVP in medium, 3% PVP in medium, and injection medium with no PVP.

3. Oocytes for injection are placed in the medium drop with no PVP.

4. Donor cells are mixed in the drop with 3% PVP.

5. The drop containing 10% PVP is used for washing the pipette. Wash the tip of the pipette several times in the drop containing 10% PVP.

6. Aspirate donor cells in medium with 3% PVP and pipette several times to ensure lysis of the membrane and removal of the cytoplasm. Aspirate several nuclei with convenient spacing between them to allow one to be injected at a time. Draw the last nucleus up the pipette away from the tip.

7. Use several piezo pulses to penetrate the zona a second time. Advance the tip of the pipette near the opposite side of the oocyte from where it is held by the holding pipette. The diameter of the opening of the holding pipette should be around 25 μm to permit a small pocket to be formed by drawing a portion of the ooplasm into the holding pipette.

8. While the tip of the injection pipette is being advanced toward the holding pipette, expel liquid from the pipette to bring the nucleus to the tip. When the nucleus reaches the tip, stop the outward flow of medium and use a single low-energy piezo pulse to penetrate the membrane.

9. Place the tip of the injection pipette at the opening of the pocket formed by the holding pipette and expel the nucleus. Withdraw the injection pipette quickly but without damaging the oocyte (see Fig. 3).

Electrofusion. This is the preferred method for larger donor nuclei. Removal of the SCC is done as described earlier, except that any remaining polar body or its debris is removed at the time of SCC removal.

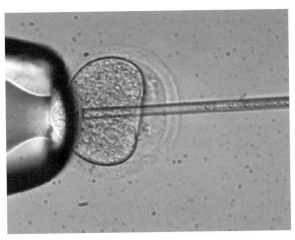

FIG. 3. Injecting a nucleus. Deposit the nucleus near the holding pipette, as shown. Additional nuclei can be seen in the injection pipette.

1. Resuspend donor cells in the same drop of injection medium as the oocytes themselves.

2. Use a larger-bore (25 to 30 μm) injection pipette to make an opening in the zona pellucida.

3. Aspirate an intact donor cell and reinsert the pipette through this opening.

4. Expel the donor cell at a location far enough away from the opening to prevent it from exiting the perivitelline space when the pipette is withdrawn. A smaller-diameter pipette may be used, but take care to avoid damage to the donor cell so that it does not lyse before fusion can be achieved.

5. After the donor cell is inserted into the perivitelline space, electrofusion can be performed, using a fusion chamber connected to a suitable pulse generator such as the BTX ECM 2001.

6. Use fusion chambers with an approximately 1-mm distance between electrodes.

7. Wash oocytes and equilibrate in the electrofusion medium before placing them between the electrodes.

8. A low-voltage (2 to 10 V) AC current can be applied to assist in aligning the cloned constructs within the electric field prior to fusion (membranes at point of contact between cells must be parallel to electrodes). Successful rotation should occur within a few seconds.

9. Fuse oocytes using a single 90-V pulse (900 V/cm) delivered for 10 μs. Wash oocytes after pulsing to remove the electrofusion medium and then incubate for about 1 h in CZB medium.

10. Subject unfused constructs remaining after a 30-min incubation to a second pulse. Although additional pulses can be given, in our experience constructs that do not fuse after two pulses generally will not be viable.

Oocyte Activation

After either injection or electrofusion, incubate the constructs for a period of time before activation. The optimum length of time before activation for most donor cell types tested is about 1 h.

Activation is performed in Ca^{2+}-free CZB-G medium supplemented with 10 mM $SrCl_2$ and 5 μg/ml cytochalasin B. The $SrCl_2$ has a tendency to precipitate so it should be added only after the culture medium is fully equilibrated at 37° and 5% CO_2.

1. Prepare drops under oil quickly and return to the incubator. Care must be taken to ensure that no precipitate forms before or during the activation.

2. Maintain oocytes in this medium for 5 to 6 h. The exact length of the activation period required may vary with donor cell type. For example,

we find that a 6-h activation works well for cumulus donor nuclei, whereas a 5-h activation works well for myoblast donor nuclei.

3. After activation, examine the cloned constructs for the presence of pseudopronuclei, wash the successfully activated constructs extensively in KSOM, and then culture the reconstructed embryos in KSOM at 37° under 5% CO_2 in air until they develop to blastocysts.

Derivation of ES Cells from Cloned Blastocysts

The first mouse ES cell lines were established in 1981 (Evans and Kaufman, 1981; Martin, 1981). Since then, various procedures have been used to establish ES cell lines from the ICM of fertilized mouse blastocysts (e.g., Nagy and Vintersten, 2006) and from single blastomeres (Chung et al., 2006a). Most lines have been derived from the 129/Sv strain in normal embryos; however, other strains, including B6D2F1, C57BL/6, and EGFP transgenic CD-1 mice, have also been used successfully to derive ES cells after nuclear transfer (Wakayama, 2006; Wakayama et al., 2001). The following procedure has been used for generating ES cell lines from cloned embryos, as well as from normal fertilized blastocysts and parthenogenetic embryos.

1. Day 1: perform the SCNT as described earlier, using cumulus cells or tail-tip cell nuclei.

2. Day 2: check for cloned embryo cleavage and remove uncleaved ones.

3. Day 3: prepare several four-well dishes with mitomycin C-treated mouse embryonic fibroblast feeder cells (MEF). Culture the MEFs in 0.5 ml DMEM supplemented with 10% FBS.

4. Day 4: By day 4 the cloned embryos will have developed to the blastocyst stage. Cloned embryos tend to be retarded by half a day compared to fertilized embryos and may require more time to develop to expanded blastocysts. Replace the MEF culture medium with ES growth medium (DMEM supplemented with glutamine, MEM nonessential amino acids, BME, antibiotics, 15% FBS, and LIF)[1] and then transfer the d4 blastocysts after removing the zona pellucidae by brief exposure to acidic Tyrode's solution (pH 2.5), adding no more than three to each well. Traditional immunosurgery to remove trophoblasts is not necessary, but removing the zona pellucida is helpful in facilitating attachment of embryos to the MEFs.

---

[1] Note that it is important to use a good batch of FBS. We have found ES qualified FBS from Hyclone gives good results, but some batches are better than others. Unfortunately, the only way to test a batch of serum for this application is to use it for this application. Batches tested and found suitable for growing established lines are not necessarily optimal for this derivation protocol.

Dishes with embryos are returned to a 37° incubator and left undisturbed for 5 days. Over this period the embryos will attach, the trophoblast will spread out and form a monolayer, and the ICM will grow and form a distinct mound of cells on the trophoblast monolayer.

5. At the end of 5 days, individual ICMs are picked by mouth pipetting with a pulled Pasteur pipette and disaggregated by treating for 5 min with 0.5% trypsin and 2 mM EDTA solution plus 1% chicken serum, after three to four washes in PBS to remove any trace of serum. When cells in the ICM start to lose contact with each other, break up the clump into smaller clusters of cells and single cells by gentle pipetting with a small-bore mouth pipette (50–100 μm in diameter). In case the clumps are not loosened even after the trypsin treatment, use a pair of syringe needles to break them up. Crossing two sharp needles (26 gauge) with clumps in the center will break the clumps into small pieces easily. Fill the mouth pipette with ES medium and blow over the cell suspension before picking up, to minimize contamination with the trypsin solution. Pick up the entire cell suspension with the minimum amount of remaining trypsin solution and transfer to a fresh well of a four-well dish that already contains the MEF feeder layer prepared 1 day earlier. Return the culture dishes to the incubator and culture for 2 days without disturbance.

6. After 2 days, check wells for the presence of ICM cells and whether they have started to form colonies. Initially these primary explants do not give rise to ES cells alone, but may include other cell types. The most prominent contaminating cell type is primitive endodermal cells, which are bigger and have more distinct cell membranes than real ES cells. The putative ES cells form tight round colonies that have smooth edges, which can be differentiated easily from other cell types. Typically individual ES cells are difficult to distinguish, but their nuclei can be recognized and contain one or two prominent nucleoli. In many instances, however, they differentiate and cease to proliferate over time. Therefore careful inspection should be made to check whether a colony continues to proliferate without differentiation. In most cases, true ES colonies are apparent within 10 days after plating the disaggregated ICM. In some cases, however, they form much later so it is important to keep the original dishes for a while even if there is initially no sign of ES cell colonies.

7. Pick the clumps of ES cells and transfer to a fresh well of a four-well dish containing mitomycin C-treated MEFs prepared 1 day earlier. In this way, putative ES cells can be selected away from other cell types that may contaminate the original culture well. When the colonies become large and numerous enough, they can be passaged with trypsin as usual for ES cell lines. Putative ES cell lines should be characterized as to their pluripotency as described in Becker and Chung (2006).

The process of producing new ES cell lines from NT embryos is time-consuming and frustrating because of the much versus lower success rate than when normal fertilized embryos are used (<10% versus 35%). This is of course partly due to epigenetic abnormalities in the cloned embryos. (It should be noted, however, that the success rate for establishment of ES cell lines from cloned embryos is far higher than the success rate for reproductive cloning [Wakayama *et al.*, 2006].) Other limitations may arise from the smaller number of cells in NT embryos (Chung *et al.*, 2002). However, a good batch of FBS and use of early passage MEFs can maximize the success rate. Overall, ES cell lines can be established at a frequency of 1 to 7% depending on the strains of nuclear donor, provided that BDF1 or BCF1 ooplasts are used for nuclear transfer.

References

Becker, S., and Chung, Y. (2006). Embryonic stem cells from a single blastomere. *Methods Enzymol.* **418**(this volume).

Blelloch, R., Wang, Z., Meissner, A., Pollard, S., Smith, A., and Jaenisch, R. (2006). Reprogramming efficiency following somatic cell nuclear transfer is influenced by the differentiation and methylation state of the donor nucleus. *Stem Cells* **24**(9), 2007–2013.

Boiani, M., Eckardt, S., Schöler, H. R., and McLaughlin, K. J. (2002). Oct4 distribution and level in mouse clones: Consequences for pluripotency. *Genes. Dev.* **16**, 1209–1219.

Chatot, C. L., Ziomek, C. A., Bavister, B. D., Lewis, J. L., and Torres, I. (1989). An improved culture medium supports development of random-bred one-cell mouse embryos *in vitro*. *J. Reprod. Fertil.* **86**, 679–688.

Chung, Y., Klimanskaya, I., Becker, S., Marh, J., Lu, S. J., Johnson, J., Meisner, L., and Lanza, R. (2006a). Embryonic and extraembryonic stem cell lines derived from single mouse blastomeres. *Nature* **439**, 216–219.

Chung, Y. G., Mann, M. R., Bartolomei, M. S., and Latham, K. E. (2002). Nuclear-cytoplasmic 'Tug-of War' during cloning: Effects of somatic cell nuclei on culture medium preferences in the preimplantation cloned mouse embryo. *Biol. Reprod.* **66**, 1178–1184.

Chung, Y. G., Ratnam, S., Chaillet, J. R., and Latham, K. E. (2003). Abnormal post-transcriptional gene regulation controlling DNA methyltransferase expression in cloned mouse embryos. *Biol. Reprod.* **69**, 146–153.

Chung, Y. G., Gao, S., and Latham, K. (2006b). Optimization of procedures for cloning by somatic cell nuclear transfer in mice. *Methods Mol. Biol.* **348**, 111–124.

Evans, M., and Kaufman, M. (1981). Establishment in culture of pluripotential cells from mouse embryos. *Nature* **292**, 154–156.

Gao, S., Chung, Y. G., Williams, J. W., Riley, J., Moley, K., and Latham, K. E. (2003). Somatic cell-like features of cloned mouse embryos prepared with cultured myoblast nuclei. *Biol. Reprod.* **69**, 48–56.

Hogan, B. F., Costantini, F., and Lacy, E. (1986). *In* "Manipulating the Mouse Embryo," pp. 109–110. Cold Spring Harbor Laboratory, Cold Spring Harbor, NY.

Kuretake, S., Kimura, Y., Hoshi, K., and Yanagimachi, R. (1996). Fertilization of mouse oocytes injected with isolated sperm heads. *Biol. Reprod.* **55**, 789–795.

Martin, G. (1981). Isolation of a pluripotent cell line from early mouse embryos cultured in medium conditioned by teratocarcinoma stem cells. *Proc. Natl. Acad. Sci. USA* **78,** 7634–7638.

Nagy, A., and Vintersten, K. (2006). Murine embryonic stem cells. *Methods Enzymol.* **418** (this volume).

Wakayama, S., Jakt, L. M., Suzuki, M., Araki, R., Hikichi, T., Kishigami, S., Ohta, H., van Thuan, N., Mizutani, E., Sadaide, Y., Senda, S., Tanaka, S., Okada, M., Miyake, M., Abe, M., Nishikawa, S. I., Shiota, K., and Wakayama, T. (2006). Equivalency of nuclear transfer-derived embryonic stem cells to those derived from fertilized mouse blastocysts. *Stem Cells.* **24**(9), 2023–2333.

Wakayama, T. (2006). Establishment of ES cell lines from adult somatic cells by nuclear transfer. *In* "Cell Biology: A Laboratory Handbook" 3rd Ed. (J. Celis, ed.) 3rd Ed., Vol. 1, pp. 87–95. Academic Press, San Diego.

Wakayama, T., Perry, A. C. F., Zuccotti, M., Johnson, K. R, and Yanagimachi, R. (1998). Full-term development of mice from enucleated oocytes injected with cumulus cell nuclei. *Nature* **394,** 369–374.

Wakayama, T., and Yanagimachi, R. (1999). Cloning the laboratory mouse. *Semin. Cell Dev. Biol.* **10,** 253–258.

Wakayama, T., Tabar, V., Rodriquez, I., Perry, A., Studer, L., and Mombaerts, P. (2001). Differentiation of embryonic stem cell lines generated from adult somatic cells by nuclear transfer. *Science* **292,** 740–743.

Section II

Differentiation of Embryonic Stem Cells

[10] Neural Stem Cells, Neurons, and Glia

By STEVEN M. POLLARD, ALEX BENCHOUA, and SALLY LOWELL

Abstract

Embryonic stem (ES) cells are a unique resource, providing in principle access to unlimited quantities of every cell type *in vitro*. They constitute an accessible system for modeling fundamental developmental processes, such as cell fate choice, commitment, and differentiation. Furthermore, the pluripotency of ES cells opens up opportunities for use of human ES cells as a source of material for pharmaceutical screening and cell-based transplantation therapies. Widespread application of ES cell-based technologies in both basic biology and medicine necessitates development of robust and reliable protocols for controlling self-renewal and differentiation in the laboratory. This chapter describes protocols that enable the conversion of mouse ES cells in simple adherent conditions to either terminally differentiated neurons and glia or self-renewing but lineage-restricted neural stem cell lines. It also reports on the current status in transfer of these approaches to human ES cells.

Introduction

Embryonic stem (ES) cells provide a valuable and convenient source of neural cells (Gottlieb and Huettner, 1999). However, despite progress in recent years, we do not as yet have full command over these cells, and it is difficult to direct differentiation of the entire population of ES cells into neural progenitors. Studying neural specification *in vitro* will help us to further improve the efficiency and predictability with which we can generate neural cells from ES cells. It could also improve our understanding of mammalian development as neural differentiation of ES proceeds through a sequence of differentiation steps that appear to closely recapitulate neural development *in vivo* (Billon *et al.*, 2002; Conti *et al.*, 2005; Ying *et al.*, 2003b). This is particularly significant if we are to understand such events in human embryogenesis, as the earliest developmental stages are not accessible. Furthermore, ES cells have several advantages over *in vivo* or primary culture systems, in particular with regard to their immortality, which provides an unlimited cellular resource for routine biochemical analysis, genetic manipulations, and small molecule screening.

METHODS IN ENZYMOLOGY, VOL. 418   0076-6879/06 $35.00
                                                              DOI: 10.1016/S0076-6879(06)18010-6

This chapter describes a protocol for neural conversion of ES cells that is designed not only for high efficiency of neural conversion, but also for tractability as an experimental system. It goes on to describe how this neural conversion protocol can be adapted for human ES cells. These protocols provide a means to study the process of neural induction and commitment.

Once neural cells are generated, they provide a resource for investigating the subsequent self-renewal and differentiation of neural progenitors. Progress in this area has been hampered by an inability to access homogeneous populations of true neural stem cells. However, we have established conditions for isolating and expanding ES cell-derived neural stem (NS) cell lines (Conti *et al.*, 2005). Such somatic stem cells provide a parallel system to ES cells in which to elucidate molecular details of self-renewal and differentiation. Protocols for the derivation and differentiation of mouse NS cells are described in later sections. A key future challenge will be to direct these NS cells to produce specific neuronal and glial subtypes.

Protocols

Conversion of Mouse ES Cells to Neural Progenitors in Adherent Monolayer

Overview

Several different approaches have been developed for generating neural cells from ES cells (Stavridis and Smith, 2003). Of these, the monolayer differentiation protocol described here has several key features that make it especially useful as an experimental system for investigating the mechanisms that regulate neural specification (Ying *et al.*, 2003b). This section outlines some of these features.

The Monolayer Protocol Is Not Based on Cell Selection

No current protocol is completely efficient in converting all ES cells into neural progenitors. Some neural differentiation protocols partially overcome this problem by relying on preferential survival of neural cells in minimal culture media (Li *et al.*, 2001; Tropepe *et al.*, 2001). Alternatively, targeted lineage selection can be applied to eliminate nonneural cells (Li *et al.*, 1998). These selective approaches are a useful means to obtain enriched populations of neural cells, but are less useful as an experimental model for studying the mechanisms that underlie neural fate choice.

The monolayer differentiation protocol is not selective; rather it brings about neural conversion with only modest cell death. We can therefore follow the fate of effectively all cells within a population, including that minority of cells that do not enter the neural lineage. This allows us to address several important questions. What proportion of cells initially resists neural conversion? Are they simply delayed in differentiation or do they irreversibly commit to an alternative fate? What alternative differentiated fate do they follow? What is their distribution in relation to the neural cells within the culture? By studying those cells that resist neural specification, rather than eliminating them through negative selection, we can gain important insights into the mechanisms that regulate this fate decision.

The Monolayer Protocol Does Not Rely on Undefined Media Components or Heterologous Cell Interactions

The monolayer protocol uses only well-characterized components. This contrasts with classical neural differentiation protocols, which rely on multicellular aggregation in serum-containing medium, usually in combination with the pleiotropic inducer retinoic acid (Bain *et al.*, 1995). Other protocols depend on coculture of ES cells with stromal cells, such as the PA6 cell line (Kawasaki *et al.*, 2000). The molecular nature of the differentiation-inducing activity delivered by PA6 cells remains obscure, and this activity can vary with cell batch or passage number. Furthermore, the rate of neural differentiation is delayed significantly in ES cells cocultured with PA6 feeders in comparison with feeder-free cultures in the same culture media (unpublished observations), which suggests that PA6 cells may deliver a complex mixture of prodifferentiation and antidifferentiation signals.

The Monolayer Protocol Allows Direct Observation and Analysis of the Differentiating Population

Cells attached to the dish in a two-dimensional monolayer can be observed readily throughout the differentiation process. This contrasts with floating aggregate cultures where the cells are not accessible to live microscopy at single-cell resolution. The absence of stromal cells also facilitates observation and simplifies population level analyses of mRNA and protein samples by avoiding the need to first separate the ES cell progeny from stromal cells.

Protocol: Conversion of Mouse ES Cells to Neural Cells in Adherent Monolayer

Media and reagents for all protocols described are described in the final section.

1. The ES cells used in this protocol should be maintained in feeder-free culture as described (Ying and Smith, 2003). If ES cells have been maintained on a feeder layer, they must be adapted to feeder-free conditions for several passages before initiating the monolayer differentiation protocol.

2. One day before initiating the protocol, trypsinize a near-confluent ES cell culture and replate at 30 to 40% confluence so that ES cells will reach 70 to 80% confluence on the following day. Both this trypsinization step itself and the relatively high ES cell density help maximize the efficiency and consistency of neural conversion by ensuring that the starting material is a substantially pure population of undifferentiated ES cells.

3. The next day, prepare dishes by coating with 0.1% gelatin for at least 1 h at room temperature. Remove gelatin and allow plates to dry. This protocol works best in either 9-cm dishes or six-well plates.

4. Wash ES cultures with phosphate-buffered saline (PBS) and then trypsinize for 5 min.

5. Harvest cells in serum-free medium without growth factors and spin for 3 min at 1000 rpm to pellet cells. Remove medium and repeat. Resuspend the washed cells in N2B27 complete medium. Unsupplemented neurobasal or DMEM-F12 is sufficient for these wash steps. Quenching of trypsin with serum medium is not necessary as long as care is taken to remove all traces of supernatant. Some workers prefer to quench with serum during the first wash: in this case, care must be taken to remove all traces using a second serum-free wash. Even very small amounts of either residual trypsin or serum can compromise neural differentiation.

6. Count cells carefully. The efficiency of neural differentiation is highly dependent on correct plating density. If the density is too low, cell viability is compromised, whereas relatively small increases in cell density can reduce the efficiency of neural differentiation dramatically, with more cells remaining as undifferentiated ES cells.

7. Plate cells onto the gelatin-coated dishes in complete N2B27 medium. The optimal plating density is generally around 10^4 cells/cm^2, but this can differ between cell lines and should be optimized in each case. Even the same cell line can differ in its density-sensitivity depending on culture conditions, so it is advisable to plate cells over a range of densities (e.g., 8×10^3/cm^2, 10^4/cm^2, and 1.2×10^4/cm^2) to maximize the probability of obtaining optimal differentiation.

8. Incubate at 37° 5% CO$_2$.

9. Replace medium every 1 to 2 days.

10. (Optional) For optimal neuronal differentiation and survival, cells should be replated onto laminin-coated dishes from the seventh day of differentiation.

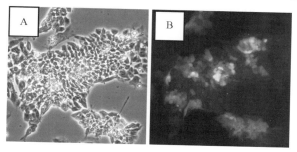

Fig. 1. Conversion of mouse ES cells (46C) to neural progenitors in adherent monolayer. (A) Typical morphology on day 3 of differentiation. (B) Sox1 GFP expression.

Monitoring Neural and Neuronal Differentiation

The Sox1-GFP-reporter cell line is convenient for monitoring and quantitating the transition from ES cell to neural precursor. This cell line, called 46C, was generated by gene targeting of E14Tg2a cells (Aubert et al., 2003). The open reading frame of Sox1 is replaced with the coding sequence for eGFP linked through an internal ribosomal entry site to a puromycin-resistant gene. Sox1 is an early marker of neuroepithelial cells throughout the extent of the developing neural tube (Pevny et al., 1998). The GFP reporter faithfully recapitulates the expression of Sox1 both in vivo and in vitro (Aubert et al., 2003) (Fig. 1).

This protocol has also been applied successfully to many other cell lines. Where Sox1-GFP-reporter cells are not used, the transition from ES cell to neural progenitor can instead be observed in live cells by a distinctive change in morphology. The nucleus, which is large and distinct in ES cells, becomes opaque and barely visible. Cells slightly elongate and pack closely together, often into rosette structures, with more distinct cell boundaries than ES cells. Commercially available antibodies can be used to monitor the loss of the ES cell marker Oct4 (Santa Cruz; C10, used at 1:200) and acquisition of the neural progenitor marker nestin (DSHB, Iowa; Rat401, used at 1:20).

Neuronal differentiation becomes detectable from around the fifth day, and the number of neurons increases progressively over the subsequent 2 to 3 days. Neurons can be identified easily by their characteristic small cell bodies with long very thin projections and by immunostaining for early neuronal markers, such as βIII-tubulin (TuJ1 antibody, Covance), or by use of a reporter cell line such as TK23, which carries an insertion of GFP into the neuronal τ locus (Ying et al., 2003a). Proneural genes such as Mash1, Neurogenin2, and Olig2 become detectable in a subset of cells shortly before the onset of neuronal differentiation. Early markers of multipotent neural precursors such as brain lipid binding protein (BLBP) and RC2 also become detectable at around this time. Note that these cultures contain a heterogeneous mixture of cells at different stages of differentiation, with Sox1, BLBP, and TuJ1 generally marking three distinct nonoverlapping subpopulations and with mosaic expression of regional identity genes in contrast to the neural stem cells described later.

Troubleshooting

POOR PLATING EFFICIENCY, SIGNIFICANT NUMBERS OF DEAD OR FLOATING CELLS 24 H AFTER PLATING. Ensure that the starting population of ES cells is healthy and that all traces of trypsin are removed carefully during the two wash steps. Coat dishes for at least 1 h with gelatin. Make sure that gelatin solution is then removed completely and that dishes are allowed to dry before plating cells.

PROGRESSIVE INCREASE IN CELL DEATH DURING THE PROTOCOL. Some cell death is normal during this protocol, but renewing the medium should reveal underlying healthy cultures throughout. However, cells sometimes begin to die in large numbers at around the third day of the protocol. The most common reasons for this are use of inappropriate basal medium or too low an initial plating density. Another common cause is use of medium or N-2 that has been stored for too long. In our experience, N-2 can deteriorate after 3 weeks even when stored at $-20°$, and N2B27 should be stored at $4°$ for a maximum of 1 week. It is also important to maintain the cells throughout in a stable $37°$ 5% CO_2 environment, avoiding prolonged periods out of the incubator.

LARGE NUMBERS OF UNDIFFERENTIATED ES CELLS PERSIST BEYOND THE FIRST 3 DAYS. A small proportion of cells, typically 10 to 15%, resist differentiation even after a week or more under this protocol. This is most likely due to local autocrine secretion of antidifferentiation factors. If significantly more than 15% of cells remain undifferentiated beyond the first 4 to 5 days, it is likely that the initial plating density was too high. The majority of residual undifferentiated ES cells will generally differentiate when replated into a fresh culture. If not, this may indicate the presence of chromosomally abnormal ES cells with impaired differentiation capacity.

NONNEURAL DIFFERENTIATION. A small subpopulation of cells, typically around 10%, differentiates into nonneural lineages under these monolayer conditions. These are readily apparent as larger flatter cells around the edge of neural colonies. High proportions of nonneural cells could indicate that the plating density is too high, that the medium is not being changed frequently enough, or that small residual traces of serum or BMP remained at the initial plating step due to incomplete washing.

It is also important to ensure that the starting population of ES cells is healthy and contains only few, or no differentiating cells. Trypsinization and replating of the ES cells 24 h prior to initiating the monolayer protocol can help eliminate predifferentiated cells.

Note also that substrates such as laminin or fibronectin, which are often used to support primary cultures taken directly from neural tissue, are not suitable for this protocol because they direct ES cells to differentiate into nonneural cell types.

VARIABILITY BETWEEN CELL LINES. Most ES cells tested undergo neural differentiation efficiently under this protocol. However, because there is line-to-line variability, especially in the optimal plating density, it is crucial to optimize this for each new cell line. Occasional cell lines may be resistant to neural differentiation using this protocol, possibly associated with genetic or epigenetic variation.

A Protocol for Neural Induction of Human ES Cells in Adherent Monolayer

Overview

The protocol described earlier for neural differentiation of mouse ES cells has been successfully adapted to three different human ES (hES) cell lines— Hs181, Hs238, and EDI1— with certain modifications as described later. The 181 and 237 cell hES lines were derived from supernumerary human blastocysts as described previously (Hovatta *et al.*, 2003). The EDI-1 cell line was derived following the same protocol (J. Nichols, unpublished result).

Protocol: Conversion of Human ES Cells to Neural Cells in Adherent Monolayer

1. hES cells are routinely cultivated in six-well plates on a feeder layer of commercially available human foreskin fibroblasts (ATCC, lines HFF-1 and HFF-2) in N2B27 medium supplemented with leukemia inhibitory factor (LIF) (10 ng/ml), BMP-4 (3 ng/ml, R&D Systems), and FGF-2 (10 ng/ml, R&D Systems). Cells are passaged at a split ratio of 1:2 each week using collagenase IV (1 mg/ml).

2. When the undifferentiated ES cell culture reaches 60 to 70% confluence (about 100 colonies in a well with 5000/10,000 cells per colony), detach the colonies using collagenase IV and further dissociate into small clumps by gentle trituration. Feeder cells will also become detached by the collagenase treatment. These can be removed by incubating the cell mixture for 6 h in a gelatin-coated flask in fresh medium without LIF and BMP-4. hES cells do not attach to gelatin, whereas feeder cells do.

3. Collect the medium containing the ES cell clumps and spin down at low speed (250g) before resuspending into N2B27 medium + FGF-2 in the absence of LIF and BMP. Plate the ES cells in another six-well plate coated previously with Matrigel growth factor-reduced matrix (BD Biosciences). Exchange medium fully every second day. Note that in contrast to the mouse ES cell protocol, where the addition of exogenous FGF-2 is not required, FGF-2 is necessary to maintain survival of both human ES cells and neural precursors.

Monitoring Neural Induction

The first morphological changes appear after about 4 days of culture. The cells adopt an elongated morphology similar to that described earlier for mouse neural progenitors and form rosettes and neural tube-like structures (Fig. 2). We have analyzed expression for neural lineage markers at these stages. Pax6 is one of the earliest markers, becoming detectable after around 4 days. Sox1 becomes detectable after around 1 week. After 10 days, nestin-positive cells start to migrate radially from the rosettes.

Terminal Differentiation

As for mouse cells, neuronal differentiation can be induced by passaging the neural precursors onto laminin-precoated dishes. Matrigel has an inhibitory effect on neuronal differentiation. Culture on laminin also strongly reduces attachment, survival, and proliferation of undifferentiated ES cells.

Troubleshooting: Human ES Cells

CELL DEATH. The most common reason for cell death during this protocol is due to dissociation of the hES cells into single cells rather than

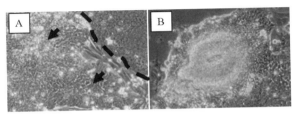

FIG. 2. Conversion of human ES cells to neural cells in adherent monolayer. (A) Typical rosette structures at day 4 (arrows). The region above the dotted line contains cells with an undifferentiated ES-like morphology. (B) Neural tube-like structures at day 7.

maintaining them throughout as small clumps. It is also important to store media and FGF-2 at 4° for no more than 1 week.

PLATING DENSITY. Because hES cells are dissociated into clumps rather than into single cells, it is not possible to quantify the plating density. The optimal plating density may also differ between different cell lines. It is therefore a good idea to test different plating densities in order to get a feel for the optimal density for each cell line. As a general guideline, the plating density should be such that the hES cells cover around 50% of the surface of the culture dish on the first day of differentiation. As with mouse ES cells, if the plating density of the hES clumps is too high or if clumps are too large they tend to resist differentiation.

Converting Mouse Pluripotent ES Cells to Tissue-Specific NS Cells

Overview

Embryonic stem cells divide symmetrically, generating a seemingly homogeneous population of stem cells without accompanying differentiation. ES cell lines enable insightful experiments to determine stem cell self-renewal and differentiation mechanisms. Studies of neural induction, for example, have revealed similarities between those mechanisms operating *in vitro* and *in vivo* (Ying *et al.*, 2003b). We investigated whether transient neural precursors generated from mouse ES cells using adherent monolayer differentiation could be captured and expanded, without use of genetic immortalization strategies, by elimination of heterologous cell types and exposure to growth factors (Conti *et al.*, 2005). We found that neural precursors are expanded readily in adherent conditions using a combination of epidermal growth factor (EGF) and fibroblast growth factor (FGF-2) in serum-free medium. Importantly, these conditions do not sustain nonneural cell types or differentiating astrocytes and thus give pure neural stem cell populations. These cells are remarkably homogeneous and show similarities to radial glia, with all cells expressing Sox2 and BLBP (Fabp7). These Sox1$^-$/Sox2$^+$/BLBP$^+$ cultures stably retain neuronal and glial differentiation potential after prolonged culture, both *in vitro* and *in vivo* following transplantation, and can be clonally expanded as cell lines. This self-renewal capacity, immortality, and symmetrical stem cell division is reminiscent of ES cells. We have termed these cell lines NS cells. It is now evident that BLBP-positive cells, which arise from early Sox1 neuroepithelial cells around 10.5 dpc, function as dividing precursors *in vivo* capable of generating neurons, astrocytes, and oligodendrocytes (Anthony *et al.*, 2004; Merkle *et al.*, 2004; Noctor *et al.*, 2002). Thus, conversion of ES cells to NS cells both temporally and in terms

TABLE I

SIMILARITIES AND DIFFERENCES BETWEEN ES CELLS AND NS CELLS

	ES cells	NS cells
Species	Rodent and primate	Rodent and primate
Source	Blastocyst	ES cells, fetal and adult CNS
Growth factor dependence	LIF plus BMP2/4 (serum free)	EGF plus FGF-2[a] (serum free)
Coculture with feeders	Unnecessary	Unnecessary
Expansion in vitro	Immortal	Immortal
Clonogenic?	Yes	Yes
Doubling time	~12 h	~25 h
Stem cell divisions	Symmetrical	Symmetrical
Karyotype	Stable diploid	Stable diploid
Niche dependence	None	None
In vivo counterpart	Similarities to inner cell mass	Similarities to radial glia
Differentiation capacity	Stable (in vitro and in vivo)	Stable (in vitro and in vivo)
Potency	Pluripotent	Multipotent
Genetic manipulation	Yes	Yes
Germ-line colonization	Yes	No

[a] EGF is necessary for derivation and maintenance of NS cells. Addition of FGF may only be required for NS cell derivation (Pollard et al., 2006b).

of lineage pathways reflects events in normal development, although it is not clear that self-renewal really occurs in the developing as opposed to adult central nervous system (CNS) (Pollard et al., 2006b).

A comparison of the salient features of both ES and NS is given in Table I. Potential applications for NS cells, both clinical and in basic biology, have been discussed elsewhere (Pollard et al., 2006a). NS cells can also be derived from fetal and adult CNS using similar protocols to that described here (Conti et al., 2005; Pollard et al., 2006b). Protocols for the derivation, maintenance, and differentiation of NS cells from mouse ES cells are described next.

Protocol: NS Cell Derivation from ES Cells

Feeder-dependent ES cells are adapted to feeder-free conditions and expanded as described earlier. We have found that NS cells can be derived following initial neural induction of ES cells using either "classic" embryoid body/retinoic acid protocols (Bain et al., 1995) or adherent monolayer differentiation (Ying et al., 2003b). The following protocol is based on the latter, as this is the preferred neural induction protocol of our laboratory, for those reasons outlined earlier.

Derivation of NS Cells Using Sox1-Lineage Selection

NS cell derivation can be achieved readily using neural differentiation of 46C Sox1-GFP ES cells. Upon neural induction the endogenous Sox1 promoter is activated and cells express both GFP and puromycin resistance. Through transient exposure of differentiated cultures to puromycin, one can enrich for Sox1-expressing neuroepithelial precursors cells. These can be replated and expanded adherently as NS cell lines following exposure to the growth factors EGF and FGF-2.

1. Differentiate 46C ES cells in adherent monolayer as described earlier to induce neural lineage commitment. Set up differentiations in 9-cm dishes (Iwaki). Efficient neural commitment can be estimated by monitoring GFP expression. N2B27 medium should be exchanged every alternate day to remove debris.

2. At day 7, the majority of cells should express GFP and have characteristic neural precursor morphology, with a proportion of cells already commencing overt neuronal differentiation. To enrich for Sox1-GFP-expressing cells, add 0.5 μg/ml of puromycin (Sigma) in N2B27 medium, replacing medium every 24 h to remove those dead/dying Sox1-negative cells.

3. Following 3 days exposure to puromycin, remove medium and wash twice with 1× PBS (Sigma) to remove debris. Add 1 ml of trypsin and incubate at 37° for 2 to 3 min. Flood the plate with 10 ml of N2B27 medium, and immediately detach and dissociate cells by smoothly pipetting several times against the culture surface. Remove the cell suspension to a 30-ml universal tube, and centrifuge cells at 300g (1300 rpm in Eppendorf 5702) for 3 min. Resuspend the cell pellet in N2B27, plus 10 ng/ml of EGF and FGF-2 (Peprotech), *without* puromycin, and seed 2 to 3 × 10^6 cells into a gelatinized T25 tissue culture flask (Iwaki).

4. These GFP$^+$ cells should readily attach and undergo expansion to near confluence within 2 to 3 days. Cells are typically split 1:3 to 1:4 every 2 to 3 days. Cells gradually extinguish expression of GFP and acquire the NS cell phenotype within 3 to 5 passages. This phenotype is then stable and can be maintained indefinitely (at least 100 passages). At this point NS cells can be transferred from N2B27 medium to NS cell expansion medium for optimal growth.

5. Once established, NS cells can be characterized using immunocytochemistry for markers such as Rat401/nestin, vimentin, and RC2 (DSHB, Iowa) and reverse transcriptase-polymerase chain reaction (RT-PCR) for expression of genes that characterize the NS cell state, such as Sox2, Olig2,

and BLBP. The cultures should be uniformly negative for astrocyte (GFAP, Sigma) and neuronal (TuJ1, Covance) markers.

Derivation of NS Cells from Any ES Cell Line

It is difficult to achieve NS cell populations from 46C ES cells using N2B27 medium in the absence of Sox1-lineage selection, as contaminating ES cells and nonneural cells persist during early passages. This is perhaps not surprising as N2B27 medium was optimized for both ES and NS cell survival/expansion. However, modifications to the NS cell derivation protocol will enable derivation of pure NS cells from any ES cell line. Central to this protocol is the replating of early neural precursors transiently into suspension culture in a medium favoring neural lineages, thus eliminating residual nonneural cells.

1. Differentiate ES cells in adherent monolayer in 9-cm dishes to induce neural lineage commitment, as described for 46C.

2. On day 7, replate 2 to 5×10^6 cells (typically half the population of a 9-cm monolayer differentiation) into a *non*gelatinized T75 flask (Iwaki) in NS medium (see media and reagents section) supplemented with EGF and FGF-2 (10 ng/ml).

3. After 24 h cultures should contain many thousands of cell aggregates in suspension culture, together with a minor population of adherent differentiated cells, and a substantial amount of cell debris. The proportion of cells in suspension varies considerably depending on the initial efficiency of neural induction, but is not a limiting factor in NS derivation. Two to 3 days later harvest the large aggregates by centrifugation at 700 rpm for 30s and resuspend in 10 ml of fresh medium in a fresh gelatinized T75 flask.

4. The majority of aggregates will settle and attach over the course of 3 to 7 days. NS cells outgrow with characteristic morphology (Fig. 3). Once the flask has reached over 50% confluence, trypsinize and split cells 1:2 to 1:3 into a fresh flask. We term these cultures passage 1 NS cells.

5. (Optional) Clonal lines can be generated once an NS cell population has been generated. Plate single cells into Terasaki microwells (Nunc) using limiting dilution. Add 10 μl of cell suspension to each well using a repeat pipette. Score wells containing a single cell after 1 to 2 h. Flood the plate with 5 ml of NS medium to exchange medium and then aspirate excess medium to leave at least 20 μl per well. Colonies appear after 7 to 10 days and can then be passaged into a well of a six-well plate. Add 20 μl of trypsin to each well, incubating for 1 to 2 min, and then pipette gently with a P20 to dissociate cells. Transfer 20 μl of cell suspension into 2 ml of NS medium in a six-well plate, + EGF + FGF-2. Upon attachment after $\sim$2 h exchange for fresh NS medium.

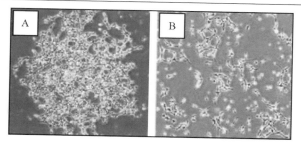

FIG. 3. NS cell derivation from mouse ES cells IB10 (129/Ola). (A) Attachment after 4 days of a cell aggregate formed following replating of a monolayer differentiation into NS expansion media. (B) Adherent NS cell lines derived following dissociation and replating of cells from A (passage 2).

Protocol: Expansion, Freezing, and Thawing of NS Cells

We routinely expand NS cells in T75 gelatinized tissue culture flasks (Iwaki). Cultures are passaged or frozen upon reaching 70 to 90% confluence ($\sim$5 to 7 $\times$ 10^6 cells), as described next.

1. Remove medium from flask and add 1 ml of trypsin solution. Note that washing of NS cells with 1$\times$ PBS solution is not required at this point as cell death is minimal and medium is serum free. Place flask at 37° for no longer than 1 to 2 min and firmly tap the side of flask against the bench to dislodge cells. All cells should detach readily. Add 10 ml of NS medium to the flask and pipette up and down, washing against the culture surface, several times to promote single cell suspension.

2. Transfer cells to a universal tube and spin for 3 min at 300 g. Aspirate the supernatant carefully, ensuring all residual trypsin is removed, and resuspend the pellet into fresh NS basal medium. We routinely split cells 1:2 to 1:5. Add the appropriate number of NS cells to a fresh gelatinized T75 flask, diluting with medium to a final volume of 10 ml. Add EGF and FGF-2 (10 ng/ml of each) to the flask for final NS cell expansion medium.

3. If NS cells are to be frozen, then following trypsinization and centrifugation resuspend the cell pellet in NS basal medium + 10% dimethyl sulfoxide (DMSO) (premixed). Typically, for a highly confluent T75 flask, resuspend in 2 ml and then aliquot 0.5 ml into four cryotubes. Transfer immediately to −80°, where NS cells can be stored and are recoverable for at least 6 months. For long-term storage, frozen vials can be transferred to liquid nitrogen following at least 24 h at −80°.

4. To recover NS cells from frozen stocks, thaw the cryotube rapidly in a 37° water bath. Immediately, but gently, transfer the 0.5 ml of cell solution into 10 ml of prewarmed NS expansion medium in a universal using a

plugged glass pipette and mix gently. Centrifuge cells at 250g for 3 min. It is important that each of these steps is performed as quickly as possible to minimize DMSO exposure.

5. Aspirate the supernatant and gently resuspend the cell pellet into 10 ml of prewarmed NS medium. Transfer this cell suspension to a fresh gelatinized T25 flask and add EGF plus FGF-2 (10 ng/ml). The following morning exchange medium or passage cells into a T75 flask. NS cell survival following thawing should be extremely high (>95%).

Differentiation of NS Cells

For NS cell differentiation to astrocytes we expose cells to 1% serum. BMP4 or LIF (10 ng/ml) also induce astrocyte differentiation with high efficiency, as assessed by GFAP expression. Seed 1×10^5 NS cells into a well of a gelatinized four-well plate (Nunc) in NS basal medium supplemented with 1% fetal calf serum, but without EGF or FGF-2. More than 95% of cells will exit the cell cycle and acquire a characteristic morphology within 24 to 48 h. There is minimal neuronal differentiation in these conditions.

NS cells maintain a capacity to generate neurons even after 100 passages. Neurons generated are largely GABAergic, as determined by GAD67 and GABA immunoreactivity (Conti et al., 2005). The following protocol results in 10 to 40% TuJ1-immunoreactive neurons.

1. NS cells are harvested using trypsin and 5×10^4 cells are plated onto laminin-coated four-well plates in NS basal medium + EGF + FGF-2 (10 ng/ml).

2. The following morning exchange the medium fully with NS basal medium supplemented with FGF2 (5 ng/ml) + $1\times$ B27 supplement (GIBCO). Subsequently replace half the medium with fresh every 3 to 4 days.

3. Following 1 week in these conditions exchange medium to NS basal medium mixed with neurobasal medium (1:1) and supplement with $1\times$ B27 (GIBCO). Cells with neuronal morphology should emerge over the next 3 to 7 days. Replace half the medium with fresh every 3 to 4 days.

4. In order to maintain neurons for longer periods, from day 14 of differentiation switch medium to neurobasal medium supplemented with B27 + BDNF (10 ng/ml), but without N2. Such conditions should enable neuronal survival for a further 3 weeks.

Troubleshooting: Mouse NS Cells

Unsurprisingly, we find that the efficiency of initial neural lineage commitment correlates with the proportion of cells that can subsequently be expanded as NS cells. However, as the NS cells expand rapidly in low-density conditions, even poor efficiency of initial neural induction provides enough founding cells to generate cell lines. Nonneural and residual ES cells are reduced/eliminated through a combination of cell death and differentiation and are not able to proliferate. A common variable is the timing of attachment of the aggregates/clusters upon replating into NS cell medium. Patience is required, as apparently nonadherent cell aggregates should eventually attach. Attachment can be promoted by replating aggregates into fresh flasks if there are large amounts of debris. Once attached and outgrown these cells will remain adherent in subsequent passages.

In optimal conditions, NS cultures are extremely healthy with minimal cell death or differentiation. If NS cultures start to show a general decline in viability over the course of passaging with a reduction in cell division, this is likely due to one of two factors. The N-2 supplement can show reduced potency if stored for long periods (>3 weeks) at $-20°$ or as part of complete NS medium (>3 weeks) at $4°$. Overtrypsinization during passage can also result in unhealthy cultures. This can be circumvented by diluting trypsin 1:10 or through use of Accutase (Sigma) or PBS to detach/dissociate cells or reducing exposure time (<2 min). NS cells can be expanded on laminin. Here cell adhesion is strong, but so long as EGF and FGF-2 are supplied there is no cell differentiation and cells continue to self-renew. For clonal differentiations, NS cells can be plated at low density (1×10^4 cells in a 9-cm plate) in NS expansion medium; colonies of several hundred cells emerge after 7 to 10 days and can be isolated and expanded or induced to differentiate through growth factor removal and/or serum exposure as described earlier.

Media and Reagents

ES Cell Medium

>   500 ml GMEM medium
>   50 ml fetal calf serum
>   5.5 ml MEM nonessential amio acids 100 × (GIBCO: final concentration
>   1×)
>   5.5 ml sodium pyruvate 100 mM (final concentration 1 mM)
>   5.5 ml L-glutamine 200 mM (final concentration 2 mM)
>   550 μl β-mercaptoethanol (final concentration 0.1 mM)
>   100 units/ml LIF (see later)

Alternatively, for serum-free culture, ES cells can be maintained in N2B27 medium supplemented with LIF (100 units/ml, see later) and 10 ng/ml BMP-4 (3 ng/ml, R&D Systems) (Ying et al., 2003a).

N2B27 Medium

    200 ml neurobasal medium
    200 ml DMEM/F12 medium
    4 ml B27 supplement (final concentration 0.5×)
    2 ml N-2 supplement (final concentration 0.5×)
    400 μl β-mercaptoethanol (final concentration 0.1 mM)
    1 ml glutamate (final concentration 0.2 mM)
    N2B27 should be stored at 4° and used within 1 week.

N-2 Supplement

Note that the N-2 supplement used in N2B27 is modified from the original formulation N-2, with higher insulin and addition of bovine serum albumin (BSA). This formulation increases attachment and survival of neural cells. Batches of N-2 can be stored in aliquots at −20° for no longer than 3 weeks.

Stock solutions are:

    Insulin 25 mg/ml (Sigma), dissolve 100 mg/4 ml 0.01 M sterile filtered HCl. Insulin should be resuspended overnight at 4°.
    Apo-transferrin 100 mg/ml (Sigma), dissolve 500 mg/5 ml sterile filtered H_2O
    BSA 75 mg/ml, dissolve in sterile PBS
    Progesterone 0.6 mg/ml (Sigma), dissolve 6 mg/10 ml ethanol and then filter sterilize
    Putrescine 160 mg/ml (Sigma), dissolve 1.6 g/10 ml H_2O and then filter sterilize
    Sodium selenite 3 mM (Sigma), dissolve 2.59 mg/5 ml H_2O and then filter sterilize
    DMEM:F12 (-glutamine) (GIBCO)

These stocks are stored at −20°. We routinely use these stocks to prepare 40 ml of N-2 supplement. To 27.5 ml of DMEM:F12 add 4 ml BSA stock, 4 ml of insulin stock (add 200 μl at a time to prevent precipitation), 4 ml apo-transferrin, 40 μl sodium selenite, 400 μl putrescine, and 132 μl progesterone.

    β-Mercaptoethanol: 200 μl of β-mercaptoethanol (14.3 M) is mixed with 28.2 ml UHP water and stored at 4° in aliquots. Final concentration 0.1 M.
    Glutamate/pyruvate: 100 mM sodium pyruvate + 5.5 ml of 200 mM L-glutamine 200 mM. Store in aliquots at −20°.
    Trypsin (1×): Add 0.186 g EDTA to 500 ml UHP water and filter sterilize. Add 5 ml chick serum and 5 ml concentrated trypsin (2.5%). Store in aliquots at −20°.

Gelatin: 1% stock solution is prepared in UHP water, autoclaved, and stored in aliquots at 4°. To prepare the 0.1% working solution, 1% gelatin is warmed to 37° until it liquefies and then it is diluted 1:10 in sterile PBS; 0.1% gelatin can be stored for up to 2 weeks at 4°.

For culture and differentiation of both ES and NS cells, plates and flasks are coated with a 0.1% gelatin solution for at least 10 min at room temperature. Gelatin is aspirated prior to use. Washing with PBS is not necessary.

NS basal medium: Euromed-N (formerly NS-A) is a basal medium with a formulation similar to DMEM (Euroclone). We supplement Euromed-N with N-2 supplement (see later) and 2 mM L-glutamine (Invitrogen). Growth factors are added directly to flasks/plates immediately prior to use to achieve the final NS cell expansion medium.

EGF (Peprotech) and FGF-2 (Peprotech) are resuspended in PBS and 20-μl aliquots are stored at $-20°$. EGF and FGF-2 are added directly to flasks and plates when required. Once thawed each is used within 1 week stored at 4°.

LIF recombinant: LIF is produced readily by transient transfection of Cos cells. The supernatant from these cultures is collected and the concentration of LIF is assayed using CP1 indicator cells. The supernatant is then diluted in 1× PBS to give a 1000× stock concentration of 100,000 units/ml and stored at $-20°$ in 0.5-ml aliquots.

Poly-L-ornithine/laminin plates: A 0.01% solution of poly-L-ornithine (Sigma) is added to plates and flasks for at least 20 min. The solution is removed and plates/flasks are washed three times with 1 × PBS. Replace PBS with a 10-μl/ml solution of laminin in PBS (Sigma) and incubate at 37° for at least 3 h (preferably overnight).

Summary

Neural differentiation of mouse and human ES cells can be achieved with high efficiency. Such protocols provide an accessible and genetically tractable model system with which to elucidate molecular mechanisms of lineage choice and differentiation. Initial studies have revealed a close similarity between those mechanisms responsible for ES cell differentiation *in vitro* and those operating during embryogenesis, suggesting that studies of stem cell mechanisms *in vitro* will be directly relevant to understanding embryogenesis, and vice versa. Further, understanding mechanisms of ES cell and NS cell differentiation provides a foundation for clinical applications.

Acknowledgments

We thank Luciano Conti for contributions to NS protocols. Austin Smith made helpful comments on the manuscript and provided support and guidance throughout these studies. This research is supported by the Biotechnology and Biological Sciences Research Council and the Medical Research Council of the United Kingdom, the Wellcome Trust, INSERM, and the Framework VI Integrated Project EuroStemCell.

References

Anthony, T. E., Klein, C., Fishell, G., and Heintz, N. (2004). Radial glia serve as neuronal progenitors in all regions of the central nervous system. *Neuron* **41**, 881–890.

Aubert, J., Stavridis, M. P., Tweedie, S., O'Reilly, M., Vierlinger, K., Li, M., Ghazal, P., Pratt, T., Mason, J. O., Roy, D., and Smith, A. (2003). Screening for mammalian neural genes via fluorescence-activated cell sorter purification of neural precursors from Sox1-gfp knock-in mice. *Proc. Natl. Acad. Sci. USA* **100**(Suppl. 1), 11836–11841.

Bain, G., Kitchens, D., Yao, M., Huettner, J. E., and Gottlieb, D. I. (1995). Embryonic stem cells express neuronal properties *in vitro*. *Dev. Biol.* **168**, 342–357.

Billon, N., Jolicoeur, C., Ying, Q. L., Smith, A., and Raff, M. (2002). Normal timing of oligodendrocyte development from genetically engineered, lineage-selectable mouse ES cells. *J. Cell Sci.* **115**, 3657–3665.

Conti, L., Pollard, S. M., Gorba, T., Reitano, E., Toselli, M., Biella, G., Sun, Y., Sanzone, S., Ying, Q. L., Cattaneo, E., and Smith, A. (2005). Niche-independent symmetrical self-renewal of a mammalian tissue stem cell. *PLoS Biol.* **3**, e283.

Gottlieb, D. I., and Huettner, J. E. (1999). An *in vitro* pathway from embryonic stem cells to neurons and glia. *Cells Tissues Organs* **165**, 165–172.

Hovatta, O., Mikkola, M., Gertow, K., Stromberg, A. M., Inzunza, J., Hreinsson, J., Rozell, B., Blennow, E., Andang, M., and Ahrlund-Richter, L. (2003). A culture system using human foreskin fibroblasts as feeder cells allows production of human embryonic stem cells. *Hum. Reprod.* **18**, 1404–1409.

Kawasaki, H., Mizuseki, K., Nishikawa, S., Kaneko, S., Kuwana, Y., Nakanishi, S., Nishikawa, S. I., and Sasai, Y. (2000). Induction of midbrain dopaminergic neurons from ES cells by stromal cell-derived inducing activity. *Neuron* **28**, 31–40.

Li, M., Pevny, L., Lovell-Badge, R., and Smith, A. (1998). Generation of purified neural precursors from embryonic stem cells by lineage selection. *Curr. Biol.* **8**(17), 971–974.

Li, M., Price, D., and Smith, A. (2001). Lineage selection and isolation of neural precursors from embryonic stem cells. *Symp. Soc. Exp. Biol.* **2**, 9–42.

Merkle, F. T., Tramontin, A. D., Garcia-Verdugo, J. M., and Alvarez-Buylla, A. (2004). Radial glia give rise to adult neural stem cells in the subventricular zone. *Proc. Natl. Acad. Sci. USA* **101**, 17528–17532.

Noctor, S. C., Flint, A. C., Weissman, T. A., Wong, W. S., Clinton, B. K., and Kriegstein, A. R. (2002). Dividing precursor cells of the embryonic cortical ventricular zone have morphological and molecular characteristics of radial glia. *J. Neurosci.* **22**, 3161–3173.

Pevny, L. H., Sockanathan, S., Placzek, M., and Lovell-Badge, R. (1998). A role for SOX1 in neural determination. *Development* **125**, 1967–1978.

Pollard, S. M., Conti, L., and Smith, A. (2006a). Exploitation of adherent neural stem cells in basic and applied neurobiology. *Regener. Med.* **1**(1), 111–118.

Pollard, S. M., Conti, L., Sun, Y., Goffredo, D., and Smith, A. (2006b). Adherent neural stem (NS) cells from foetal and adult forebrain. *Cerebral Cortex* **16**(Suppl. 1), 112–120.

Stavridis, M. P., and Smith, A. G. (2003). Neural differentiation of mouse embryonic stem cells. *Biochem. Soc. Trans.* **31**, 45–49.

Tropepe, V., Hitoshi, S., Sirard, C., Mak, T. W., Rossant, J., and van der Kooy, D. (2001). Direct neural fate specification from embryonic stem cells: A primitive mammalian neural stem cell stage acquired through a default mechanism. *Neuron* **30,** 65–78.

Ying, Q. L., Nichols, J., Chambers, I., and Smith, A. (2003a). BMP induction of Id proteins suppresses differentiation and sustains embryonic stem cell self-renewal in collaboration with STAT3. *Cell* **115,** 281–292.

Ying, Q. L., and Smith, A. G. (2003). Defined conditions for neural commitment and differentiation. *Methods Enzymol.* **365,** 327–341.

Ying, Q. L., Stavridis, M., Griffiths, D., Li, M., and Smith, A. (2003b). Conversion of embryonic stem cells into neuroectodermal precursors in adherent monoculture. *Nature Biotechnol.* **21,** 183–186.

[11] Retinal Pigment Epithelium

By Irina Klimanskaya

Abstract

Retinal pigment epithelium (RPE) arises from neuroectoderm and plays a key role in support of photoreceptor functions. Several degenerative eye diseases, such as macular degeneration or *retinitis pigmentosa*, are associated with impaired RPE function that may lead to photoreceptor loss and blindness. RPE derived from human embryonic stem (hES) cells can be an important source of this tissue for transplantation to cure such degenerative diseases. This chapter describes differentiation of hES cells to RPE, its subsequent isolation, maintenance in culture, and characterization.

Introduction

Human embryonic stem (hES) cells bear a promise for cellular therapy of many ailments because of their unique ability to differentiate into the derivatives of all three germ layers. It is considered that in the absence of other inductive cues, embryonic stem (ES) cells choose a "default" neural pathway for differentiation (Smukler *et al.*, 2006; Ying *et al.*, 2003; reviewed by Muñoz-Sanjuán and Brivanlou, 2002), and to date different types of derivatives of this lineage have been isolated (reviewed by Kania *et al.*, 2004; Wei *et al.*, 2005; Teramoto *et al.*, 2005; Olsen *et al.*, 2006; Peschle and Condorelli, 2005; Ben-Hur, 2006; Reubinoff *et al.*, 2001). Among them is retinal pigment epithelium (RPE), a derivative of neuroectoderm, which progenitor it shares with neuronal retina in early development. Transplantation of RPE has been studied extensively in animal models (reviewed by Lund *et al.*, 2001) and in a few human trials (Binder *et al.*, 2004; van Meurs

METHODS IN ENZYMOLOGY, VOL. 418
0076-6879/06 $35.00
DOI: 10.1016/S0076-6879(06)18011-8

et al., 2004; Radtke *et al.*, 2004; Weisz *et al.*, 1999) as a potential treatment for retinal degenerative diseases, such as macular degeneration or *retinitis pigmentosa*. Several cell sources have been considered for such therapy: fetal RPE (Radtke *et al.*, 2004; Weisz *et al.*, 1999), autologous RPE (Binder *et al.*, 2004; van Meurs *et al.*, 2004), or established RPE cell lines (Lund *et al.*, 2001). However, each source is not perfect. With all human donor tissue there is batch-to-batch variation and safety issues; in addition, fetal tissue as a cell source raises ethical concerns. Autologous RPE may already have an impaired function due to the developing disease. Cell lines such as ARPE-19 and h1RPE7 were used by Lund and coauthors (2001) in the Royal College of Surgeons (RCS) rat model of retinal dystrophy and showed preservation of the photoreceptor. These or similar cell lines could be a good source, if they can prove to maintain stable karyotype and RPE functions over multiple passages. However, it could be challenging to generate multiple lines that meet these criteria with minimal batch-to-batch variation if donor tissue is used as a source. Generation of RPE from hES cells has numerous advantages, as it can be done from pathogen-free cell lines under good manufacturing practices (GMP) conditions and with minimal variation among batches. Such cells can be characterized extensively prior to preclinical studies or for clinical applications, and large numbers of cells can be generated from a virtually unlimited supply of each hES cell line. With the future development of technologies such as somatic cell nuclear transfer- or parthenote-generated ES cells, banks of RPE cell lines can be established for future selection of cell lines more closely immune matched with a patient.

There are currently several reports on producing RPE from ES cells. In 2002, in the same experiments when primate ES cells differentiated into dopamine neurons, RPE was also observed and isolated in the same cultures (Kawasaki *et al.*, 2002), and more extensive characterization of such cells was done 2 years later (Haruta *et al.*, 2004), which showed that these cells express mRNA for RPE-specific markers RPE65 (Redmond *et al.*, 1998) and CRALBP (Saari *et al.*, 2001), perform phagocytosis with latex beads, and attenuate the loss of visual function after transplantation into the subretinal space of RCS rats. Mouse ES cells were differentiated into RPE (among other retinal structures) in the experiments of another group (Hirano *et al.*, 2003). In all these experiments, coculture with mouse skull stromal cell line PA6 was used, and this differentiation was attributed to the stromal cell-derived inducing activity.

Experiments with differentiating hES cells showed that no such coculture is required for efficient and reliable differentiation of RPE. In the model system used in our laboratory, such differentiation occurs spontaneously when overgrown hES cells are maintained in the same plates, with or without feeders, until clusters of pigmented epithelia appear and can be harvested (Klimanskaya *et al.*, 2004). Alternatively, embryoid bodies (EB) that are produced from hES cells and cultured for 6 to 8 weeks show pigmented areas on the surface; such EBs can be plated for outgrowth and produce primary cultures of RPE. Such differentiation was observed in medium supplemented with fetal bovine serum (FBS) or Serum Replacement (Invitrogen), with or without basic fibroblast growth factor (bFGF), on feeder cells, mouse embryonic fibroblasts (MEF), and in feeder-free systems: on MEF-derived extracellular matrix (Klimanskaya *et al.*, 2005), on fibronectin, collagens I and IV, and laminin. The differentiating cultures of hES cells are diverse, showing the presence of various cell types, and the sequence of events leading to the formation of RPE is still unclear. One of the possible models is that the earliest step is "default" neural lineage commitments of ES cells and formation of neuroectoderm or similar retinal progenitor cells. In eye formation during early mammalian development, the dorsal part of the optic vesicle adjacent to the mesoderm receives RPE-inductive signals, such as activin A expressed by extraocular mesenchyme (Chow and Lang, 2001; Feijen *et al.*, 1994; Fuhrmann *et al.*, 2000), which promote RPE formation, while the distal part receiving the FGF signals from surrounding ectoderm becomes neural retina. It is likely that similar events leading to RPE specification occur in differentiating hES cultures in response to and as a result of cues produced by the differentiating derivatives of hES cells that surround clusters of neuroectoderm.

Differentiation of hES Cells to RPE

Our method of producing RPE from hES cells mostly relies on long-term spontaneous differentiation of hES cells in serum-free medium. The cells are grown (Fig. 1A) on mitomycin C-treated primary mouse embryonic fibroblasts (PMEF) until they "overgrow" (usually 7 to 10 days; Fig. 1B), and then the medium is replaced with FGF-free differentiation medium. As an alternative, embryoid body cultures can be set up. Our experiments isolated RPE cells from 15 different hES cell lines: 6 lines derived at Harvard University in the laboratory of Dr. Douglas Melton (Cowan *et al.*, 2004), 3 lines from Wicell (Thomson *et al.*, 1998), and 6 lines established at Advanced Cell Technology, Inc.

After 5 to 7 days in culture, we usually see signs of differentiation, when the typical ES cell morphology is lost and various differentiated cell types appear (Fig. 1B and C). Most colonies usually show signs of neural lineage commitment, including cells that stain positive for tubulin βIII, pax6, and GFAP. These observations are in agreement with numerous observations in the literature that ES cells in culture select the neuronal pathway of differentiation most readily, which could be chosen by default (Smukler

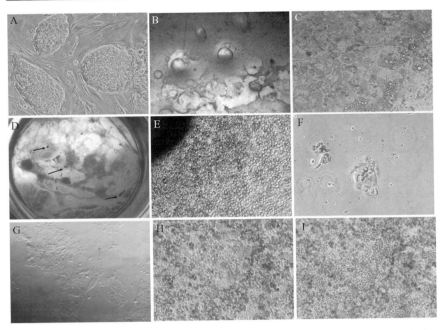

Fig. 1. Differentiation of hES cells to RPE. (A) Undifferentiated hES cell colonies. (B and C) Differentiating three-dimensional structures and various cell types from hES cells. (D and E) Appearance of pigmented cell clusters in long-term differentiating cultures of hES cells. (F and G) Growth of isolated RPE cells. (F) The next day after isolation. (G) Five days after isolation. (H and I) Mature hES–RPE culture, the same field is shown with (H) HMC and (I) phase contrast. Original magnification: A, C, E, F, H, I, 200×; B, 50×; D, 7.5×; G, 50×.

et al., 2006; Tropepe *et al.*, 2001), in response to the activity of FGF (Bouhon *et al.*, 2005; Ying *et al.*, 2003), or as a result of elimination of other inductive signals (Ying *et al.*, 2003). The plates are then cultured until clusters of pigmented epithelial cells begin to appear, which usually happens in 6 to 8 weeks (Fig. 1D and E). Such clusters keep slowly increasing in size, while new clusters continue to emerge. The same process can be initiated in conventional embryoid body culture (EB), in which case pigmented epithelial cells would appear on the surface of EBs and then this transition of nonpigmented cells to pigmented epithelium would slowly take over the whole EB. The cell lines were used at various passages, and the visible efficiency of RPE formation was higher at earlier passages. While on average the first pigmented epithelial clusters appeared around 6 to 8 weeks after the cells were subcultured, at early passages such pigmented cells were observed after 3 weeks. This is possibly happening because the lines were either passaged only mechanically or just adapted to trypsinization and had not undergone multiple passaging with trypsin, which could be removing cell surface molecules and thus causing a certain degree of selection of such cells.

Of note, clusters of cells stained positively for neural lineage markers Pax6 and/or tubulin βIII, often in close conjunction with pigmented epithelium, were found in such differentiating systems. Cells of various types, still unidentified, are also found in the same differentiating cultures of hES cells, surrounding the clusters of RPE and their presumptive progenitors. It is possible that cells producing signals promoting RPE specification in clusters of Pax6-positive progenitors, similar to the signaling of ocular mesoderm in patterning ocular tissues, could be found among such differentiated cells next to Pax6-positive clusters.

These weeks-old cultures are composed of several layers of cells with a lot of extracellular matrix deposition, which makes it difficult to disperse them into a single cell suspension to select the desired cell type using FACS or magnetic beads. Instead, we use an approach in which the multilayer of cells is loosened with trypsin or collagenase and the pigmented cells are picked under the dissecting microscope using a glass capillary. Collected cells are plated on laminin or gelatin in RPE culture medium containing Serum Replacement and FBS with optional bFGF; within 24 to 48 h, clusters of cells begin to proliferate. Proliferating cells lose pigment and acquire a fibroblastic phenotype (Fig. 1F, G, Fig. 2A and B), strongly resembling the transdifferentiated RPE, which dedifferentiate as they proliferate and return to typical RPE morphology after they establish a monolayer (Fig. 1H and I), which usually takes 2 to 3 weeks (Chen *et al.*, 2003; Reh *et al.*, 1987; Sakaguchi *et al.*, 1997; Vinores *et al.*, 1995). Such RPE transdifferentiation has been shown to result in the formation of neuronal, amacrine, and photoreceptor cells (Zhao *et al.*, 1995), glia (Sakaguchi *et al.*, 1997), neural retina (Galy *et al.*, 2002), and neuronal progenitors (Opas and Dziak, 1994). bFGF accelerates transdifferentiation and RPE proliferation (Fig. 2), thus allowing the cells to reach confluence and begin to revert to the RPE phenotype much sooner. hES-derived RPE (hES–RPE) in the transdifferentiated state express the neural lineage markers Pax6 and tubulin βIII (Fig. 3), strongly resembling immature neural cells, and our comparative gene expression profiling showed their similarity to neural stem cells (Klimanskaya *et al.*, 2004).

Materials and Equipment

Unless the manufacturer and model are specified, most brands are acceptable.

−bFGF   +bFGF

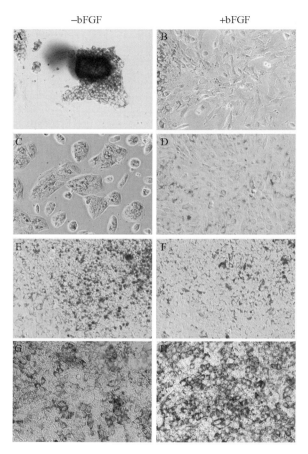

Fig. 2. Proliferation of hES–RPE in culture and transdifferentiation. (Left) No FGF; (right) 10 ng/ml bFGF. (A and B) Initial outgrowth after isolation of RPE from hES cells, 5 days. (C and D) Three days after passaging of hES–RPE. (E and F) Seven days after passaging. (G and H) Twenty-five days after passaging. Note that in the presence of bFGF transdifferentiation is more prominent and the monolayer is established faster (C and D). Cells in mature cultures of hES–RPE of the same age are more pigmented (G and H). Original magnification: A–F, 200×; G, H, 400×.

Equipment and Cell Culture Disposables

Stereomicroscope for microdissection (we use Nikon SMZ-1500)
Inverted microscope (we use Nikon TE 300 and TS 100) with phase (4, 10, 20, 40×) and Hoffman modulation Optics (HMC, 20, 40×) objectives

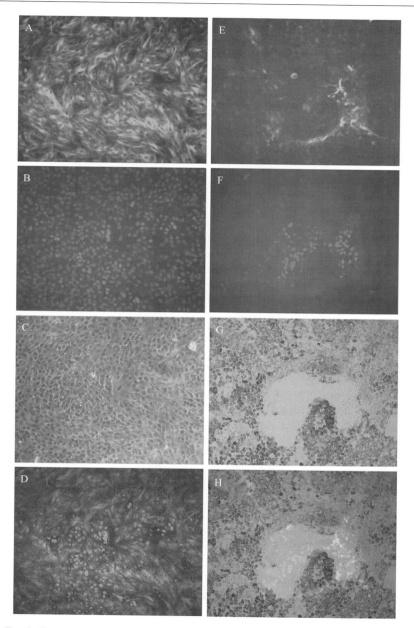

Fɪɢ. 3. Transdifferentiated (A–D) and "mature" (E–H) hES–RPE stained with antibodies to tubulin βIII (A and E) and Pax6 (B and F). (C and G) Phase contrast of the same fields. (D and H) Merged images of the first figures. Original magnification: 200×. Reproduced from Klimanskaya *et al.* (2004) with permission.

Bench-top biological safety cabinet (Terra Universal, Anaheim, CA) or micromanipulation workstation (MidAtlantic Diagnostics, Mount Laurel, NJ). The dissecting microscope is set up in this biosafety cabinet/workstation.

Biosafety cabinet (laminar flow hood) for cell culture

CO_2 incubator

Cell culture centrifuge

Automatic pipettors P1000, P200, P20

Six-well tissue culture plates

Four-well tissue culture plates

100-mm tissue culture dishes

Ultralow attachment cell culture plates or flasks (Corning)

Tissue culture flasks

15- and 50-ml conical centrifuge tubes

Glass capillaries flame pulled from Pasteur pipettes. *Note:* Pasteur pipettes need to be autoclaved or otherwise sterilized using biological indicators for quality control. We use spore strips from Steris (Mentor, OH).

Pipette-aid, cell culture disposables, etc.

Media Components and Other Reagents

Knockout Dulbecco's modified Eagle's medium (DMEM) (Invitrogen)

DMEM high glucose (Invitrogen)

Serum Replacement (Invitrogen)[1]

Plasmanate (Bayer)[1]

FBS (Hyclone)

β-Mercaptoethanol, 1000× solution (Invitrogen)

Nonessential amino acids (NEAA), 100× solution (Invitrogen)

Penicillin/streptomycin, 100× solution (Invitrogen)

Glutamax-I, 100× solution (Invitrogen)

bFGF (Invitrogen)

Human LIF (Chemicon International)

0.05% trypsin/0.53 mM EDTA (Invitrogen)

Collagenase type IV (Invitrogen)

Gelatin from porcine skin (Sigma)

Laminin from human placenta (Sigma)

PBS, Ca^{2+}, Mg^{2+} free (Invitrogen)

---

[1] Each lot of these reagents needs to be tested for quality before it is used for hES cell medium preparation. For more detailed quality control protocols, see Klimanskaya and McMahon (2004).

Normocin (Invivogen, San Diego, CA): combination of three antibiotics active against mycoplasma, both positive and negative Gram bacteria and fungi. Usually well tolerated by the cells.

Mitomycin C (Sigma): the mitomycin C water solution normally has a deep purple color. Some batches have an insoluble precipitate forming the next day, which is reflected in the color of the solution becoming very light, and the mitomycin becomes less effective in arresting the cell division.

Primary Mouse Embryo Fibroblast (PMEF) Feeders[1]

PMEF are prepared from E12.5 fetuses of CD1 mice, expanded 1:5, and frozen at passage one; passage two is treated with 10 μg/ml mitomycin C for 3 h at 37° (for more detailed procedures, see Klimanskaya and McMahon [2004, 2005]). PMEF are plated at a density of 50,000 to 60,000 cells/cm^2 and used within 3 days.

hES Cell Culture and Differentiation Culture Setup

Basal medium (BM): knockout DMEM, supplemented with 1:100 NEAA, 1:100 penicillin/streptomycin, 1:100 glutamax, 1:1000 β-mercaptoethanol

hES cell growth medium (GM): BM, supplemented with 8% Serum Replacement, 8% plasmanate, 8 ng/ml human bFGF, 10 ng/ml human LIF

Differentiation medium (DM): BM, supplemented with 15% Serum Replacement

Note: to prevent contamination in long-term cultures, Normocin can be added.

hES cells are grown in GM, and are passaged routinely every 4 to 6 days using trypsin or mechanical dispersion (for more detailed procedures, see Klimanskaya and McMahon [2004] and Cowan et al. [2006]). For adherent differentiation the cells are allowed to overgrow on MEF until dome-like structures begin to appear, usually 7 to 10 days. The medium is then changed to DM and replaced every day or every other day, depending on the volume of the medium per well. Note: because medium containing NEAA and Serum Replacement generally looks more yellowish than orange/pink, a slight color change toward yellow is acceptable before the medium needs to be changed. The frequency of medium change usually depends on the number of the cells per well and the well size/shape. Usually, four-well plates that can only hold 0.5 to 0.8 ml of medium need change every day, whereas six-well plates with 5 to 7 ml of medium can be changed every 36 to 48 h. These are approximate guidelines, and the frequency of medium change for other sizes of tissue culture dishes needs to be established empirically.

For differentiation as embryoid bodies, hES cultures are treated with 2 to 4 mg/ml collagenase IV in GM for 5 to 10 min or until the colonies begin to detach from PMEF. The colonies are then collected by gentle pipetting and centrifuged at 1000 rpm in a standard cell culture centrifuge for 5 min (about $160g$). The medium is aspirated, and the cell clumps are plated into ultralow attachment plates/flasks in DM. The medium is changed as required (see earlier discussion) by careful aspiration from the top, leaving a layer of the medium on the bottom to prevent disturbing of the clumps/EBs. If the EBs become hollow and rise to the surface, a cell culture pipette is used to remove the medium carefully. *Note 1:* spreading the EB to multiple wells reduces the frequency of medium change. *Note 2:* Ultralow attachment plates/flasks cannot be substituted with nontissue culture Petri dishes because many EBs will eventually attach and begin to grow out.

Pigmented clusters usually become visible within 6 to 8 weeks and will continue to grow slowly, and more of them may appear (Fig. 1D and E). To harvest more cells, the cultures need to continue, usually 2 to 3 months. In some experiments, we had 9-month-old cultures of EBs that produced passageable RPE cultures after being plated on gelatin. *Note:* when cells are cultured for such a long time, the medium is replaced only partially, and the wells are filled almost to the top; extra care needs to be exercised to prevent contamination. Normocin can be used as a wide-spectrum agent (fights Gram-positive and Gram-negative bacteria, fungi, and mycoplasma without any noticeable effect on hES cell performance). The plates need to be stacked carefully and be aligned properly to prevent accidental sliding and medium spills. If this happens, the spills need to be aspirated immediately. Setting up long-term cultures in flasks may reduce the risk of contamination.

Observe the cultures under a stereoscope at low power and at higher power under an inverted microscope (preferably using HMC objectives) for the appearance of pigmented cell clusters with cobblestone morphology. Usually when they appear, they can be seen clearly without any microscope by simply putting a dish against a white surface, appearing as little "freckles" (Fig. 1D) on the bottom of the plate or on EBs. However, to confirm that these are the anticipated cells, microscopic observation is required. *Note:* HMC observation is highly desirable for seeing both pigment and three-dimensional cell shapes (Fig. 1H and I), but if it is unavailable, using the "wrong" phase match for the regular phase-contrast objective allows one to see pigmentation better and adds some depth to the picture.

Harvesting RPE Cells

After several weeks of differentiation, a lot of extracellular matrix is deposited by the cells and it becomes very difficult to dissociate them into single cells to collect RPE by FACS or another cell-sorting method. However, its unique appearance allows one to handpick the cells of the right phenotype. We use two approaches: (1) handpicking pigmented cells under the stereomicroscope after the monolayer has been loosened with collagenase IV or trypsin or (2) outgrowth of pigmented EB.

Media and Reagents

RPE growth medium (RPE-GM)
BM supplemented with 7% Serum Replacement and 4% FBS. A concentration 10 ng/ml bFGF is optional.
0.05% trypsin/0.53 mM EDTA (Invitrogen)
PBS, Ca^{2+}, Mg^{2+} free
Collagenase IV, 20 mg/ml in DMEM stock solution (sterilize by 0.22-μm filtration and keep frozen in 1- to 2-ml aliquots)
Gelatin 0.01% (1 mg/ml) solution in PBS, sterilize by 0.22-mm filtration
Laminin from human placenta, 10 μg/ml solution in PBS, sterilize by 0.22-mm filtration

Coating Tissue Culture Plates with Gelatin or Laminin

Add solution of gelatin (0.1 to 0.2 mg/ml) or laminin (5 to 10 μg/cm^2) to cell culture plates.

Method 1. Handpicking Pigmented Cells

Method 1 can be performed on multiwell plates or tissue culture dishes. Cell culture flasks with a detachable side can also be used.
Trypsin is only used to assist removing the cells; it could be done without any enzymes, if the cells detach easily, so the cultures need to be probed first. Collagenase IV at a concentration of 5 to 10 mg/ml can be used instead of trypsin.
RPE-GM used for culture of isolated cells can be used with or without bFGF: in the presence of bFGF cells will grow faster, so the monolayer will be reached faster and reacquisition of the RPE phenotype after transdifferentiation will happen faster (Fig. 2).

1. Rinse the plate with PBS two or three times, add 0.05% trypsin/0.53 mM EDTA. Incubate for a few minutes, checking frequently under the

microscope and probing the culture with a flame-pulled capillary. If using collagenase, expect to wait longer, that is, 1 to 2 h.

2. The technique is simple gentle "scraping" off of the pigmented clusters and aspirating the removed cells; this should be done as soon as the monolayer has loosened enough to allow the cells to be collected easily. Keep scraping the pigmented clusters gently and transfer the collected cells to another plate or tube with RPE medium. This should be done quickly because the cells remain in trypsin and cell damage will occur after prolonged exposure. Using collagenase allows more time for the procedure.

3. Rinse RPE cells either by transferring them through two to three changes of medium in four-well plates (for larger clumps) or by centrifugation at 1000 rpm for 5 min in 5 to 10 ml RPE medium.

4. If large clumps of cells are collected, they can be transferred into a conical centrifuge tube, washed with PBS by centrifugation, and treated with 0.05 trypsin/0.53 mM EDTA for several minutes in a water bath. *Note:* agitate the tube by gently tapping it frequently and observe under a dissecting microscope; add RPE medium to quench trypsin immediately after large clumps are broken into desired smaller-sized clumps and single cells.

5. Plate into one or two wells of a four-well plate on gelatin or laminin in RPE medium.

Note: because this technique is designed for isolation of very small numbers of RPE cells, one or two wells of a four-well plate are recommended. Usually, even several hundred cells will fill such wells in 2 to 3 weeks and can be passaged after that by regular methods. See Fig. 1F and G for initial stages of growth of such manually isolated RPE cells.

Method 2. Outgrowth of RPE from EBs

Note: Use for EB cultures that show pronounced RPE areas on the surface and for large clumps of RPE cells that sometimes spontaneously detach from the cellular multilayer in adherent differentiating cultures of hES cells.

1. Plate EBs or RPE clumps onto cell culture dishes coated with gelatin or laminin in RPE medium.

2. After 2 to 3 days outgrowth of RPE cells should be visible.

3. Collect any large clusters of RPE cells that remain loosely attached after 1 to 2 weeks using a glass capillary or a P20/P200 automatic pipette under a dissecting microscope, wash by transferring large clumps through two to three wells of a four-well plate filled with PBS, and incubate in 0.05% trypsin/0.53 mM EDTA (in a drop or in another well) for several minutes, checking frequently under the microscope. Quench trypsin with

RPE medium when the desired small cell clump/single cell suspension is obtained.

4. Collect the cells in a centrifuge tube, centrifuge at 1000 rpm for 5 min, remove the supernatant, resuspend the cells in RPE growth medium, and plate onto gelatin- or laminin-coated tissue culture plates.

Note: proliferating cells lose their RPE morphology, turning into lightly or nonpigmented elongated cells. After the confluent monolayer is established, they will begin to revert to RPE morphology. Adding bFGF to growth medium after isolation will accelerate formation of the confluent monolayer so the cells will reacquire the RPE morphology faster.

Culture and Properties of hES-Derived RPE

Proliferating RPE will transdifferentiate and then begin to redifferentiate upon formation of the confluent monolayer (Fig. 2). The full cycle usually takes about 2 to 3 weeks, but even after that RPE will continue to "mature," becoming more pigmented. However, because very "mature" cells do not survive trypsinization and freezing well, we prefer to subculture them every 2 to 4 weeks. Passaging RPE too soon before they can fully reacquire RPE morphology results in a reduced life span of such cultures: the cells do not revert to RPE morphology and stop growing. RPE are relatively "slow" cells: even in the presence of bFGF, which accelerates their transdifferentiation and proliferation, it may take up to 2 to 3 weeks at each passage at a 1:3 ratio before they "mature" and regain the RPE phenotype. Such cell behavior requires slow propagation, and two or three confluent wells of a four-well plate can be produced in 3 to 4 weeks from several clusters of RPE cells usually found in one 35-mm plate of differentiating ES cells (each cluster usually has several hundred cells; some large older ones may have several thousand). After that the cells are usually subcultured at a 1:3 to 1:6 ratio at 2- to 3-week intervals. However, high numbers of RPE can be obtained from large-scale differentiating cultures of hES cells.

RPE Culture Protocols

Media and Reagents

    RPE-GM, with and without bFGF
    0.05% trypsin/0.53 mM EDTA (Invitrogen)
    Laminin- or gelatin-coated tissue culture plates
    For passaging, 0.05% trypsin/0.53 mM EDTA is usually sufficient but
        may require incubation for several minutes at 37°.

Medium Change

For freshly isolated or passaged cultures it is best to use RPE growth medium supplemented with 10 ng/ml bFGF, which we replace with bFGF-free medium after the monolayer is established. Because they differentiate faster in the presence of PEDF secreted by RPE, it may be beneficial to only replace two-thirds to one-half of the medium in a well. On average, the medium is changed once or twice a week. The monolayer is usually established within the first week after a 1:3 to 1:6 split, and the RPE phenotype is regained in 2 to 3 weeks. After that the cells can be maintained without significant loss of properties for several months; however, after 6 to 8 weeks it would require extended time in trypsin to passage them, so the viability can be decreased. On average, our usual passaging time is between 2 and 8 weeks.

Characterization of hES–RPE Cells

After the culture of hES–RPE is established, the next important step is to characterize the cells at the molecular and functional level. The RPE markers used in our studies are bestrophin, a 68-kDa product of the Best vitelliform macular dystrophy gene (Marmorstein *et al.*, 2000), CRALBP, a water-soluble 36-kDa cellular retinaldehyde-binding protein (CRALBP), which is found in apical microvilli of RPE and in Muller glia (Bunt-Milam and Saari, 1983; Saari *et al.*, 2001), RPE65, a 65-kDa cytoplasmic protein involved in retinoid metabolism (Hamel *et al.*, 1993; Ma *et al.*, 2001; Redmond *et al.*, 1998), and PEDF (Jablonski *et al.*, 2000; Karakousis *et al.*, 2001; Steele *et al.*, 1993). Pax6, although seen by some authors as a molecular marker of RPE (Kawasaki *et al.*, 2002), is normally downregulated in mature RPE, so it could rather indicate the presence of immature cells. CRALBP, PEDF, and bestrophin can be detected by Western blot or immunofluorescence (see Fig. 4 for localization of bestrophin and CRALBP in differentiated hES–RPE by immunostaining), and PEDF can be measured conveniently by ELISA in conditioned medium and/or lysed cells. Translationally controlled RPE65 has been reported previously to be absent from cultured RPE at the protein level (Nicoletti *et al.*, 1995), although real-time RT-PCR has detected high levels of RPE65 mRNA, and we found the same thing happening with our hES–RPE cells. Interestingly, the level of its expression correlated with the differentiation: in more mature cultures its expression was several times higher than in recently passaged cells (Klimanskaya *et al.*, 2004). Therefore, RT-PCR can be used for RPE65 detection.

Because every culture of hES–RPE has variable proportions of more and less "mature" RPE cells, for more comprehensive characterization of

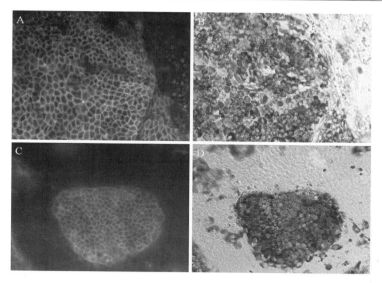

F<small>IG</small>. 4. Localization of bestrophin (A and B) and CRALBP (C and D) in hES–RPE. Original magnification: A, B, 400×; C, D, 200×. Parts of the figure are reproduced from Klimanskaya *et al.* (2004) with permission.

different populations/batches of RPE, real-time quantitative PCR can be used. Additionally, gene expression profiling allows a thorough comparison of hES–RPE with their *in vivo* counterparts, such as primary cultures of fetal and adult RPE, retinal tissues, or established lines (Klimanskaya *et al.*, 2004).

Immunofluorescence and Western Blot

The following antibodies are used.

Anti-bestrophin antibody from Novus Biologicals (Littleton, CO)
CRALBP antibody (we used a gift from Dr. John Sari, University of Washington; antibody is now available commercially)
PEDF ELISA kit (Chemicon)
Pax6 (Chemicon AB 5409)
Secondary antibodies conjugated with FITC, rhodamine red, or biotin from Jackson Immunoresearch (West Grove, PA)
Streptavidin-FITC/streptavidin-Texas red from Amersham
Blocking solution: PBS with 10% goat serum, 10% donkey serum (Jackson Immunoresearch)
Mounting medium Vectashield with DAPI (Vector Laboratories, Burlingame, CA)

For immunofluorescence staining, we fix the cells with 2% freshly made PBS-buffered paraformaldehyde for 10 to 20 min at room temperature and then permeabilize with 0.1% PBS-buffered NP-40 for 10 to 15 min at room temperature. *Note:* we do not have any particular preference for manufacturer of paraformaldehyde or NP-40; however, it is *crucial* (as with any cell/tissue sections staining) that freshly prepared paraformaldehyde is used. Freshly made PBS-buffered 4% paraformaldehyde can be stored in frozen aliquots. See Appendix 1 for a simple procedure for making the paraformaldehyde solution.

Fixed (but not yet permeabilized) cells can be stored for several weeks at 4° before staining, if necessary. Blocking is done for 1 h at room temperature, primary antibodies are added overnight at 4°, and secondary antibodies are added for 1 h at room temperature; fluorescently labeled streptavidin, if desired, is added for 20 min. The specimens are mounted in Vectashield with DAPI and observed/photographed under a fluorescent microscope.

For Western blot the cells are lysed in 4× Laemmli buffer (Laemmli, 1970) without reducing agents (proteinase inhibitors are optional), frozen, and stored at −20°. Reducing agents (β-mercaptoethanol or dithiothreitol) can be added before boiling the samples.

For PEDF ELISA, collect the medium conditioned by RPE cells for several days so it becomes yellow. The cells can also be lysed according to the instructions of the PEDF ELISA kit (Chemicon).

Primers Used in RT-PCR and Quantitative PCR (QPCR)

Gene-specific primer pairs are given for the following genes: RPE65, bestrophin, CRALBP, PEDF, Pax6, β-actin, and GADPDH:

> RPE65-F
> ATGGACTTGGCTTGAATCACTT
> RPE65-R
> GAACAGTCCATGAAAGGTGACA
> Bestrophin-F
> TAGAACCATCAGCGCCGTC
> Bestrophin-R
> TGAGTGTAGTGTGTATGTTGG
> CRALBP-F
> AAATCAATGGCTTCTGCATCATT
> CRALBP-R
> CCAAAGAGCTGCTCAGCAAC
> PEDF-F1
> TCTCGGTGTGGCGCACTTCA

PEDF-R1
GTCTTCAGTTCTCGGTCTATG
Pax6 F1
GTTTCAGCACCAGTGTCTAC
Pax6 R1
TATTGAGACATATCAGGTTCAC
β-actin-F
GCGGGAAATCGTGCGTGACA
β-actin-R
GATGGAGTTGAAGGTAGTTTCG
GADPH-F
CGATGCTGGCGCTGAGTAC
GADPH-R
CCACCACTGACACGTTGGC

For RNA isolation and RT-PCR/real-time QPCR conditions, see Appendix 2. For more details on gene expression profiling of hES RPE, including data analysis, see Klimanskaya et al. (2004) and Hipp and Atala (2006).

Functional Tests

One of the major RPE functions in supporting the photoreceptor is phagocytosis of the shed photoreceptor fragments, and *in vitro* assays for phagocytosis include assays with latex beads and, more specific for RPE, with labeled photoreceptor segments. Briefly, a monolayer of differentiated RPE grown in plastic chamber slides is incubated with 10^8/ml latex beads for up to 24 h, fixed with 2.5% glutaraldehyde in PBS for 30 min, rinsed three times with PBS, and then processed for electron microscopy as described elsewhere. Electron microscopy shows the presence of latex beads in the cytoplasm of the RPE cells. RPE-specific phagocytosis of rod outer segments is described in detail by Finnemann and coauthors (1997; http://www.pnas. org/cgi/content/full/94/24/12932).

Freezing ES–RPE

We use basic freezing medium and techniques. Briefly, trypsinized cells are centrifuged in DMEM with 10% FBS, the pellet is resuspended in cold (stored on ice or at 4°) 90% FBS, 10% DMSO, and the suspension is dispensed in prelabeled cryovials. For larger quantities (more than four to five vials) the vials are kept on ice during the cell-dispensing process. The vials are then sandwiched between two Styrofoam racks from 15-ml centrifuge tubes, and the racks are taped together and put into a −80° freezer overnight. The vials are transferred to liquid nitrogen storage next morning or within 1 to 2 weeks. *Note*: because strongly pigmented cells do not recover well after thaw, we prefer to freeze hES–RPE before they become too pigmented, usually between 1 and 2 weeks after passaging.

Concluding Remarks

In our experiments, all hES cell lines ever handled reliably produced RPE cells (the total number of such hES cell lines is 18) in multiple experiments (over 80). We are currently investigating the *in vivo* performance of some of these hES–RPE cell lines in animal models (such as RCS rat), and preliminary data indicate that there can be a variation between lines or cultures of hES–RPE due to their transdifferentiation abilities and thus differences in the proportion of more and less differentiated cells across the cultures at any time point. This unique ability of RPE to transdifferentiate into cells of the neural lineage creates certain challenges for research and production of FDA-compliant cells for therapeutic applications. At the very minimum, a quantitative evaluation of markers of differentiated RPE versus neural lineage needs to be performed by real-time PCR and by FACS with different populations of cells used for experiments. For any preclinical studies or phase I clinical trials, batch-to-batch variation has to be minimized and the optimal level of differentiation needs to be found. Approaches to generating a suitable RPE cell population are currently being developed in our laboratory and are beyond the scope of this chapter.

Acknowledgments

I appreciate the hard work of Rebeca Ramos-Kelsey on optimization of hES–RPE culture and PCR/QPCR conditions and preparing RNA isolation and PCR protocols. My deep gratitude goes out to Sandy Becker for critical reading of the manuscript. I thank Advanced Cell Technology, Inc. for supporting this work.

Appendix 1

Preparation of 4% Paraformaldehyde Solution

1. Weigh 2 g of paraformaldehyde into a 50-ml centrifuge tube.
2. Add 3.5 ml of Milli-Q or similar quality water.
3. Add 10 μl of 10 N KOH.
4. Bring to a boiling point in a microwave.

Because microwaves vary, times will be different for each microwave. Watch the tube carefully *all the time* to avoid the paraformaldehyde solution boiling over and producing toxic fumes! Ideally, do it under a fume hood; if this is not available, follow these steps:

1. Close the cap on the tube tightly; mix the contents by vortexing or shaking.
2. Unscrew the cap slightly, put the tube into a small (100 to 200 ml) glass beaker or plastic container, and place in the middle of the rotary table in the microwave.
3. Start the microwave and *watch the tube closely* until the liquid *begins* to boil and rise in the tube.
4. *Before* it reaches the top of the tube, press the door button to open. *Note:* when the liquid begins to boil and rise, it takes 1 s or less for it to overflow, so watch carefully and open the door promptly!
5. Screw on the cap tightly, shake the tube to mix the remaining particles (if everything is already dissolved, skip this step), unscrew the cap slightly, put the tube back, and repeat steps 3 and 4. Usually, after that everything is dissolved; if not, repeat step 5.
6. Prepare a 15-ml centrifuge tube and 5-ml syringe with a syringe filter (acrodisc), pour contents into the syringe, and filter into the prepared tube. Do this under the fume hood; if a fume hood is not available, immediately close the tube and rinse the syringe, filter, and original tube with plenty of *cold* running water.
7. Dilute the paraformaldehyde solution with PBS 1:10. This produces approximately 3.7 to 4% paraformaldehyde solution. Make aliquots and freeze for storage or use fresh.
8. Discard the tubes, syringe, and filter according to the chemical waste disposal procedures used in your laboratory.

Appendix 2

RT-PCR and Quantitative Real-Time PCR

Total RNA Isolation Using the RNeasy Minikit

1. Grow cells in a four- or six-well plate to almost confluency. Remove media and add 350 or 600 μl of buffer RLT with 1% BME.
2. Transfer sample to a Qiashredder column and spin for 2 min at maximum speed.
3. Toss the column and add an equal volume of 70% ETOH to the sample and transfer all to a RNeasy column. Spin for 15 to 30 s at maximum speed.
4. Add 350 μl of RW1 buffer. Spin for 15 to 30 s at maximum speed.

5. Add 70 μl of RRD buffer to 10 μl of reconstituted DNase and mix gently. Add 80 μl of diluted DNase to column. Incubate for 15 min.
6. Add 350 μl of RW1 buffer. Spin for 15 to 30 s at maximum speed.
7. Change column to a new collection tube. Add 500 μl of buffer RPE and spin for 15 to 30 s at maximum speed. Do this wash twice.
8. Spin for 1 min at maximum speed to dry the column.
9. Put column in a microcentrifuge tube and add 50 μl of RNase-free water. Spin for 1 min at maximum speed.

Note: for highly differentiated (pigmented) cells it is advisable to start with the Trizol purification step.

RT-PCR Using Qiagen One-Step RT-PCR Kit

Setup for a 25-μl Reaction

> 5 μl of RT-PCR buffer
> 5 μl of Q buffer
> 1.5 μl of 10 μM forward primer
> 1.5 μl of 10 μM reverse primer
> 1 μl of dNTP mix
> 1 μl of enzyme mix
> 5 units of RNase inhibitor
> 10 ng^{-1} μg of template
> RNase-free water up to 25 μl

RT-PCR Program

> 50° for 30 min/reverse transcription
> 95° for 15 min/inactivation of RT/activation of *Taq*
> Cycles (25–40×)
> 94° for 30 s
> 55 to 60° for 30 s
> 72° for 1 min/kb
> 72° for 10 min/final extension

First-Strand cDNA Synthesis (Promega Kit)

1. Incubate 2 or 10 μl of RNA (~1 μg) at 70° for 10 min.
2. Spin down sample and place on ice.
3. To 8.25 or 41.25 μl of RNase-free dH$_2$O add 4 or 20 μl 25 mM MgCl$_2$, 2 or 10 μl 10× RT buffer, 2 or 10 μl 10 mM dNTP, 0.5 or 2.5 μl RNase inhibitor, 0.75 or 3.75 μl AMV RT (15 μg), 0.5 or 2.5 μl oligo(dT) (0.5 μg), and 2 or 10 μl of denatured RNA.
4. Incubate at 42° for 15 to 60 min.

5. Heat samples at 95° for 5 min and place on ice for 5 min.
6. Dilute sample with 80 or 400 μl of water.
7. Store sample at $-20°$. Use 2 to 5 μl for PCR.

PCR Using Amplitaq Gold

Setup for a 25-μl Reaction

2.5 μl of 10× buffer II
0.5 μl of 10 μM forward primer
0.5 μl of 10 μM reverse primer
2 μl of 2.5 mM each dNTP mix
1.5 μl of 25 mM MgCl$_2$
0.25 μl Amplitaq Gold
1 μl of DNA (100 to 200 ng)
16.75 μl of water

Final Concentrations

1× PCR buffer II
0.2 μM forward primer
0.2 μM reverse primer
200 μM dNTP mix
1.5 mM MgCl$_2$
1.25 units Amplitaq Gold

PCR Program

95° for 10 min/activation of Amplitaq Gold
Cycles (25 to 35×)
94° for 30 s
55 to 60° for 30 s
72° for 1 min/per kb
72° for 5 min/final extension

Quantitative Real-Time PCR

Primers for RPE65, Bestrophin, PEDF, CRALBP, Pax6, and β-actin are optimized using various concentrations of 50, 100, 150, 200, 250, and 300 nM. The optimal concentration that gives the lowest Ct value and the highest fluorescence value is used to generate standard curves using various concentrations of the positive control. Optimal standard curves should fall between 90 and 110% efficiency. Except for Bestrophin, whose highest efficiency achieved was 86.8%, all other efficiencies fall within the acceptable limits

Once standard curves are generated, comparative quantification is possible using β-actin as the normalizer gene. Amplification plots should display Ct values that fall within the acceptable range of 15 to 30 cycles. All amplification plots generated fall within this acceptable range.

Real-Time PCR Using SYBR Green

Dilute the passive reference dye 1:500 and keep protected from light. Thaw the master mix and keep on ice protected from light.

For 150 nM Final Primer Concentration

12.5 μl of 2× master mix
0.375 μl of 10 μM forward primer
0.375 μl of 10 μM reverse primer
0.375 μl of diluted reference dye
2 to 5 μl of cDNA
Water up to 25 μl

For 200 nM Final Primer Concentration

12.5 μl of 2× master mix
0.5 μl of 10 μM forward primer
0.5 μl of 10 μM reverse primer
0.375 μl of diluted reference dye
2 to 5 μl of cDNA
Water up to 25 μl

For 250 nM Final Primer Concentration

12.5 μl of 2× master mix
0.625 μl of 10 μM forward primer
0.625 μl of 10 μM reverse primer
0.375 μl of diluted reference dye
2 to 5 μl of cDNA
Water up to 25 μl
Mix gently and spin down plate.

Real-Time PCR Program

Amplification Curve

95° for 10 min
Cycles (40×)
95° for 30 s
55 to 60° for 1 min
72° for 1 min

Dissociation Curve

95° for 1 min
Ramping down to 55°
Ramp up from 55° to 95° at 0.2°/s.

References

Akutsu, H., Cowan, C., and Melton, D. (2006). Human embryonic stem cells. *Methods Enzymol.* **418**(this volume).

Ben-Hur, T. (2006). Human embryonic stem cells for neuronal repair. *Isr. Med. Assoc. J.* **8**(2), 122–126.

Binder, S., Krebs, I., Hilgers, R. D., Abri, A., Stolba, U., Assadoulina, A., Kellner, L., Stanzel, B. V., Jahn, C., and Feichtinger, H. (2004). Outcome of transplantation of autologous retinal pigment epithelium in age-related macular degeneration: A prospective trial. *Invest. Ophthalmol. Vis. Sci.* **45**(11), 4151–4160.

Bouhon, I. A., Kato, H., Chandran, S., and Allen, N. D. (2005). Neural differentiation of mouse embryonic stem cells in chemically defined medium. *Brain Res. Bull.* **68**(1-2), 62–75.

Bunt-Milam, A. H., and Saari, J. C. (1983). Immunocytochemical localization of two retinoid-binding proteins in vertebrate retina. *J. Cell Biol.* **97**(3), 703–712.

Chen, S., Samuel, W., Fariss, R. N., Duncan, T., Kutty, R. K., and Wiggert, B. (2003). Differentiation of human retinal pigment epithelial cells into neuronal phenotype by N-(4-hydroxyphenyl) retinamide. *J. Neurochem.* **84**(5), 972–981.

Chow, R. L., and Lang, R. A. (2001). Early eye development in vertebrates. *Annu. Rev. Cell Dev. Biol.* **17**, 255–296.

Cowan, C. A., Klimanskaya, I., McMahon, J., Atienza, J., Witmyer, J., Zucker, J. P., Wang, S., Morton, C. C., McMahon, A. P., Powers, D., and Melton, D. A. (2004). Derivation of embryonic stem-cell lines from human blastocysts. *N. Engl. J. Med.* **350**(13), 1353–1356.

Feijen, A., Goumans, M. J., and van den Eijnden-van Raaij, A. J. (1994). Expression of activin subunits, activin receptors and follistatin in postimplantation mouse embryos suggests specific developmental functions for different activins. *Development* **120**(12), 3621–3637.

Finnemann, S. C., Bonilha, V. L., Marmorstein, A. D., and Rodriguez-Boulan, E. (1997). Phagocytosis of rod outer segments by retinal pigment epithelial cells requires alpha(v)-beta5 integrin for binding but not for internalization. *Proc. Natl. Acad. Sci. USA* **94**(24), 12932–12937.

Fuhrmann, S., Levine, E. M., and Reh, T. A. (2000). Extraocular mesenchyme patterns the optic vesicle during early eye development in the embryonic chick. *Development* **127**(21), 4599–4609.

Galy, A., Neron, B., Planque, N., Saule, S., and Eychene, A. (2002). Activated MAPK/ERK kinase (MEK-1) induces transdifferentiation of pigmented epithelium into neural retina. *Dev. Biol.* **248**(2), 251–264.

Hamel, C. P., Tsilou, E., Pfeffer, B. A., Hooks, J. J., Detrick, B., and Redmond, T. M. (1993). Molecular cloning and expression of RPE65, a novel retinal pigment epithelium-specific microsomal protein that is post-transcriptionally regulated *in vitro. J. Biol. Chem.* **268**(21), 15751–15757.

Haruta, M., Sasai, Y., Kawasaki, H., Amemiya, K., Ooto, S., Kitada, M., Suemori, H., Nakatsuji, N., Ide, C., Honda, Y., and Takahashi, M. (2004). *In vitro* and *in vivo* characterization of pigment epithelial cells differentiated from primate embryonic stem cells. *Invest. Ophthalmol. Vis. Sci.* **45**(3), 1020–1025.

Hirano, M., Yamamoto, A., Yoshimura, N., Tokunaga, T., Motohashi, T., Ishizaki, K., Yoshida, H., Okazaki, K., Yamazaki, H., Hayashi, S., and Kunisada, T. (2003). Generation of structures formed by lens and retinal cells differentiating from embryonic stem cells. *Dev. Dyn.* **228**(4), 664–671.

Jablonski, M. M., Tombran-Tink, J., Mrazek, D. A., and Iannaccone, A. (2000). Pigment epithelium-derived factor supports normal development of photoreceptor neurons and opsin expression after retinal pigment epithelium removal. *J. Neurosci.* **20**(19), 7149–7157.

Kania, G., Blyszczuk, P., and Wobus, A. M. (2004). The generation of insulin-producing cells from embryonic stem cells—a discussion of controversial findings. *Int. J. Dev. Biol.* **48**(10), 1061–1064.

Karakousis, P. C., John, S. K., Behling, K. C., Surace, E. M., Smith, J. E., Hendrickson, A., Tang, W. X., Bennett, J., and Milam, A. H. (2001). Localization of pigment epithelium derived factor (PEDF) in developing and adult human ocular tissues. *Mol. Vis.* **7**, 154–163.

Kawasaki, H., Suemori, H., Mizuseki, K., Watanabe, K., Urano, F., Ichinose, H., Haruta, M., Takahashi, M., Yoshikawa, K., Nishikawa, S., Nakatsuji, N., and Sasai, Y. (2002). Generation of dopaminergic neurons and pigmented epithelia from primate ES cells by stromal cell-derived inducing activity. *Proc. Natl. Acad. Sci. USA* **99**(3), 1580–1585.

Klimanskaya, I., Chung, Y., Meisner, L., Johnson, J., West, M. D., and Lanza, R. (2005). Human embryonic stem cells derived without feeder cells. *Lancet* **365**(9471), 1636–1641.

Klimanskaya, I., Hipp, J., Rezai, K. A., West, M., Atala, A., and Lanza, R. (2004). Derivation and comparative assessment of retinal pigment epithelium from human embryonic stem cells using transcriptomics. *Cloning Stem Cells* **6**(3), 217–245.

Klimanskaya, I., and McMahon, J. (2004). Approaches for derivation and maintenance of human ES cells: Detailed procedures and alternatives. *In* "Handbook of Stem Cells" (R. Lanza, *et al.*, eds.),. Academic Press, San Diego.

Laemmli, U. K. (1970). Cleavage of structural proteins during the assembly of the head of bacteriophage T4. *Nature* **227**(5259), 680–685.

Lund, R. D., Adamson, P., Sauve, Y., Keegan, D. J., Girman, S. V., Wang, S., Winton, H., Kanuga, N., Kwan, A. S., Beauchene, L., Zerbib, A., Hetherington, L., Couraud, P. O., Coffey, P., and Greenwood, J. (2001). Subretinal transplantation of genetically modified human cell lines attenuates loss of visual function in dystrophic rats. *Proc. Natl. Acad. Sci. USA* **98**(17), 9942–9947.

Lund, R. D., Kwan, A. S., Keegan, D. J., Sauve, Y., Coffey, P. J., and Lawrence, J. M. (2001). Cell transplantation as a treatment for retinal disease. *Prog. Retin. Eye Res.* **20**(4), 415–449.

Ma, J., Zhang, J., Othersen, K. L., Moiseyev, G., Ablonczy, Z., Redmond, T. M., Chen, Y., and Crouch, R. K. (2001). Expression, purification, and MALDI analysis of RPE65. *Invest. Ophthalmol. Vis. Sci.* **42**(7), 1429–1435.

Marmorstein, A. D., Marmorstein, L. Y., Rayborn, M., Wang, X., Hollyfield, J. G., and Petrukhin, K. (2000). Bestrophin, the product of the Best vitelliform macular dystrophy gene (VMD2), localizes to the basolateral plasma membrane of the retinal pigment epithelium. *Proc. Natl. Acad. Sci. USA* **97**(23), 12758–12763.

Muñoz-Sanjuán, I., and Brivanlou, A. H. (2002). Neural induction, the default model and embryonic stem cells. *Nature Rev. Neurosci.* **3**(4), 271–280.

Nicoletti, A., Wong, D. J., Kawase, K., Gibson, L. H., Yang-Feng, T. L., Richards, J. E., and Thompson, D. A. (1995). Molecular characterization of the human gene encoding an abundant 61 kDa protein specific to the retinal pigment epithelium. *Hum. Mol. Genet.* **4**(4), 641–649.

Olsen, A. L., Stachura, D. L., and Weiss, M. J. (2006). Designer blood: Creating hematopoietic lineages from embryonic stem cells. *Blood* **107**(4), 1265–1275.

Opas, M., and Dziak, E. (1994). bFGF-induced transdifferentiation of RPE to neuronal progenitors is regulated by the mechanical properties of the substratum. *Dev. Biol.* **161**(2), 440–454.

Peschle, C., and Condorelli, G. (2005). Stem cells for cardiomyocyte regeneration: State of the art. *Ann. NY Acad. Sci.* **1047**, 376–385.

Radtke, N. D., Aramant, R. B., Seiler, M. J., Petry, H. M., and Pidwell, D. (2004). Vision change after sheet transplant of fetal retina with retinal pigment epithelium to a patient with retinitis pigmentosa. *Arch. Ophthalmol.* **122**(8), 1159–1165.

Redmond, T. M., Yu, S., Lee, E., Bok, D., Hamasaki, D., Chen, N., Goletz, P., Ma, J. X., Crouch, R. K., and Pfeifer, K. (1998). Rpe65 is necessary for production of 11-cis-vitamin A in the retinal visual cycle. *Nature Genet.* **20**(4), 344–351.

Reh, T. A., Nagy, T., and Gretton, H. (1987). Retinal pigmented epithelial cells induced to transdifferentiate to neurons by laminin. *Nature* **330**(6143), 68–71.

Reubinoff, B. E., Itsykson, P., Turetsky, T., Pera, M. F., Reinhartz, E., Itzik, A., and Ben-Hur, T. (2001). Neural progenitors from human embryonic stem cells. *Nature Biotechnol.* **19**(12), 1134–1140.

Saari, J. C., Nawrot, M., Kennedy, B. N., Garwin, G. G., Hurley, J. B., Huang, J., Possin, D. E., and Crabb, J. W. (2001). Visual cycle impairment in cellular retinaldehyde binding protein (CRALBP) knockout mice results in delayed dark adaptation. *Neuron* **29**(3), 739–748.

Sakaguchi, D. S., Janick, L. M., and Reh, T. A. (1997). Basic fibroblast growth factor (FGF-2) induced transdifferentiation of retinal pigment epithelium: Generation of retinal neurons and glia. *Dev. Dyn.* **209**(4), 387–398.

Smukler, S. R., Runciman, S. B., Xu, S., and van der Kooy, D. (2006). Embryonic stem cells assume a primitive neural stem cell fate in the absence of extrinsic influences. *J. Cell Biol.* **172**(1), 79–90.

Steele, F. R., Chader, G. J., Johnson, L. V., and Tombran-Tink, J. (1993). Pigment epithelium-derived factor: Neurotrophic activity and identification as a member of the serine protease inhibitor gene family. *Proc. Natl. Acad. Sci. USA* **90**(4), 1526–1530.

Teramoto, K., Asahina, K., Kumashiro, Y., Kakinuma, S., Chinzei, R., Shimizu-Saito, K., Tanaka, Y., Teraoka, H., and Arii, S. (2005). Hepatocyte differentiation from embryonic stem cells and umbilical cord blood cells. *J. Hepatobiliary Pancreat. Surg.* **12**(3), 196–202.

Thomson, J. A., Itskovitz-Eldor, J., Shapiro, S. S., Waknitz, M. A., Swiergiel, J. J., Marshall, V. S., and Jones, J. M. (1998). Embryonic stem cell lines derived from human blastocysts. *Science* **282**(5391), 1145–117. Erratum in *Science* **282**(5395), 1827 (1998).

Tropepe, V., Hitoshi, S., Sirard, C., Mak, T. W., Rossant, J., and van der Kooy, D. (2001). Direct neural fate specification from embryonic stem cells: A primitive mammalian neural stem cell stage acquired through a default mechanism. *Neuron* **30**(1), 65–78.

van Meurs, J. C., ter Averst, E., Hofland, L. J., van Hagen, P. M., Mooy, C. M., Baarsma, G. S., Kuijpers, R. W., Boks, T., and Stalmans, P. (2004). Autologous peripheral retinal pigment epithelium translocation in patients with subfoveal neovascular membranes. *Br. J. Ophthalmol.* **88**(1), 110–113.

Vinores, S. A., Derevjanik, N. L., Mahlow, J., Hackett, S. F., Haller, J. A., deJuan, E., Frankfurter, A., and Campochiaro, P. A. (1995). Class III beta-tubulin in human retinal pigment epithelial cells in culture and in epiretinal membranes. *Exp. Eye Res.* **60**(4), 385–400.

Wei, H., Juhasz, O., Li, J., Tarasova, Y. S., and Boheler, K. R. (2005). Embryonic stem cells and cardiomyocyte differentiation: Phenotypic and molecular analyses. *J. Cell. Mol. Med.* **9**(4), 804–817.

Weisz, J. M., Humayun, M. S., De Juan, E., Jr., Del Cerro, M., Sunness, J. S., Dagnelie, G., Soylu, M., Rizzo, L., and Nussenblatt, R. B. (1999). Allogenic fetal retinal pigment epithelial cell transplant in a patient with geographic atrophy. *Retina* **19**(6), 540–545.

Ying, Q. L., Stavridis, M., Griffiths, D., Li, M., and Smith, A. (2003). Conversion of embryonic stem cells into neuroectodermal precursors in adherent monoculture. *Nature Biotechnol.* **21**(2), 183–186.

Zhao, S., Thornquist, S. C., and Barnstable, C. J. (1995). *In vitro* transdifferentiation of embryonic rat retinal pigment epithelium to neural retina. *Brain Res.* **677**(2), 300–310.

[12] Mesenchymal Cells

By TIZIANO BARBERI and LORENZ STUDER

Abstract

Human embryonic stem cells (hESC) provide a potentially unlimited source of specialized cell types for regenerative medicine. Nonetheless, one of the key requirements used to fulfill this potential is the ability to direct the differentiation of hESC to selective fates *in vitro*. Studies have reported the development of culture strategies to derive multipotent mesenchymal precursors from hESCs *in vitro*. This chapter reviews the techniques that allow the selective derivation of such precursors and their differentiation toward various mesenchymal cell types. It also discusses current limitations and future perspectives on the use of hESC-derived mesenchymal tissues.

Introduction

The isolation of human embryonic stem cells (Thomson *et al.*, 1998) has led to renewed focus on developing *in vitro* differentiation strategies to control embryonic stem cell (ESC) fate. Such strategies may yield specialized cell types suitable for cell therapy in degenerative diseases. Understanding the differentiation steps directing human ESCs (hESCs) toward mesenchymal fates will also provide an important tool to study developmental biology and the molecular mechanism controlling mesodermal and mesenchymal fate specification in humans. It is established that spontaneous differentiation of hESCs in immunocompromised hosts *in vivo* leads to the formation of teratomas, tumor masses that typically include a wide variety of mesenchymal tissues, including cartilage, bone, and striate muscle in addition to ectodermal and endodermal derivatives (Reubinoff *et al.*, 2000; Thomson *et al.*, 1998). These data suggest that hESCs have the intrinsic ability to generate mesenchymal tissues and lead

METHODS IN ENZYMOLOGY, VOL. 418
0076-6879/06 $35.00
DOI: 10.1016/S0076-6879(06)18012-X

to the question of how to efficiently harness this potential for the selective derivation of these tissues.

The availability of unlimited numbers of mesenchymal tissues will have important implications for tissue engineering and regenerative medicine. Cartilage or bone precursors derived from hESCs could be used, for example, in cosmetic and reconstructive surgery. Patients with head and neck tumors often have large pieces of cartilage or bone removed and there may be an insufficient supply of suitable replacement cartilage or bone structures for grafting. One of the clinically most relevant applications may be the use of skeletal muscle cells. While there have been attempts of using fetal skeletal myoblasts for transplantation, the lack of sufficient donor cells has been a major hurdle to developing this approach for therapeutic use.

Developmental Considerations

Experience from various ESCs differentiation studies has revealed that undifferentiated ESCs recapitulate key developmental events during *in vitro* differentiation. For example, differentiation of mouse ESCs toward neural fates retains the normal developmental sequence and timing of gene expression as compared to neural development *in vivo* (Barberi *et al.*, 2003). Induction of mesodermal fates from mouse ESCs also follows a highly reproducible and stereotypic temporal time frame. Hematopoietic and cardiac mesoderm appear in sequential waves during embryoid body differentiation in mouse ESCs (Kouskoff *et al.*, 2005) and the *in vitro* developmental progression closely recapitulates the sequential appearance of hematopoietic and cardiac mesoderm *in vivo* (Kinder *et al.*, 1999).

Unsegmented mesoderm is specified during gastrulation and gives rise to the notochord, paraxial, intermediate, and lateral plate mesoderm. In vertebrate embryos, the paraxial mesoderm is subdivided into somites through progressive epithelialization. This process is controlled by an intrinsic oscillating "clock" mechanism that switches the presomitic mesoderm between permissive and nonpermissive states. The oscillating mechanism of somite formation is highly correlated with periodic waves of Notch activation (Pourquie, 2003).

Interestingly, epithelialization of somites appears not to be an essential feature for the fate specification within the somite. Mouse mutants for the bHLH factor *paraxis* lack epithelialization of the somites but generate largely normal muscle and sclerotome derivatives (Burgess *et al.*, 1995). This suggests that somite formation and epithelialization may not be an essential step in the formation of somatic-derived tissues from ESCs.

During development the somite gets subdivided into a ventral half, forming the sclerotome while the residual dorsal half generates the dermomyotome. The sclerotome will give rise to various mesenchymal structures, including the vertebral column, ribs, tendons, and the meninges. The dermatomyotome yields vertebral muscles and muscles of the limbs, as well as endothelial cells, dermis, and the cartilage of the scapula blade (for review, see Christ et al., 2004). Various molecular markers can be used to identify subpopulations within somite-derived tissues (for review, see Kalcheim and Ben-Yair, 2005; Pownall et al., 2002). These include Pax1 and Pax9 in progenitors of the sclerotome, Dermo, Sim1, and En-1 expression in dermal and lateral somite lineages, Pax3, Pax7, Dash2, Eya2, and Six1 in the dermomyotome, and Myf5 and MyoD as earliest markers of committed myoblasts.

Adult Mesenchymal Stem Cells (MSCs)

Multipotent mesenchymal precursors were initially described in the guinea pig bone marrow as colony-forming unit fibroblasts in the early 1970s (Friedenstein et al., 1970). Nevertheless, their developmental potential was discovered only years later and only recently have these precursor cells been more fully characterized. Typically, MSCs are isolated from the adult bone marrow and can be differentiated in vitro and in vivo into adipocytes, chondrocytes, and osteoblasts (Pittenger et al., 1999). Such mesenchymal precursors have been also isolated and characterized from a variety of other connective tissues, including adipose tissue (Zuk et al., 2001) and dermis and other connective tissues (Young et al., 2001). MSCs are also found in different regions of the developing embryo, such as the fetal liver (Campagnoli et al., 2001) or from the umbilical cord (Erices et al., 2000). Their isolation is based mainly on the combined expression of certain surface markers such as CD73, CD105, and CD166. Despite significant efforts to define MSCs at the molecular level, little is known about the specific molecular requirements of these cells. The identification of MSCs is still based on surface marker expression and on characterizing differentiation potentiation. The ability to differentiate into bone and cartilage has made MSCs highly relevant for clinical applications in skeletal disease. However, only few small-scale clinical trials have been performed since the first isolation of MSCs in the 1970s. The first patients receiving in vitro MSCs were treated with autologous cells (Lazarus et al., 1995). Subsequent trials were extended to patients receiving MSCs from matched sibling donors (Koc et al., 2000). In both trials, MSC infusions were well tolerated without any obvious safetyconcerns. The hypothesis pursued by these early studies is improved recovery of patients with hematopoietic malignancies after myeloablative therapy by facilitating engraftment of hematopoietic stem cells.

Other applications for MSCs that have been tested in limited clinical trials are stimulating growth in children with osteogenesis imperfecta (Horwitz *et al.*, 2002); improving cardiac function after myocardial infarct (Chen *et al.*, 2004; Katritsis *et al.*, 2005; Perin *et al.*, 2003, 2004; Strauer *et al.*, 2002, 2005), presumably by reducing scar formation; use in cartilage repair (see www.osiristx.com); and treatment of graft versus host disease (Le *et al.*, 2004) due to unique immunological features exhibited by MSCs (for review, see Barry *et al.*, 2005).

One major factor impeding clinical translation is the limited *in vitro* proliferative potential. In addition to limited *in vitro* expansion, MSCs also exhibit decreasing differentiation potential over time. These limitations impact not only the clinical use of MSCs, but also their application in basic biology. While there is a great need for mesenchymal tissues in regenerative medicine, adult MSCs have never truly fulfilled this potential. Another example demonstrating the limitations of adult MSCs for cell replacement is their limited potential for skeletal muscle differentiation.

Mouse ESC-Derived Mesenchymal Tissues

One of the first strategies to derive differentiated tissues from mouse ES cells is the formation of embryoid bodies (EBs). Embryoid bodies are spherical structures composed of aggregated ESCs. Aggregation induces ESC differentiation and the formation of derivatives of the three germ layers (for review, see Keller, 1995). Differentiation methods based on EB formation were developed in the 1960s to characterize the differentiation properties of teratocarcinoma cells (Pierce *et al.*, 1960), then used to study the differentiation of clonal lines derived from teratocarcinomas (Martin and Evans, 1975), and eventually for inducing differentiation of mouse ES cells (Keller, 1995; Risau *et al.*, 1988).

While spontaneous *in vitro* differentiation of mouse ES cells toward adipocytic fates has been observed for many years (Field *et al.*, 1992), more directed differentiation approaches were developed based on mouse ES-derived EBs exposed to retinoic acid (RA) (Dani *et al.*, 1997). Later work in mouse ES cells revealed that there is an early RA-dependent phase of adipocytic differentiation and a later phase dependent on peroxisome proliferator-activated receptor γ (PPARγ) expression. Treatment of ES-derived progeny during this later phase with specific agonists of PPARγ increases the number of adipocytes greatly, an effect blocked in PPARγ $-/-$ ES cells (Rosen *et al.*, 1999). Adipocytic differentiation of mouse ES cells serves as an important model system to develop pharmacological agents affecting adipogenesis (for review, see Phillips *et al.*, 2003).

Embryoid body-based protocols were also developed for the generation of chondrocytic progeny. Most of these early protocols were EB based, such as the hanging drop methods, and demonstrated an increased formation of cartilage cells in the presence of BMP2 or BMP4 (Kramer *et al.*, 2000). Using similar approaches for the generation of mouse ES-derived chondrocytes, it was suggested that chondrocytic progeny may have significant plasticity and that these conditions would allow for the generation of a wide range of mesenchymal derivatives, including bone cells and skeletal muscle (Hegert *et al.*, 2002).

More recent work was aimed at defining some of the key factors that drive mouse ES into a wide range of specific mesenchymal fates. The approach is based on the classic RA protocol for adipocytic differentiation but RA treatment is followed by BMP or transforming growth factor (TGFβ3) exposure that bias toward bone and cartilage differentiation, respectively (Kawaguchi *et al.*, 2005). However, the mouse system is still lacking defined conditions for the isolation of pure mesenchymal precursor populations from ESCs. Such conditions should greatly facilitate future mechanistic studies on mesenchymal fate specification.

Mesenchymal Cells Derived from Human ESCs

The very first studies describing the isolation of hESC cell lines reported on the presence of mesenchymal tissues in hESC-derived teratomas *in vivo* (Reubinoff *et al.*, 2000; Thomson *et al.*, 1998). However, despite clear *in vivo* evidence of mesenchymal differentiation, surprisingly little work was performed on directing mesenchymal fates from hESCs *in vitro*. There have been reports on the derivation of specific mesenchymal cell types from hESCs. For example, hESCs could be differentiated into mineralizing bone. Differentiation was induced using embryoid body-based protocols in the presence of various bone-inducing factors such as ascorbic acid, β-glycerophosphate, and dexamethasone (Bielby *et al.*, 2004; Sottile *et al.*, 2003). These studies also showed evidence of *in vivo* survival of hESC-derived osteocytes. More recently, EB-free conditions were reported that demonstrated a further improvement in osteogenic differentiation of hESCs *in vitro* (Karp *et al.*, 2006). Adipocytic differentiation was achieved (Xiong *et al.*, 2005) using a protocol based on mouse ESC studies (Dani *et al.*, 1997). Under these conditions hESCs are differentiated in EB culture in serum-containing medium followed by replating and exposure to the PPARγ agonist rosiglitazone (Xiong *et al.*, 2005).

Interest in generating mesenchymal derivatives from hESCs stemmed in part from the need for human-derived feeders for hESC maintenance or differentiation. Such feeders could replace the need for mouse stroma and could provide a potentially autologous cell source for hESC expansion and differentiation. The feasibility of this approach was demonstrated via immortalization of fibroblast-like populations derived from hESCs (Xu et al., 2004). In this study, immortalization was performed by ectopic expression of human telomerase reverse transcriptase. Conditioned medium from these hESC-derived fibroblast-like populations was capable of maintaining growth of undifferentiated hESCs. In addition to fibroblast morphologies, these cells could also be induced to express markers of osteogenic lineage (Xu et al., 2004).

Our own work on deriving mesenchymal precursors from hESCs (Barberi et al., 2005) pursued a strategy distinct from previous work. The goal of our study was to initially isolate multipotent mesenchymal precursor cell populations rather than specific mesenchymal derivatives. We demonstrated that pure populations of mesenchymal precursors can be isolated using FACS-based isolation of $CD73^+$ progeny. This approach offers the advantage of working with defined intermediate cell populations that can be further expanded as mesenchymal precursors (hESMPCs) or differentiated into various mesenchymal derivatives, including skeletal muscle cells. A detailed discussion of our approach is provided.

A strategy similar to our study has been pursued by isolating mesenchymal precursor populations from hESCs but under conditions that do not require EB formation or the use of feeder layers (Olivier et al., 2006). These hESC-derived mesenchymal precursors express markers characteristic of adult MSCs and are capable of adipocytic and osteogenic differentiation in vitro. Similar to the study by Xu and colleagues (2004), cells isolated under these conditions can also be used for maintaining undifferentiated hESCs (Olivier et al., 2006).

Derivation, Isolation, and Differentiation of hESC-Derived Mesenchymal Precursors

Our system for obtaining pure mesenchymal precursors from hESC is based on combining two main strategies. The first strategy is the induction of mesodermal fates from hESCs using a coculture approach with a murine bone marrow-derived stromal cell line called OP9. OP9 cells were derived from the osteopetrotic op/op mouse, a spontaneous mouse mutant deficient for macrophage colony-stimulating factor (Kodama et al., 1994). OP9 cells were used previously to induce mesoderm differentiation from mouse ESCs (Nakano et al., 1994). The second step is based on FACS-mediated isolation of mesenchymal precursors after the stromal induction phase. Cells are isolated based on the expression of CD73, a marker used routinely for the isolation of adult bone marrow-derived MSCs. The resulting $CD73^+$ population of mesenchymal precursors can be proliferated extensively *in vitro* or differentiated into various mesenchymal lineages, such as fat, cartilage, bone, and skeletal myoblasts (see Fig. 1).

Induction of Mesodermal and Mesenchymal Fates

Undifferentiated hESCs can be maintained on mitotically inactivated mouse embryonic fibroblasts (Zhang *et al.*, 2001), human fibroblasts (Amit *et al.*, 2003; Richards *et al.*, 2002), or under feeder-free conditions (Xu *et al.*, 2001). Human ESCs are routinely tested at regular intervals to ensure

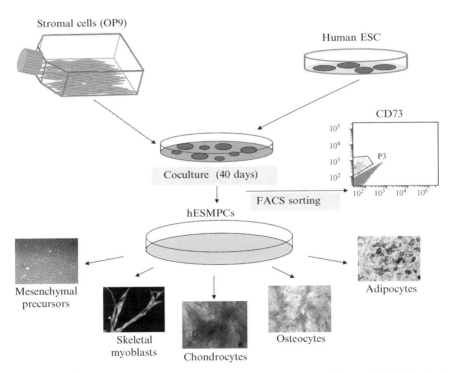

FIG. 1. Schematic illustration of the derivation and differentiation of hESC-derived mesenchymal precursors (hESMPCs). hESCs are dissociated and replated onto confluent OP9 cells. After 40 days of coculture, hESMPCs are isolated and purified using FACS for CD73. hESMPCs can be further expanded and maintained as mesenchymal precursors or selectively differentiated into specific mesenchymal derivatives, including fat, cartilage, bone, and skeletal muscle cells. (See color insert.)

normal karyotype before starting an experiment. OP9 cells are maintained in alpha-minimal essential medium (αMEM) containing 20% fetal bovine serum (FBS) and 2 mM L-glutamine. Mesenchymal differentiation is induced upon plating hESCs at low densities (10 to 25 × 10³ cells/cm²) on a monolayer of OP9 cells. During induction, hESC/OP9 cocultures are also maintained in αMEM in the presence of 20% heat-inactivated FBS (Fig. 2).

FACS-Mediated Isolation of hESMPCs

Isolation of mesenchymal precursors can be performed between days 30 and 50 of differentiation. Typically longer coculture periods yield increased proportions of mesenchymal precursors ranging from 1% to about 20% of the total cell population. Isolation of CD73$^+$ populations is performed using FACS sorting (MoFlo; Cytomation). Good results are routinely obtained using a CD73-PE-conjugated antibody (Pharmingen) after trypsinization of the mixture of differentiating hESCs and OP9 cells at day 40 of coculture. Under these conditions about 5% of all the cells are CD73$^+$ and yield polyclonal lines of hESC-derived mesenchymal precursors. hESMPCs can be further proliferated and differentiated into various mesenchymal derivatives.

Characterization of hESMPCs

Characterization involves FACS analysis for the expression of a set of surface markers commonly used to define adult bone marrow-derived MSCs, including CD105(SH2), STRO-1, VCAM (CD106), CD29 (integrin β1), CD44, ICAM-1(CD54), ALCAM(CD166), vimentin, and α smooth muscle actin. The hESMPC populations should be negative for hematopoietic

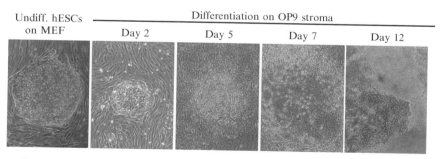

Undiff. hESCs on MEF	Differentiation on OP9 stroma			
	Day 2	Day 5	Day 7	Day 12

FIG. 2. *In vitro* progression of hESCs toward hESMPCs. Representative images are shown illustrating the early transition of hESCs toward mesenchymal fates. Note that cultures can contain contaminating extraembryonic yolk sac-like structures (see day 12) in parallel to early hESMPCs.

markers such as CD34, CD45, and CD14 and lack markers of neuroecto-dermal, epithelial, and mature muscle cells such as nestin, pancytokeratin, and desmin. The antibodies used for isolation and characterization are species specific and will not cross-react with OP9 feeders. The human identity of hESMPCs can be further confirmed using human-specific markers such as human nuclear antigen (Chemicon).

Interestingly, genome-wide expression studies using the Affymetrix platform showed a remarkable overlap on global gene expression profiles between hESMPC derived from various hESC lines and adult bone marrow-derived MSCs (Barberi *et al.*, 2005). As expected, among the genes shared between hESMPCs and primary adult MSC are many of the known MSC markers such as the *mesenchymal stem cell protein* DSC54, *neuropilin 1*, *hepatocyte growth factor*, *forkhead box D1*, and *notch homolog 2* (Barberi *et al.*, 2005). Despite these striking similarities in gene expression between hESMPCs and primary adult MSC, it will be interesting to analyze the genes that are regulated differently between the two cell types. Such differences may provide fundamental insights into the molecular mechanisms responsible for the wider proliferation and differentiation potential of hESMPCs compared with adult MSCs.

Differentiation of hESMPCs into Specialized Mesenchymal Derivatives

hESMPCs can be differentiated readily into the classic mesenchymal fates obtained from adult MSCs, including fat, cartilage, and bone.

Adipocytic differentiation of hESMPCs is induced at high cell densities (100% confluent) via exposure to 1 mM dexamethasone, 10 μg/ml insulin, and 0.5 mM isobutylxanthine (all Sigma) in αMEM medium containing 10% FBS for 2 to 4 weeks. This protocol has been widely used for adipocytic differentiation of primary adult MSCs (Pittenger *et al.*, 1999). In hESMPC cultures subjected to adipocytic differentiation, cells harboring fat granules are first observed 7 to 12 days after induction. After 3 weeks typically at least 70% of all cells contain Oil Red O+ fat granules. At that stage *PPARγ*, a marker of adipocytic differentiation, can be detected readily by gene expression analysis.

Chondrocytic differentiation also follows the strategies initially developed for the differentiation of adult MSCs (Pittenger *et al.*, 1999). The first step involves growth of hESMPCs as pellet cultures followed by exposure to 10 ng/ml TGFß-3 (R&D Systems) and 200 μM ascorbic acid (AA; Sigma) in αMEM containing 10% FBS. Using this protocol, more than 50% of the cells exhibit robust staining for Alcian blue after 4 weeks of induction. Alcian blue is a specific dye used to detect extracellular matrix proteoglycans.

Additional markers that confirm the chondrocyte identity of the hESMPC derivatives are *collagen II* and *aggrecan*. These are two other components of the extracellular matrix that are selectively produced by chondrocytes.

Osteogenic differentiation of hESMPCs is induced at low cell density (1 to 2.5×10^3 cells/ cm^2) on tissue culture-treated dishes. Osteogenic induction is carried out in αMEM containing 10% FBS supplemented with 10 mM β-glycerol phosphate (Sigma), 0.1 μM dexamethasone, and 200 μM AA. After 3 to 4 weeks of induction, cells become positive to von Kossa and Alizarin red stainings for the detection of calcium deposition in the matrix. Osteocytes derived from hESMPCs express specific markers of osteogenesis, such as *bone-specific alkaline phosphatase* and *bone sialoprotein.* Typically about 70% of all cells at day 28 of treatment are stained positive for Alizarin red.

Skeletal Muscle Differentiation

In addition to a greater proliferative potential, hESMPCs differ from adult MSCs because of their ability to efficiently generate skeletal muscle cells. Few reports have described the differentiation of MSCs into skeletal muscle cells. Typically the derivation of skeletal muscle cells from MSCs requires genetic manipulation such as overexpression of the Notch receptor (Dezawa *et al.*, 2005). Other MSC-based protocols yield cells expressing skeletal muscle markers but only after coculture with committed myoblasts such as C2C12 (Shi *et al.*, 2004).

While generation of true skeletal muscle cells from MSCs remains a controversial issue, hESMPCs should have a much greater potential for skeletal muscle generation due to their developmental state. Conditions based on 5-azacytidine (5-Aza) exposure, described previously for skeletal muscle differentiation of adult MSCs (Wakitani *et al.*, 1995), did not yield substantial numbers of skeletal myoblasts. However, long-term (3 to 4 weeks) cultures in αMEM containing 20% heat-inactivated FBS can readily induce the expression of specific skeletal muscle markers in most hESMPC progeny. Expression of skeletal muscle markers such as MyoD and fast-switch myosin is not dependent on 5-Aza exposure. More rapid myogenic differentiation can be obtained in the presence of medium conditioned for 24 h in the murine myoblastic cell line C2C12. C2C12 cells are known to spontaneously differentiate and form myotubes under serum-free conditions (Yaffe and Saxel, 1977). Interestingly, coculture of hESMPCs with C2C12 induces more mature myogenic markers and leads to the formation of hESMPC-derived myotubes. hESMPC-derived myotubes can be visualized readily by expression of human-specific markers such as human nuclear antigen.

The percentage of hESMPC-derived myotubes in C2C12 cocultures can be increased further by altering the initial plating ratio of hESMPCs to C2C12 cells. Gene expression studies using human-specific primers for muscle-related transcripts is another strategy used to confirm the skeletal muscle identity in hESMPC progeny. Such transcripts include *MyoD, myosin heavy chain IIa*, and *myogenin* (Barberi *et al.*, 2005).

Summary and Perspectives for the Future

In summary, the methods described clearly demonstrate that hESCs can be differentiated in mesenchymal precursors and a variety of specialized mesenchymal derivatives. One concern for the clinical application of hESC-derived progeny in regenerative medicine is the risk of teratoma formation due to the potential presence of residual undifferentiated ES cells among the differentiated progeny. Barberi *et al.* (2005) showed that hESMPCs do not express detectable levels of ES-specific markers such as *Nanog* (Chambers *et al.*, 2003) or *Oct-4* (Scholer *et al.*, 1991) by reverse transcription-polymerase chain reaction (RT-PCR) or immunocytochemistry. Additional work in SCID/Beige mice suggests that hESMPCs do not form teratomas *in vivo* (Barberi *et al.*, unpublished data).

However, long-term studies in large groups of animals are required to confirm the safety of these cells *in vivo*. While the current protocol has been tested extensively in two hESC lines (WA-01 and WA-09), it will be important to compare derivation and differentiation potential of hESMPCs among a larger numbers of hESC lines.

Future studies also need to revisit the requirement of OP9 for the induction of mesenchymal fates from hESCs. The contamination of human cells with xenogenic cells or cell products makes our strategy not suitable for clinical translation. However, studies have shown the presence of mesenchymal derivatives in hESC cultures under feeder-free conditions (Olivier *et al.*, 2006; Xu *et al.*, 2004). Ongoing studies will address whether our FACS sorting and differentiation strategies may be applicable to alternative hESMPC induction protocols. Another important question relates to the multipotentiality of individual hESMPCs. This issue can be addressed readily by systematically comparing the differentiation potential of clonal hESMPC derivatives isolated after various *in vitro* expansion periods. hESMPs will also provide a unique tool for basic developmental studies aimed at unraveling the molecular signals that control mesenchymal subtype specification into fat, cartilage, bone, and muscle cells. Finally, hESMPC function needs to be explored *in vivo* in various animal models of musculoskeletal disease.

References

Amit, M., Margulets, V., Segev, H., Shariki, K., Laevsky, I., Coleman, R., and Itskovitz-Eldor, J. (2003). Human feeder layers for human embryonic stem cells. *Biol. Reprod.* **68**(6), 2150–2156.

Barberi, T., Klivenyi, P., Calingasan, N. Y., Lee, H., Kawamata, H., Loonam, K., Perrier, A. L., Bruses, J., Rubio, M. E., Topf, N., Tabar, V., Harrison, N. L., Beal, M. F., Moore, M. A., and Studer, L. (2003). Neural subtype specification of fertilization and nuclear transfer embryonic stem cells and application in parkinsonian mice. *Nature Biotechnol.* **21**, 1200–1207.

Barberi, T., Willis, L., Socci, N. D., and Studer, L. (2005). Derivation of multipotent mesenchymal precursors from human embryonic stem cells. *Plos Med.* **2**, e161.

Barry, F. P., Murphy, J. M., English, K., and Mahon, B. P. (2005). Immunogenicity of adult mesenchymal stem cells: Lessons from the fetal allograft. *Stem Cells Dev.* **14**, 252–265.

Bielby, R. C., Boccaccini, A. R., Polak, J. M., and Buttery, L. D. (2004). *In vitro* differentiation and *in vivo* mineralization of osteogenic cells derived from human embryonic stem cells. *Tissue Eng.* **10**, 1518–1525.

Burgess, R., Cserjesi, P., Ligon, K. L., and Olson, E. N. (1995). Paraxis: A basic helix-loop-helix protein expressed in paraxial mesoderm and developing somites. *Dev. Biol.* **168**, 296–306.

Campagnoli, C., Roberts, I. A., Kumar, S., Bennett, P. R., Bellantuono, I., and Fisk, N. M. (2001). Identification of mesenchymal stem/progenitor cells in human first-trimester fetal blood, liver, and bone marrow. *Blood* **98**, 2396–2402.

Chambers, I., Colby, D., Robertson, M., Nichols, J., Lee, S., Tweedie, S., and Smith, A. (2003). Functional expression cloning of Nanog, a pluripotency sustaining factor in embryonic stem cells. *Cell* **113**, 643–655.

Chen, S. L., Fang, W. W., Ye, F., Liu, Y. H., Qian, J., Shan, S. J., Zhang, J. J., Chunhua, R. Z., Liao, L. M., Lin, S., and Sun, J. P. (2004). Effect on left ventricular function of intracoronary transplantation of autologous bone marrow mesenchymal stem cell in patients with acute myocardial infarction. *Am. J. Cardiol.* **94**, 92–95.

Christ, B., Huang, R., and Scaal, M. (2004). Formation and differentiation of the avian sclerotome. *Anat. Embryol. (Berl.)* **208**, 333–350.

Dani, C., Smith, A. G., Dessolin, S., Leroy, P., Staccini, L., Villageois, P., Darimont, C., and Ailhaud, G. (1997). Differentiation of embryonic stem cells into adipocytes *in vitro.* *J. Cell Sci.* **110**(Pt. 11), 1279–1285.

Dezawa, M., Ishikawa, H., Itokazu, Y., Yoshihara, T., Hoshino, M., Takeda, S., Ide, C., and Nabeshima, Y. (2005). Bone marrow stromal cells generate muscle cells and repair muscle degeneration. *Science* **309**, 314–317.

Erices, A., Conget, P., and Minguell, J. J. (2000). Mesenchymal progenitor cells in human umbilical cord blood. *Br. J. Haematol.* **109**, 235–242.

Field, S. J., Johnson, R. S., Mortensen, R. M., Papaioannou, V. E., Spiegelman, B. M., and Greenberg, M. E. (1992). Growth and differentiation of embryonic stem cells that lack an intact c-fos gene. *Proc. Natl. Acad. Sci. USA* **89**, 9306–9310.

Friedenstein, A. J., Chailakhjan, R. K., and Lalykina, K. S. (1970). The development of fibroblast colonies in monolayer cultures of guinea-pig bone marrow and spleen cells. *Cell Tissue Kinet.* **3**, 393–403.

Hegert, C., Kramer, J., Hargus, G., Muller, J., Guan, K., Wobus, A. M., Muller, P. K., and Rohwedel, J. (2002). Differentiation plasticity of chondrocytes derived from mouse embryonic stem cells. *J. Cell Sci.* **115**, 4617–4628.

Horwitz, E. M., Gordon, P. L., Koo, W. K., Marx, J. C., Neel, M. D., McNall, R. Y., Muul, L., and Hofmann, T. (2002). Isolated allogeneic bone marrow-derived mesenchymal cells engraft and stimulate growth in children with osteogenesis imperfecta: Implications for cell therapy of bone. *Proc. Natl. Acad. Sci. USA* **99,** 8932–8937.

Kalcheim, C., and Ben-Yair, R. (2005). Cell rearrangements during development of the somite and its derivatives. *Curr. Opin. Genet. Dev.* **15,** 371–380.

Karp, J. M., Ferreira, L. S., Khademhosseini, A., Kwon, A. H., Yeh, J., and Langer, R. S. (2006). Cultivation of human embryonic stem cells without the embryoid body step enhances osteogenesis *in vitro. Stem Cells* **24,** 835–843.

Katritsis, D. G., Sotiropoulou, P. A., Karvouni, E., Karabinos, I., Korovesis, S., Perez, S. A., Voridis, E. M., and Papamichail, M. (2005). Transcoronary transplantation of autologous mesenchymal stem cells and endothelial progenitors into infarcted human myocardium. *Catheter. Cardiovasc. Interv.* **65,** 321–329.

Kawaguchi, J., Mee, P. J., and Smith, A. G. (2005). Osteogenic and chondrogenic differentiation of embryonic stem cells in response to specific growth factors. *Bone* **36,** 758–769.

Keller, G. M. (1995). *In-vitro* differentiation of embryonic stem-cells. *Curr. Opin. Cell Biol.* **7,** 862–869.

Kinder, S. J., Tsang, T. E., Quinlan, G. A., Hadjantonakis, A. K., Nagy, A., and Tam, P. P. (1999). The orderly allocation of mesodermal cells to the extraembryonic structures and the anteroposterior axis during gastrulation of the mouse embryo. *Development* **126,** 4691–4701.

Koc, O. N., Gerson, S. L., Cooper, B. W., Dyhouse, S. M., Haynesworth, S. E., Caplan, A. I., and Lazarus, H. M. (2000). Rapid hematopoietic recovery after coinfusion of autologous-blood stem cells and culture-expanded marrow mesenchymal stem cells in advanced breast cancer patients receiving high-dose chemotherapy. *J. Clin. Oncol.* **18,** 307–316.

Kodama, H., Nose, M., Niida, S., and Nishikawa, S. (1994). Involvement of the c-kit receptor in the adhesion of hematopoietic stem cells to stromal cells. *Exp. Hematol.* **22,** 979–984.

Kouskoff, V., Lacaud, G., Schwantz, S., Fehling, H. J., and Keller, G. (2005). Sequential development of hematopoietic and cardiac mesoderm during embryonic stem cell differentiation. *Proc. Natl. Acad. Sci. USA* **102,** 13170–13175.

Kramer, J., Hegert, C., Guan, K., Wobus, A. M., Muller, P. K., and Rohwedel, J. (2000). Embryonic stem cell-derived chondrogenic differentiation *in vitro:* Activation by BMP-2 and BMP-4. *Mech. Dev.* **92,** 193–205.

Lazarus, H. M., Haynesworth, S. E., Gerson, S. L., Rosenthal, N. S., and Caplan, A. I. (1995). *Ex vivo* expansion and subsequent infusion of human bone marrow-derived stromal progenitor cells (mesenchymal progenitor cells): Implications for therapeutic use. *Bone Marrow Transplant.* **16,** 557–564.

Le, B. K., Rasmusson, I., Sundberg, B., Gotherstrom, C., Hassan, M., Uzunel, M., and Ringden, O. (2004). Treatment of severe acute graft-versus-host disease with third party haploidentical mesenchymal stem cells. *Lancet* **363,** 1439–1441.

Martin, G. R., and Evans, M. J. (1975). Differentiation of clonal lines of teratocarcinoma cells: Formation of embryoid bodies *in vitro. Proc. Natl. Acad. Sci. USA* **72,** 1441–1445.

Nakano, T., Kodama, H., and Honjo, T. (1994). Generation of lymphohematopoietic cells from embryonic stem cells in culture. *Science* **265,** 1098–1101.

Olivier, E. N., Rybicki, A. C., and Bouhassira, E. E. (2006). Differentiation of human embryonic stem cells into bipotent mesenchymal stem cells. *Stem Cells* **24**(8), 1914–1922.

Perin, E. C., Dohmann, H. F., Borojevic, R., Silva, S. A., Sousa, A. L., Mesquita, C. T., Rossi, M. I., Carvalho, A. C., Dutra, H. S., Dohmann, H. J., Silva, G. V., Belem, L., Vivacqua, R., Rangel, F. O., Esporcatte, R., Geng, Y. J., Vaughn, W. K., Assad, J. A., Mesquita, E. T., and Willerson, J. T. (2003). Transendocardial, autologous bone marrow cell transplantation for severe, chronic ischemic heart failure. *Circulation* **107,** 2294–2302.

Perin, E. C., Dohmann, H. F., Borojevic, R., Silva, S. A., Sousa, A. L., Silva, G. V., Mesquita, C. T., Belem, L., Vaughn, W. K., Rangel, F. O., Assad, J. A., Carvalho, A. C., Branco, R. V., Rossi, M. I., Dohmann, H. J., and Willerson, J. T. (2004). Improved exercise capacity and ischemia 6 and 12 months after transendocardial injection of autologous bone marrow mononuclear cells for ischemic cardiomyopathy. *Circulation* **110**, II213–II218.

Phillips, B. W., Vernochet, C., and Dani, C. (2003). Differentiation of embryonic stem cells for pharmacological studies on adipose cells. *Pharmacol. Res.* **47**, 263–268.

Pierce, G. B., Dixon, F. J., and Verney, E. L. (1960). Teratocarcinogenic and tissue-forming potentials of the cell types comprising neoplastic embryoid bodies. *Lab. Invest.* **9**, 583–602.

Pittenger, M. F., Mackay, A. M., Beck, S. C., Jaiswal, R. K., Douglas, R., Mosca, J. D., Moorman, M. A., Simonetti, D. W., Craig, S., and Marshak, D. R. (1999). Multilineage potential of adult human mesenchymal stem cells. *Science* **284**, 143–147.

Pourquie, O. (2003). The segmentation clock: Converting embryonic time into spatial pattern. *Science* **301**, 328–330.

Pownall, M. E., Gustafsson, M. K., and Emerson, C. P., Jr. (2002). Myogenic regulatory factors and the specification of muscle progenitors in vertebrate embryos. *Annu. Rev. Cell Dev. Biol.* **18**, 747–783.

Reubinoff, B. E., Pera, M. F., Fong, C. Y., Trounson, A., and Bongso, A. (2000). Embryonic stem cell lines from human blastocysts: Somatic differentiation *in vitro*. *Nature Biotechnol.* **18**, 399–404.

Richards, M., Fong, C. Y., Chan, W. K., Wong, P. C., and Bongso, A. (2002). Human feeders support prolonged undifferentiated growth of human inner cell masses and embryonic stem cells. *Nature Biotechnol.* **20**, 933–936.

Risau, W., Sariola, H., Zerwes, H. G., Sasse, J., Ekblom, P., Kemler, R., and Doetschman, T. (1988). Vasculogenesis and angiogenesis in embryonic-stem-cell-derived embryoid bodies. *Development* **102**, 471–478.

Rosen, E. D., Sarraf, P., Troy, A. E., Bradwin, G., Moore, K., Milstone, D. S., Spiegelman, B. M., and Mortensen, R. M. (1999). PPAR gamma is required for the differentiation of adipose tissue *in vivo* and *in vitro*. *Mol. Cell* **4**, 611–617.

Scholer, H. R., Ciesiolka, T., and Gruss, P. (1991). A nexus between Oct-4 and E1A: Implications for gene regulation in embryonic stem cells. *Cell* **66**, 291–304.

Shi, D., Reinecke, H., Murry, C. E., and Torok-Storb, B. (2004). Myogenic fusion of human bone marrow stromal cells, but not hematopoietic cells. *Blood* **104**, 290–294.

Sottile, V., Thomson, A., and McWhir, J. (2003). *In vitro* osteogenic differentiation of human ES cells. *Cloning Stem Cells* **5**, 149–155.

Strauer, B. E., Brehm, M., Zeus, T., Bartsch, T., Schannwell, C., Antke, C., Sorg, R. V., Kogler, G., Wernet, P., Muller, H. W., and Kostering, M. (2005). Regeneration of human infarcted heart muscle by intracoronary autologous bone marrow cell transplantation in chronic coronary artery disease: The IACT Study. *J. Am. Coll. Cardiol.* **46**, 1651–1658.

Strauer, B. E., Brehm, M., Zeus, T., Kostering, M., Hernandez, A., Sorg, R. V., Kogler, G., and Wernet, P. (2002). Repair of infarcted myocardium by autologous intracoronary mononuclear bone marrow cell transplantation in humans. *Circulation* **106**, 1913–1918.

Thomson, J. A., Itskovitz-Eldor, J., Shapiro, S. S., Waknitz, M. A., Swiergiel, J. J., Marshall, V. S., and Jones, J. M. (1998). Embryonic stem cell lines derived from human blastocysts. *Science* **282**, 1145–1147.

Wakitani, S., Saito, T., and Caplan, A. I. (1995). Myogenic cells derived from rat bone marrow mesenchymal stem cells exposed to 5-azacytidine. *Muscle Nerve* **18**, 1417–1426.

Xiong, C., Xie, C. Q., Zhang, L., Zhang, J., Xu, K., Fu, M., Thompson, W. E., Yang, L. J., and Chen, Y. E. (2005). Derivation of adipocytes from human embryonic stem cells. *Stem Cells Dev.* **14**, 671–675.

Xu, C., Jiang, J., Sottile, V., McWhir, J., Lebkowski, J., and Carpenter, M. K. (2004). Immortalized fibroblast-like cells derived from human embryonic stem cells support undifferentiated cell growth. *Stem Cells* **22**, 972–980.

Xu, C. H., Inokuma, M. S., Denham, J., Golds, K., Kundu, P., Gold, J. D., and Carpenter, M. K. (2001). Feeder-free growth of undifferentiated human embryonic stem cells. *Nature Biotechnol.* **19**, 971–974.

Yaffe, D., and Saxel, O. (1977). A myogenic cell line with altered serum requirements for differentiation. *Differentiation* **7**, 159–166.

Young, H. E., Steele, T. A., Bray, R. A., Hudson, J., Floyd, J. A., Hawkins, K., Thomas, K., Austin, T., Edwards, C., Cuzzourt, J., Duenzl, M., Lucas, P. A., and Black, A. C., Jr. (2001). Human reserve pluripotent mesenchymal stem cells are present in the connective tissues of skeletal muscle and dermis derived from fetal, adult, and geriatric donors. *Anat. Rec.* **264**, 51–62.

Zhang, S. C., Wernig, M., Duncan, I. D., Brustle, O., and Thomson, J. A. (2001). *In vitro* differentiation of transplantable neural precursors from human embryonic stem cells. *Nature Biotechnol.* **19**, 1129–1133.

Zuk, P. A., Zhu, M., Mizuno, H., Huang, J., Futrell, J. W., Katz, A. J., Benhaim, P., Lorenz, H. P., and Hedrick, M. H. (2001). Multilineage cells from human adipose tissue: Implications for cell-based therapies. *Tissue Eng.* **7**, 211–228.

[13] Hematopoietic Cells

By MALCOLM A. S. MOORE, JAE-HUNG SHIEH, and GABSANG LEE

Abstract

Murine embryonic stem cells (mESC) readily form embryoid bodies (EBs) that exhibit hematopoietic differentiation. Methods based on EB formation or ESC coculture with murine bone marrow stromal cell lines have revealed pathways of both primitive and definitive hematopoietic differentiation progressing from primitive mesoderm via hemangioblasts to endothelium and hematopoietic stem and progenitor cells. The addition of specific hematopoietic growth factors and morphogens to these cultures enhances the generation of neutrophils, macrophages, megakaryocyte/platelets, and hemoglobinized mature red cells. In addition, selective culture systems have been developed to support differentiation into mature T lymphocytes, natural killer cells, B cells, and dendritic cells. In most cases, culture systems have been developed that support equivalent differentiation of various human ESC (hESC). The major obstacle to translation of ESC hematopoietic cultures to clinical relevance has been the general inability to produce hematopoietic stem cells (HSC) that can engraft adult, irradiated recipients. In this context, the pattern of ES hematopoietic development mirrors the yolk sac phase of hematopoiesis that precedes the appearance of engraftable HSC in the aorta–gonad–mesonephros

METHODS IN ENZYMOLOGY, VOL. 418   0076-6879/06 $35.00
Copyright 2006, Elsevier Inc. All rights reserved.   DOI: 10.1016/S0076-6879(06)18013-1

region. Genetic manipulation of mESC hematopoietic progeny by upregulation of HOXB4 or STAT5 has led to greatly enhanced long- or short-term multilineage hematopoietic engraftment, suggesting that genetic or epigenetic manipulation of these pathways may lead to functional HSC generation from hESC.

Introduction

Murine Embryonic Stem Cell (mESC) Differentiation

Hematopoietic differentiation of ESC lines has been subject to a number of reviews covering derivations from murine, primate, and human ESC (Bhatia, 2005; Choi et al., 2005; Daley, 2003; Keller, 2005; Lengerke and Daley, 2005; Lensch and Daley, 2006; Lerou and Daley, 2005; Martin and Kaufman, 2005; Olsen et al., 2006; Priddle et al., 2006; Tian and Kaufman, 2005).

Early ESC Development

When murine ESCs are cultured in hanging drop systems or directly in semisolid media (methyl cellulose), they proliferate and differentiate to generate colonies known as embryoid bodies (EBs) (Wiles and Keller, 1991). These EBs consist of differentiated cells from a number of lineages, including those of the hematopoietic system. When EBs at 3 to 3.5 days of development were replated in methyl cellulose with vascular endothelial growth factor (VEGF) and c-kit ligand/stem cell factor (SCF), colonies of blast morphology (BL-CFC) developed. Upon replating of day 6 blast colonies, colonies of primitive erythroid cells, as well as colonies of definitive erythroid (BFU-E), multilineage (CFU-Mix), and myeloid (CFU-GM), developed (Kennedy et al., 1997). BL-CFC growth was augmented greatly by the addition of conditioned medium from an endothelial cell line derived from EBs (Choi et al., 1998). ES-derived transitional colonies express brachyury, Flk1, SCL/Tal-1, Gata-1, βH1, and β-major, reflecting the combination of mesodermal, hematopoietic, and endothelial populations (also cardiomyocyte) (Robertson et al., 2000). Replating studies demonstrated that transitional colonies contain low numbers of primitive erythroid precursors, as well as a subset of precursors associated with early-stage definitive hematopoiesis. BL-CFC (brachyury negative) contain higher numbers and a broader spectrum of definitive precursors than found in transitional colonies, and SCL$-/-$ ES form transitional but not blast colonies (Robertson et al., 2000). ES cultured on type IV collagen-coated dishes formed Flk1+ mesoderm. The hematopoietic developmental sequence is from proximal lateral mesoderm (E-cadherin-Flk1 +VE-cadherin-) to progenitors with hemoangiogenic potential (Flk1+VE-cadherin+, CD45-) and then to hemopoietic progenitor (CD45+, c-kit+)

and mature blood cells (c-kit-, CD45+ or Ter119+) (Nishikawa *et al.*, 1998). Flk1, SCL, and basic fibroblast growth factor (bFGF)-mediated signaling is critical for hemangioblast development, with activin A synergizing to increase BL-CFC (Faloon *et al.*, 2000). In mESC cultured in serum-free, chemically defined medium supplemented with BMP-2 or BMP-4, a process resembling primitive streak formation occurred, at least at the molecular level, with the formation of mesoderm and subsequently endothelial and hematopoietic cells (Wiles and Johansson, 1997). VEGF is necessary for subsequent expansion and differentiation of hematopoietic precursors with the Smad1 and Smad5 and MAP kinase pathways activated by BMP-4 and VEGF, respectively (Park *et al.*, 2004). VEGF-mediated expansion of hematopoietic and endothelial cell progenitors was inhibited by TGFβ1, but was augmented by activin A. Smad5 ($-/-$) EBs contained an elevated number of BL-CFCs and an increased frequency of high proliferative potential primitive hematopoietic progenitors (HPP-CFCs) (Liu *et al.*, 2003). These HPP-CFCs displayed enhanced self-renewal capacity and decreased sensitivity to TGFβ1 inhibition, suggesting a critical role of Smad5 in TGFβ1-mediated negative regulation of embryonic HPP-CFCs.

Runx1-deficient EBs form 10- to 15-fold fewer BL-CFCs and have a complete block in definitive hematopoiesis. Runx$-/-$ EBs and embryos generated normal numbers of primitive erythroid precursors, with the latter developing from a subset of BL-CFC that can develop in a Runx1-independent fashion (Lacud *et al.*, 2004). Runx1 heterozygosity leads to an acceleration of mesodermal commitment and specification to the BL-CFCs and to the hematopoietic lineages in EBs (Lacud *et al.*, 2004). In contrast to normal ES cells, GATA-1 null ES cells fail to generate primitive erythroid (EryP) precursors. Definitive erythroid (EryD) precursors, however, are normal in number but undergo developmental arrest and death at the proerythroblast stage (Weiss *et al.*, 1994). Flk-1$-/-$ ESCs are capable of blast colony formation (Ema *et al.*, 2003); however, in Flk1$-/-$ embryonic stem cell chimeras, Shalaby *et al.* (1997) showed that Flk1 is required cell autonomously for endothelial development. Flk1 is involved in the movement of cells from the posterior primitive streak to the yolk sac and, possibly, to the intraembryonic sites of early hematopoiesis. Flk1$-/-$ EBs showed myeloid–erythroid differentiation, but ESC in OP9 stromal coculture failed to generate hematopoietic clusters even with cytokine (Hidaka *et al.*, 1999). Thus, the requirement for Flk-1 in early hematopoietic development can be abrogated by alterations in the microenvironment. This finding is consistent with a role for Flk-1 in regulating the migration of early mesodermally derived precursors into a microenvironment that is permissive for hematopoiesis (Hidaka *et al.*, 1999). An ES-derived Tie-2+, Flk1+ cell fraction is enriched for hematopoietic and endothelial progenitors, but Tie2−/− ES cells had no defect in hematopoiesis (Hamaguchi *et al.*, 2006). Shp-2, a member of a small family of cytoplasmic Src homology 2 (SH2) domain-containing protein tyrosine phosphatases, is integrally necessary for bFGF-mediated hemangioblast production (Zou *et al.*, 2006). Hemangioblast formation and primitive and definitive hematopoietic progenitor formation were decreased significantly following transfection with Shp-2 siRNA.

Embryoid Body Differentiation Systems

Embryoid bodies can be generated by the hanging drop technique, by suspension culture, or by methyl cellulose culture (Dang *et al.*, 2002). For large-scale production of EBs in a controlled environment, an agarose encapsulation, stirred bioreactor system has been developed (Dang *et al.*, 2004). ES cells differentiated on porous three-dimensional scaffold structures developed EBs similar to those in traditional two-dimensional (2D) cultures; however, unlike 2D differentiation, these EBs integrated with the scaffold and appeared embedded in a network of extracellular matrix, exhibiting enhanced progenitor (CFC) and myeloid differentiation (Liu and Roy, 2005). Hematopoietic differentiation can proceed autonomously in developing EBs with the development of erythroid precursors by day 4 and by days 6 to 10, 40 to 85% of EBs are hematopoietic, containing visible erythropoietic cells (i.e., red with hemoglobin). BMP-4 addition in the first 4 days and VEGF for a further 3 days enhanced hematopoietic development (Nakayama *et al.*, 2000). Hematopoiesis was also increased by the addition of interleukin (IL)-11 and SCF, and the kinetics of precursor development was similar to that of the yolk sac (Keller *et al.*, 1993).

βH1 globin mRNA is detectable in EBs within 5 days of differentiation, whereas β(maj)-globin RNA appears by day 6 (Wiles and Keller, 1991). Addition of erythroid stimulating factors (Epo, SCF, IGF-1, transferrin) promoted enhanced and prolonged erythropoiesis (Carotta *et al.*, 2004). BMP-4 and VEGF synergized in generation of CD45+ myelomonocytic and Ter119+ erythroid cells. The development of macrophages is enhanced significantly by the addition of IL-3 alone or in combination with IL-1 and M-CSF or GM-CSF. When well-differentiated EBs are allowed to attach onto tissue culture plates and grown in the presence of IL-3, a long-term output of cells of the mast cell lineage is observed (Wiles and Keller, 1991).

mESC Stroma Coculture Systems

mESC in coculture with mouse stromal cells (OP9) give rise to erythroid progenitors (EryP and EryD) sequentially, with a time course similar to that seen in murine ontogeny (Nakano *et al.*, 1996, 1997). Analysis of the role of different growth factor requirements and limiting dilution analysis of precursor frequencies indicated that most EryP and EryD probably developed from different precursors by way of distinct differentiation pathways. In OP9 cocultures, a CD41+(dim) population was the immediate precursor of TER119+ EryP cells (Otani *et al.*, 2005). Coculture of mESC with the murine MS-5 stromal line, together with hematopoietic growth factors (KL, IL-3, IL-6, IL-11, G-CSF, Epo), enhanced hematopoietic differentiation and addition of Tpo induced differentiation to megakaryocytes (Berthier *et al.*, 1997). In mESC-OP9 cocultures with Tpo, small megakaryocytes were generated that rapidly produced proplatelets by day 8 and large hyperploid megakaryocytes were developed after day 12, suggesting the existence of both primitive and definitive megakaryopoiesis (Fujimoto *et al.*, 2003). From 10^4 ES cells up to 10^8 platelets could be produced. Lieber *et al.* (2004) used a three-step culture system to generate 10^7 neutrophils from 8×10^4 ES cells. In this system, day 8 EBs were cocultured on OP9 stroma with IL-6, bFGF, oncostatin M, SCF, IL-11, and LIF for 3 days and then transferred to a medium supplemented with G-CSF, GM-CSF, and IL-6 for 4 to 20 days.

Lymphoid Differentiation of ESC

B-Cell Development

mESC cocultured with OP9 stroma generated erythroid, myeloid, and B cells (Nakano *et al.*, 1994, 1996, 1997). Isolated CD34+ cells from differentiating mESC were cultured on OP9 with IL-2 and IL-7 and developed into B220+, CD34-ve B lymphocytes and CD19+ pre-B cells (Nakayama *et al.*, 2000). OP9 stroma produces both SCF and IL-7, and the addition of Flt3L led to a 10-fold increase in B-cell production [CD19+, CD45R+(B220), AA4.1+, CD24+, IgM+] with reduced erythroid and myeloid differentiation. By 4 weeks of coculture, >90% of cells were surface IgM+, IgD+ B cells that could secrete immunoglobulin upon mitogen (LPS) stimulation (Cho *et al.*, 1999).

T-Cell Development

T cells can be generated from Flk-1+, CD45-, mESC-derived cells from 5- to 6-day OP9 cultures in a fetal thymic organ culture system or in a reaggregated thymic culture (de Pooter *et al.*, 2003). Schmitt *et al.* (2004) reported a normal program of T-cell differentiation in cocultures of mESC on OP9 stroma expressing the Notch ligand Delta (DLL1). The T cells displayed a diverse antigen receptor repertoire, and CD8+ T cells proliferated and produced interferon-γ in response to T-cell receptor stimulation. ESC-derived T-cell progenitors effectively reconstituted the T-cell compartment of Rag2−/− immunodeficient mice, enabling an effective response to a viral infection.

Natural Killer (NK) Cell Development

Culture of mESC-derived CD34+ cells on OP9 with IL-2 and IL-7 produced cytotoxic lymphocytes with NK markers (NKR-P1, perforin, granzyme) (de Pooter *et al.*, 2005). Lian *et al.* (2002) generated NK cells from CD34+ isolated from mEBs by coculturing on OP9 stroma with IL-6, IL-7, SCF, and Flt3L for 5 days and then plating on fresh OP9 stroma with IL-2, IL-15, IL-18, and IL-12 for 7 days with a final expansion with cytokines without stroma. The ES-derived NK (ES-NK) cells expressed NK cell-associated proteins and were capable of killing certain tumor cell lines, as well as MHC class I-deficient lymphoblasts. They also express CD94/NKG2 heterodimers, but not Ly49 molecules.

Dendritic Cell (DC) Development

mESC cocultured with OP9 and GM-CSF generated immature DCs that share many characteristics of macrophages, but upon maturation acquire the allo-stimulatory capacity and surface phenotype of classical DCs, including expression of CD11c, major histocompatibility complex (MHC) class II and costimulatory molecules CD80, and CD86 (Fairchild *et al.*, 2000; Senju *et al.*, 2003). Upon stimulation with IL-4 plus tumor necrosis factor (TNF)-α, combined with anti-CD40 monoclonal antibody or lipopolysaccharide, ES-DCs became mature DCs, characterized by a typical morphology and higher capacity to stimulate MLR and to process and present protein antigen to T cells. Immunization with ESC-derived DCs expressing a model antigen (OVA) provided protection from OVA-expressing tumor cells more potently than immunization with OVA alone (Fukuma *et al.*, 2005; Matsuyoshi *et al.*, 2004). ESC-derived DCs may also offer prospects for reprogramming the immune system to tolerate grafted tissues. ES-derived genetically modified DC-presenting myelin oligodendrocyte glycoprotein (MOG) peptide in the context of MHC class II molecules and simultaneously expressing TRAIL or programmed death-1 ligand significantly reduced the severity of MOG-induced experimental autoimmune encephalomyelitis following pretreatment of mice (Hirata *et al.*, 2005).

In Vitro *Engraftment of mESC-Derived Hematopoietic Cells*

There are reports that mESC-derived hematopoietic cells can produce long-term lymphomyeloid reconstitution of irradiated adult mice (Burt *et al.*, 2004; Miyagi *et al.*, 2002). Others have reported that the hematopoietic potential of ES cells *in vivo* is limited to low levels of repopulation and is restricted to the lymphoid lineage (Muller and Dzierzak, 1993).

There is a limited temporal window for the derivation of multilineage repopulating hematopoietic progenitors during embryonal stem cell differentiation *in vitro*. Day 4 murine EBs can generate primitive hematopoietic progenitors, but upon transplant into irradiated mice, very low levels of CD45+ lymphoid and myeloid cells were detected by 12 weeks (Hole *et al.*, 1996). Upon transfer into lymphoid-deficient mice, mESC-derived CD45 +, B220+ (CD45R)+, AA4.1+ cells generated a single transient wave of IgM+, IgD+ B cells but failed to generate T cells (Potocnik *et al.*, 1997). In contrast, transfer of the B220-, AA4.1+ fraction achieved long-term repopulation of both T and B lymphoid compartments and restored humoral and cell-mediated immune reactions in the recipients.

Primate ESC-Derived Hematopoiesis

Primate ES cell lines have been shown to differentiate into multiple hematopoietic lineages. Rhesus ES cells cocultured on S17 stroma with hematopoietic growth factors and BMP-4 generated CFC and CD34+ cells that formed cobblestone areas on secondary replating (Honig *et al.*, 2004; Li *et al.*, 2001). CD34+ and CD34+ CD38– cells derived from Rhesus ES cells expressed embryonic ε and ζ, as well as α, β, and γ globin genes, whereas no expression of embryonic globins could be detected in the cell preparations from bone marrow (BM)-derived CD34+ cells (Lu *et al.*, 2002). Enhanced hemangioblast development and hematopoietic and CD34+ differentiation were reported in a rhesus EB culture system when Flt3L and SCF were supplemented with VEGF and Tpo (Wang *et al.*, 2005e). In addition, analysis of gene expression during hemangioblast development demonstrated that Tpo is capable of increasing the mRNA expression of the VEGF receptor and its own receptor (c-mpl). OP-9 stromal coculture with cytokines has been used to support hematopoietic differentiation of cynomolgus ES lines (Sasaki *et al.*, 2005; Umeda *et al.*, 2004). Primitive generation erythropoiesis was detected on day 8 of coculture without exogenous Epo, whereas definitive erythropoiesis appeared on day 16 and had an indispensable requirement for exogenous Epo (Umeda *et al.*, 2004). VEGF increased, in a dose-dependent manner, not only the number of floating hematopoietic cells, but also the number of adherent hematopoietic cell clusters containing CD34-positive immature progenitors. In colony assays, exogenous VEGF also had a dose-dependent stimulatory effect on the generation of primitive erythroid colonies. Hematopoietic cells generated in this manner have been injected intrahepatically after the first trimester in fetal sheep, and microchimerism was observed up to 17 months posttransplantation with cynomolgus cells detected in bone marrow (1 to 2%) and circulation (<0.1%) (Sasaki *et al.*, 2005).

Hematopoietic Differentiation of Human ESC Lines

Kaufman *et al.* (2001) first reported the generation of hematopoietic cells and erythroid (BFU-E) and myeloid (CFU-GM) progenitors following a 2- to 3-week coculture of hESC (H1 and H9) on a murine marrow stromal line (S17) or on a yolk sac endothelial line. CFCs were enriched in the CD34+ cell fraction. A greater output of CD34+ cells was seen with ESC coculture on an hTERT-immortalized human fetal liver stromal line compared to S17 (Qiu *et al.*, 2005). Human marrow stromal cells, plus a low-dose cocktail of hematopoietic cytokines, also efficiently supported the generation of KDR-positive hemangioblasts, CD34+ hematopoietic precursors, and CD45+ mature hematopoietic cells from EBs (Wang *et al.*, 2005a). The murine OP9 stromal line is superior to either MS-5 or S17 in supporting hematopoietic differentiation of hESC, with 10^7 CD34+ cells (>95% purity and 1:66 cells forming CFC) generated from a similar number of initially plated hES cells after 8 to 9 days of coculture (Vodyanik *et al.*, 2005). These CD34+ cells displayed the phenotype of primitive hematopoietic progenitors as defined by coexpression of CD90, CD117, CD164, and aldehyde dehydrogenase along with a lack of CD38 expression and possessing a verapamil-sensitive ability to efflux rhodamine 123. The OP9 coculture system was used to generate CD45+, CD33+, myeloperoxidase (MPO)+ myeloid precursors from an Oct4-EGFP knock-in human ES cell line, demonstrating that Oct4-EGFP expression was extinguished in these precursors (Yu *et al.*, 2006).

An alternative technique for the generation of hematopoietic elements involves the formation of EBs from H1 or H9 ESC lines in the presence of hematopoietic growth factors (Flt3 L, SCF, IL-3, IL-6, and G-CSF) (Chadwick *et al.*, 2003). Hematopoietic commitment was defined by the appearance of CD45+ cells in day 10 EBs with an increase to day 15. Up to 90% of cells from day 22 hEBs were hematopoietic as defined by CD45 expression and CFC potential (Wang *et al.*, 2005b). EB hematopoietic development involves a temporal sequence with few or no clonogenic progenitors earlier than day 14 and none later than day 28. hEB differentiation, like that of the mouse, begins with the emergence of semiadherent mesodermal-hematoendothelial (MHE) colonies that can generate endothelium and form organized, yolk sac-like structures that secondarily generate multipotent primitive hematopoietic stem and progenitor cells, erythroblasts, and CD13+, CD45+ macrophages (Zambidis *et al.*, 2005). A first wave of hematopoiesis follows MHE colony emergence and is predominated by primitive erythropoiesis characterized by a brilliant red hemoglobinization, CD71/CD325a (glycophorin A) expression, and exclusively embryonic/fetal hemoglobin expression. A second wave of definitive-type BFU-E, CFU-E, GM-CFC, and multilineage CFCs follows. These stages of hematopoiesis proceed spontaneously from hEB-derived cells without a requirement for supplemental growth factors. Initiation of hematopoiesis correlated with increased levels of SCL/TAL1, GATA1, GATA2, CD34, CD31, and the homeobox gene-regulating factor CDX4 (Zambidis et al., 2005). Addition of BMP-4 to the cytokine cocktail enhanced the generation of hematopoietic progenitor colonies that could undergo secondary passage. Treatment of EBs with $VEGF_{165}$ in addition to hematopoietic growth factors and BMP-4 increased the number of cells coexpressing CD34 and KDR, as well as cells expressing erythroid markers (Cerdan et al., 2004). Under serum-free conditions with SCF, Flt3 L, Tpo, and the obligate presence of BMP-4 and VEGF, EBs generated CD45+, CD34+ hematopoietic stem/progenitor cells (Tian and Kaufman, 2005; Tian et al., 2004). In a further improvement in efficiency, hESCs in serum-free medium with cytokines can be aggregated by centrifugation to foster the formation of EBs of uniform size (spin EBs) with 90% forming blood cells and CFC (Ng et al., 2005). Clonal isolation of a PECAM-1 (CD31+) population coexpressing Flk1 (KDR) and VE-cadherin and lacking CD45 (CD45-ve PFV) from day 10 EBs has defined a human bipotent precursor (hemangioblast?) with endothelial and hematopoietic capacity (Menendez et al., 2004; Wang et al., 2004, 2005a,b). These cells express the GALVR-1 receptor, permitting their efficient transduction with GALV-pseudotyped retroviral vectors (Menendez et al., 2004).

Immune Cell Generation

Zhan et al. (2004) reported a long-term culture system in which hES-derived EBs were cultured for up to 6 weeks with a cytokine cocktail that induced hematopoietic expansion (SCF, Flt3L, Tpo, IL-3) and dendritic cell differentiation (GM-CSF, IL-4). Of the leucocytes generated (2.3/106/week from an input of 40 EBs), ~25% acquired MHC class II and co-stimulatory molecule (CD80 or CD86) expression. Cells expressing CD40 (a marker for antigen-presenting cells), CD83 (a dendritic cell marker), or CD14 (a macrophage and monocyte marker) were detected confirming the findings on Wright-Giemsa staining that dendritic cells and macrophages were present. Isolated hES-derived CD34+ cells cocultured on MS-5 stroma in the presence of SCF, Flt3-L, IL-7, and IL-3 differentiated into lymphoid (B and NK cells) as well as myeloid (macrophages and granulocytes) lineages (Vodyanik et al., 2005). Woll et al. (2005) used a two-step culture method to demonstrate efficient generation of functional NK cells from hESCs. The CD56+ CD45+ hESC-derived lymphocytes express inhibitory and activating receptors typical of mature NK cells, including killer cell Ig-like receptors, natural cytotoxicity receptors, and CD16. NK cells acquire the ability to lyse human tumor cells by both direct cell-mediated cytotoxicity and antibody-dependent cellular cytotoxicity.

In Vivo *Engraftment*

CD34+ and CD34+ CD38− cells derived from hESC cocultured on OP9 have been transplanted intraperitoneally into first-trimester fetal sheep, and low levels of human hematopoietic engraftment were detected by FACS and colony assay (Narayan *et al.*, 2006). The human cells also showed secondary passaging potential. The hemogenic precursor population from H1 or H9 day 10 hEBs, purified according to their expression of PECAM-1, coexpression of Flk1+ and VE-cadherin, and lack of CD45, was transplanted by tail-vein injection into sublethally irradiated NOD/SCID mice (Wang *et al.*, 2005c). In contrast to 100% survival of recipient mice receiving a similar dose of cultured primitive somatic hematopoietic cells (cord blood CD34+), <40% of the mice transplanted with hESC-derived hematopoietic cells survived 8 weeks. Postinjection (24 h), numerous emboli were found lodged in small pulmonary capillaries and mice showed minimal engraftment. Up to 80% of ES-derived hematopoietic precursors underwent rapid aggregation when exposed to mouse serum for 2 h *in vitro*. In contrast, adult mouse serum did not cause aggregation of somatic hematopoietic cells. To bypass the circulatory system and complications associated with systemic delivery, Wang *et al.* (2005c) transplanted hESC-derived hematopoietic cells by means of intrabone marrow transplantation directly to the femur (i.f.). In contrast to intravenously transplanted mice, >90% of the i.f.-transplanted mice survived >8 weeks and most demonstrated human reconstitution, indicative of human HSC function. Human engraftment in the BM of the injected femur was confirmed by Southern blot analysis for human-specific satellite sequences, and human engraftment was detected in contralateral femurs and other bones, but at lower levels than in the injected femur. The human hematopoietic graft composition from hESC-derived HSCs was similar to that shown previously for somatic HSCs derived from cord blood and included lymphoid (CD45+/CD19+), myeloid (CD45+/CD33+), and erythroid (glycophorin A+, CD45−/human, MHC-1+) hematopoietic lineages.

Hematopoiesis and Homeobox Gene Expression in ESC

HOXB4 is expressed at the time of initiation of hematopoiesis in the yolk sac (McGrath and Palis, 1997) and ES cell differentiation models mimic embryonic hematopoiesis, with coexpression of HOX genes and their cofactors coinciding with the appearance of hematopoietic progenitor cells (Pineault *et al.*, 2002). Helgason *et al.* (1996) first reported that HOXB4 overexpression significantly increased the number of progenitors of mixed erythroid/myeloid lineage and definitive, but not primitive, erythroid colonies derived from embryoid bodies. Expression of HOXB4 in ES-derived primitive progenitors combined with culture on hematopoietic stroma induced a switch to the definitive HSC phenotype (Kyba et al., 2002). These progenitors engrafted lethally irradiated adults and contributed to long-term, multilineage hematopoiesis in primary and secondary recipients. Initial reports employing retroviral transduction of murine bone marrow with a HOXB4 retrovirus showed no disruption in hematopoiesis, but more recent data with in vivo transplantation of HOXB4-transduced ES-derived cells showed that while myeloid development was enforced, T and B lymphoid development was suppressed over a wide range of expression levels (Kyba et al., 2002). High expression levels of HOXB4 were also detrimental for erythroid development (Pilat et al., 2005). Provided that HOXB4 levels are kept within a certain therapeutic window, for example, by utilizing inducible gene expression systems, ES cells carry the potential of efficient and safe somatic gene therapy.

Cdx4 belongs to the caudal family of homeobox genes that have been implicated in anteroposterior patterning of the axial skeleton and are thought to regulate HOX gene expression. Davidson et al. (2003) reported that a Cdx4 mutation in zebra fish resulted in severe anemia with a complete absence of Runx1and decreased GATA1 expression but with normal myeloid (PU.1+) and angioblast (Flk-1+) development. HOXB7 and HOXA9 mRNA almost completely rescued the mutant. Retroviral transduction of ES-derived embryoid body cells with Cdx4 increased expression of HoxB4, HoxB3, HoxB8, and Hoxa9 (all implicated in HSC or progenitor expansion) and significantly increased CFU-GM/CFU-MIX and primitive erythroid generation (Davidson et al., 2003). Wang et al. (2005d) engineered mESC to express Cdx4 under a tetracycline-inducible system and found that the greatest effect on hematopoietic progenitor generation in EBs was when Cdx4 was expressed on days 4 and 5. Ectopic Cdx4 expression promoted hematopoietic mesoderm specification, increased blast colony and hematopoietic progenitor formation, and, together with HOXB4, enhanced long-term multilineage hematopoietic engraftment of lethally irradiated adult mice. The combination of ectopic HOXB4 and Cdx4 expression resulted in a high degree of lymphoid engraftment and thymic repopulation with both CD4 and CD8 cells and a capacity to engraft secondary recipients. It should be noted that this long-term lymphoid engraftment was associated with transcriptional silencing of the HOXB4 provirus, as the lymphoid populations lacked GFP expression (Wang et al., 2005d). In human ESC studies, Bowles et al. (2006) developed H1 ES clones stably expressing HOXB4 and showed in an EB differentiation system that this led to greatly increased production of CD45+ cells and progenitors of all lineages.

Transplantation of HOXB4-transduced cord blood Lin-/CD34+ cells into NOD/SCID mice demonstrated an enhanced increase in reconstitution capacity compared with vector-transduced human HSCs and a marked enhancement in the generation of primitive CD34+ cells (Schiedlmeier *et al.*, 2003). However, ectopic expression of HOXB4 is unable to induce hematopoietic repopulating capacity from hESCs, questioning the notion, derived from murine studies, that single genes, such as HOXB4, represent a master gene capable of conferring engraftment potential to hESC-derived hematopoietic cells. The failure to obtain robust engraftment using hESC-derived HSCs may be due to a failure to activate a molecular program similar to somatic HSCs. Gene expression patterns of CD34+ CD38− cells derived from human ESC have been compared with those of cells isolated from adult human bone marrow using microarrays (Lu *et al.*, 2004). Flt-3 gene expression was decreased markedly in cells from ESCs, whereas there was substantial Flt-3 expression in cells from adult marrow. The Flt3 gene was also undetectable in Rhesus monkey ES cell-derived CD34+ and CD34+ CD38− cells (Lu *et al.*, 2002). hESC-derived hemogenic progenitors expressed higher levels of a negative hematopoietic regulator, CD164, as well as the migratory and/or adhesion proteins CKLF-1, integrin-β3, matrix metalloproteinase 9, macrophage-inhibiting factor, and monocyte chemo-attractant protein-1. These features may account for the reduced ability of hESC-derived hematopoietic cells to migrate beyond the injected site and enter the circulation (Wang *et al.*, 2005c).

Methods

The methodology to be outlined covers methods for optimization of hematopoietic stem/progenitor production and for specific lineage differentiation, including lymphoid. Methods are restricted to human or primate ESC systems except where no human data are available and in such cases methods based on murine ESC are shown. More detailed methods for murine ESC hematopoietic and lymphoid differentiation have been presented elsewhere (Fairchild *et al.*, 2003; Kitajima *et al.*, 2003; Kyba *et al.*, 2003).

Maintenance of Human Embryonic Stem Cells

We have undertaken studies on hematopoietic differentiation using two human hESC lines, H1 and Miz-4. These are maintained on mitotically inactivated mouse embryonic fibroblast (Chemicon & Specialty Media) with Serum Replacement media consisting of 80% knockout Dulbecco's modified Eagle medium (KO-DMEM, GIBCO), 20% Serum Replacement (GIBCO), 4 ng/ml bFGF (R&D systems, MN), 1% nonessential amino acids (GIBCO), 1 mM L-glutamine (GIBCO), and 0.1 mM β-mercaptoethanol (Sigma, Canada). For propagation of undifferentiated hESC lines, hESC colonies are dissociated with 1 mg/ml collagenase IV (GIBCO) for 5 min and split every 6 to 7 days. Occasionally, hESC colonies are dissected manually during passaging. For more extensive methodology for derivation and maintenance of human ESC, see Cowan *et al.* (2006).

Hematopoietic Differentiation of hESC Coculture with Stromal Cell Lines

Kaufman *et al.* (2001) pioneered a stromal coculture system for generating hematopoietic cells from hESC. As outlined originally, undifferentiated hES cells are cultured on S17 mouse bone marrow stromal cell monolayers to derive cystic bodies containing CD34+ hematopoietic progenitor stem cells. hES cell cultures are treated with collagenase IV (1 mg/ml) for 10 min at 37° and subsequently detached from the plate by gentle scraping off of the colonies. The hES cell clusters are then transferred to irradiated (35 Gy) S17 cell layers and cultured with RPMI differentiation medium containing 15% fetal bovine serum (FBS) (HyClone, Logan, UT), 2 mM L-glutamine, 0.1 mM β-mercaptoethanol, 1% minimum essential medium (MEM)-nonessential amino acids, and 1% penicillin/streptomycin. The medium is changed every 2 to 3 days during 14 to 17 days of culture on S17 cells. After allowing adequate time for differentiation, hES cystic bodies are harvested and processed into a single cell suspension by collagenase IV treatment followed by digestion with trypsin/EDTA supplemented with 2% chick serum (Invitrogen, Carlsbad, CA) for 20 min at 37°. Cells are then washed twice with phosphate-buffered saline (PBS) and filtered through a 70-μm cell strainer to obtain a single cell suspension. To assess the levels of CD34+ cells in the bulk cell suspension, cells are labeled with PE-conjugated anti-CD34 antibody (BD Biosciences, San Jose, CA) and analyzed by FACS. CD34+ cells can be isolated using a CD34 progenitor cell isolation kit (Miltenyi Biotech, Auburn, CA) following the manufacturer's protocol. Subsequent adaptations of this method increase the serum concentration to 20% (Tian *et al.*, 2004) and add recombinant human cytokines and growth factors to the S17 stroma, including SCF, IL-3, IL-6, VEGF, G-CSF, Flt3-L, Epo, and BMP-4 (Hematti *et al.*, 2005) or a combination of BMP-4, -2, and -7 (Honig *et al.*, 2004).

Maintenance of Stromal Cell Lines

A number of murine and human stromal cell lines, developed for hematopoietic stem cell support, have been used to promote hESC hematopoietic differentiation.

S17

The mouse bone marrow stromal cell line S17 (Collins and Dorshkin, 1987) is maintained in α-modified Eagle's minimum essential medium (α-MEM) with 2 mM L-glutamine, 1.5 g/liter sodium bicarbonate, and 10% FBS. Stromal cocultures with S17 have been established using serum-free medium, either Stemline medium (Sigma) or QBSF60 medium (Quality Biologics, Gaithersburg, MD) supplemented with 4 mM L-glutamine and defined cytokines (Flt 3 L, SCF, Tpo, VEGF, and BMP-4) (Tian et al., 2004).

AFT024

A mouse fetal liver stromal cell (Moore et al., 1997) is routinely cultured in Dulbecco's modified Eagle's medium (DMEM) supplemented with 10% FBS, 5 × 105 mol/liter β-mercaptoethanol (2-ME) at 32°, 5% CO2, and 100% humidity.

MS-5

The murine stromal cell line MS-5 (Itoh et al., 1989) (provided by K. Mori, Kyoto University, Kyoto, Japan) is cultured in α-MEM (GIBCO-BRL Life Technologies, Grand Island, NY) supplemented with 10% FBS (HyClone Laboratories, Logan, UT) and passaged weekly.

OP9

Kodama et al. (1994) developed a cell line (OP9) from calvaria of newborn osteopetrotic mutant op/op mice that lack M-CSF and exhibit an osteoclast formation defect. OP9-supported hematopoiesis shows a marked reduction in macrophage production and we have observed that macrophages or their products inhibit stem cell proliferation in stromal coculture systems (Feugier et al., 2005). OP9 cells can be obtained from ATCC and are maintained in α-MEM with 2 mM L-glutamine, 1.5 g/liter sodium bicarbonate, 20% FBS, and 50 μg/ml ascorbic acid (Kitajima et al., 2003). OP9 cells can easily lose the ability to maintain lymphohematopoiesis, particularly after prolonged passage or if maintained in suboptimal medium or certain lots of FCS; consequently, it is important to screen a number of FCS batches to ensure optimal OP9 hematopoietic support function. While OP9 stroma is contact inhibited and confluent cultures can be used for some weeks, unirradiated stroma may detach after 2 weeks under coculture conditions, whereas irradiated stroma (40 Gy) can support hESC differentiation for >4 weeks. It should be noted that while OP9 can support long-term murine stem cell proliferation and differentiation, it is unable to do so with human CD34+ cells, either adult or neonatal, in the absence of additional cytokine supplementation (Feugier et al., 2005).

We have shown that it is one of the most effective cell lines supporting long-term human stem cell proliferation and differentiation, provided that it is either supplemented with recombinant thrombopoietin or transduced with an adenovector expressing Tpo (Feugier *et al.*, 2005; Gammaitoni *et al.*, 2004). We have shown prolonged hematopoiesis in OP9 stromal cocultures in serum-free medium (QBSF60) and have observed that the serum-free conditions inhibited overgrowth of ES-derived nonhematopoietic cells whose presence normally requires frequent repassaging onto fresh stroma.

Human Bone Marrow Stroma

Hematopoietic differentiation of hESC has been reported using a coculture with primary cultures of adult (Kim *et al.*, 2005) and fetal (Wang *et al.*, 2005a) bone marrow stroma. Ficoll (Sigma)-separated marrow cells (5×10^5 cells/ml) are plated in 10 ml of IMDM plus 12.5% FCS and 12.5% horse serum and 5 μM hydrocortisone in T-25 flasks. Cultures are subject to weekly demidepopulation with the addition of fresh medium. By 2 to 3 weeks a semiconfluent layer of fibroblasts with some adipocyte differentiation is apparent. The cultures at this point may be trypsinized and repassaged into fresh flasks or six-well plates. Irradiation (15 Gy) is used to eliminate residual hematopoietic cells. At this stage, hESC can be added and differentiation monitored over 2 to 5 weeks. Upon continual passage of primary stroma, hematopoietic support function is progressively lost so studies should be restricted to early passage stroma.

Chorionic Mesenchyme

Kim *et al.* (2006) reported hESC differentiation on primary chorionic mesoderm. Human chorionic plate membranes are separated from placenta, a cell suspension made by enzymatic digestion, and cells cultured in DMEM + 20% FCS. Upon reaching confluence, this stroma supports the hematopoietic differentiation of EBs cocultured in the presence of IMDM (Invitrogen) 12.5% HS and 12.5% FBS and L-glutamine.

Hematopoietic Differentiation of hESC in Coculture with Stromal Cells

Stromal-Supported hESC Differentiation

OP9, MS-5, or S17 cells are plated onto gelatinized six-well plates, 10-cm dishes, or in T-12.5 flask with vented cap (BD) in α-MEM supplemented with 20% heat-inactivated FBS, 0.1 mM β-mercaptoethanol, 1 mM L-glutamine, and 10 U/ml of penicillin and streptomycin. After formation of confluent stromal cultures on days 4 and 5, half of the medium is replaced and cells are cultured for an additional 3 to 4 days. Undifferentiated hESC are harvested by treatment with 1 mg/ml collagenase IV (Invitrogen), dispersed by scraping, and added to stromal cultures at a density of 20 colonies/20 ml per 10-cm dish, or 4 to 5 colonies/4 ml per well of a six-well plate, in α-MEM supplemented with 10% FBS (HyClone) and 100 μM MTG (Sigma). When human ES cells are dispersed as individual cells or small aggregates ($\sim$50 cells), no or only a few hematopoietic cells develop in the coculture. The larger the ES cell colony, the greater the degree of generation of hematopoietic cells. This suggests that a critical ESC mass is necessary to generate the mesodermal differentiation, leading to hemangioblast and hematopoietic cell differentiation. The hES cell/stromal cocultures are incubated at 37° in 5% CO_2 with a half-medium change on days 4, 6, and 8. When needed, single cell suspensions are prepared by treatment of the hESC/stromal cocultures with collagenase IV (Invitrogen; 1 mg/ml in α-MEM) for 20 min at 37°, followed by treatment with 0.05% trypsin/0.5 mM EDTA (Invitrogen) or cell dissociation medium (Accutase, Innovative Cell Technologies) for 15 min at 37°. Cells are washed twice with PBS/5% FBS, filtered through a 100-μm cell strainer (BD Biosciences), counted, and used for clonogenic and flow cytometric assays and gene expression analysis or for continuation of hematopoietic development, reseeded onto fresh stroma. Hematopoietic foci are detected by 2 to 3 weeks, consisting of phase dark cells beneath the stroma and phase bright clusters of cells loosely attached to the stromal surface (Fig. 1). The cells may be repassaged onto OP9 stroma or switched from OP9 stroma to MS-5 stroma. The latter provides a standard for evaluation of cobblestone area formation (CAFC) or secondary colony formation (long-term culture-initiating assay) at 2 to 5 weeks as a measure of progenitor and stem cell function (Jo et al., 2000). Both assays can be undertaken under limiting-dilution conditions in 96-well plates. It is useful to have control CD34+ cells from neonatal (umbilical cord blood) or adult (bone marrow or G-CSF mobilized peripheral blood) where progenitors and CAFC are to be evaluated. For the progenitor (CFC) assay, 5×10^4 differentiated human ES cells or 5×10^3 ES-derived CD34+ cells are plated in triplicate in 35-mm tissue culture dishes containing 1 ml assay medium consisting of IMDM, 1.2% methyl cellulose (Fisher Scientific), 30% FBS, 5×10^{-5} M β-mercaptoethanol, 2 mM L-glutamine (GIBCO), and 0.5 mM hemin (Sigma) and supplemented with SCF, Flt3L, IL-3, G-CSF (all at 20 ng/ml), and 6 U/ml Epo (Amgen). We also routinely include one dish with no cytokine as a negative control to ensure that hematopoietic colonies are specifically responding to hematopoietic cytokines. After 14 days of incubation at 37° in 5% CO_2 in air, CFU-GM, burst-forming unit erythroid (BFU-E), and mixed colonies (CFU-Mix) are

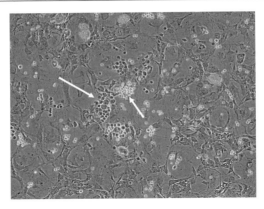

FIG. 1. Coculture of irradiated OP9 cells and human ES cells. Two human ES colonies were cocultured with irradiated OP9 cells for 3 weeks in IMDM supplemented with 20% FCS and VEGF, SCF, Flt3L, and Tpo. Half of the medium was replaced every 4 days. Foci of loosely adherent hematopoietic cells and phase dark cobblestone area-like cells are observed (arrows).

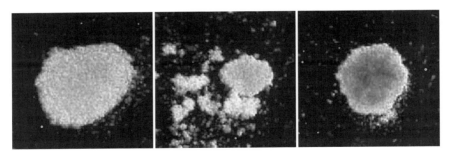

FIG. 2. Hematopoietic colonies (erythroid, mixed, and myeloid) developing at day 14 in cytokine-stimulated methyl cellulose of hESC-derived EB-CD34+ cells.

scored (Figs. 2 and 3). Between 2 and 3 weeks of coculture on various stromal lines, 0.04 to 0.15% of total cells are progenitors, predominantly granulocyte/macrophage, with a lower frequency of BFU-E and mixed colonies (Table I). Progenitor cells are almost exclusively present in the CD34+ fraction, and the cloning efficiency of this population increases markedly from 14 days to 17 to 21 days of coculture on S17 (Table II). FACS sorting can be used to demonstrate that CFC generated in the stromal coculture system are exclusively CD34+ and CD45+ by 2 to 3 weeks of culture (Fig. 4). The cloning efficiency of this CD34+, CD45+ population (0.68%) is considerably lower than that of neonatal or adult CD34+ populations (10 to 20%).

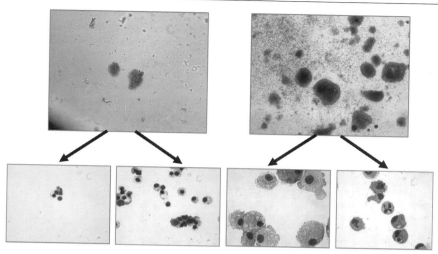

FIG. 3. Cell morphology of CFC derived from human ESC OP9 stromal cocultures with cytokine supplemention (VEGF, SCF, Flt3L, Tpo). The hematopoietic colonies were isolated by a micropipette, washed free of methyl cellulose, cytospun on a glass slide, and stained with Giemsa. Erythroid differentiation from BFU-E (left) and macrophage and neutrophil differentiation from CFU-GM/CFU-Mix (right).

TABLE I

HEMATOPOIETIC PROGENITOR CELLS DERIVED FROM COCULTURES OF H1 HES CELLS ON VARIOUS STROMAL CELL LINES[a]

	Cell # × 10^5	GM-CFC	BFU-E	CFU-Mix	Total CFC	% CFC
S17	13.0	1690	78	182	1970	0.15
AFT-024	33.0	660	0	0	660	0.02
MS-5	9.0	648	18	54	720	0.08
OP-9	15.5	589	31	62	682	0.04

[a] Cocultures of human ES cells on various mouse stromal cells with cytokine (VEGF, SCF, Flt3L, Tpo) addition were harvested at day 18 and subjected to a clonogenic (CFC) assay.

Stromal Plus Cytokine-Supplemented Cultures

Improved hematopoietic differentiation is achieved by coculturing hESC (H1) on irradiated (or nonirradiated) mouse stromal cell lines (S-17, MS-5, AFT-024, and OP-9) in the presence of cytokines, including rhVEGF (R&D System), rhSCF, (Kirin Brew Inc.), rhFlt-3/Flk2 ligand (Biosources), rhTpo (Kirin Brewery Inc.), rhEPO, (Amgen), and rhIL-3 (R&D system). Table III shows increasing production of CFC in OP9

TABLE II

FREQUENCY OF CD34+ CELLS AND THEIR CFC PRODUCTION IN hESC COCULTURE ON S17
STROMA PLUS CYTOKINE (VEGF, SCF, FLT3L, TPO) OVER 2 TO 3 WEEKS

Day	CD34+ %	CFC/10^5 unseparated	CFC/10^5 CD34+ fraction	CFC/10^5 CD34− fraction
14	4.70	8 ± 2	50 ± 14	1 ± 1
17	6.02	37 ± 5	652 ± 20	11 ± 5
21	6.72	38 ± 2	638 ± 62	9 ± 2

Cell fraction	CFC/10^5 cells
Unsorted cells	68 ± 4
CD34$^+$ cells (R3)	750 ± 110
CD34$^+$ CD45$^+$ cells (R3)	680 ± 60
CD45$^-$ cells (R14)	2 ± 0

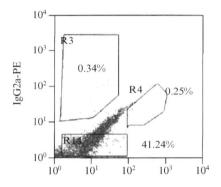 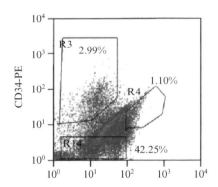

FIG. 4. Coculture of human H1 hES cells on irradiated OP9 stroma with cytokine addition (VEGF, Flt3L, SCF, Tpo). Cells were harvested at day 18 and sorted on the basis of CD45 and CD34 expression. Sorted fractions were plated for CFC and scored after 2 weeks. Results are displayed as mean ± SD (in triplicate).

coculture of hESC with the addition of various cytokine combinations, with a maximum effect seen with a combination of SCF, Flt3L, VEGF, IL-3, and Tpo, all added at 10 ng/ml. Further enhancement has been reported with the addition of BMP-4 to a comparable cytokine cocktail (Hematti *et al.*, 2005; Honig *et al.*, 2004). The concentration of cytokines and the frequency of cytokine replenishment are variables in a number of studies, with cytokine availability and cost as factors limiting continuous use at optimal plateau levels that can be as high as 300 ng/ml with SCF or Flt3L. We have developed a panel of replication-incompetent adenovectors expressing SCF,

TABLE III
EFFECT OF CYTOKINE COMBINATIONS ON INDUCTION OF hESC DIFFERENTIATION TO
HEMATOPOIETIC PROGENITORS IN OP9 COCULTURE[a]

Cytokine[b]	Cell number	BFU-E (per 10^5)	GM (per 10^5)	Mix (per 10^5)	Total CFC
S/F/T	1.0×10^6	0 ± 0	16 ± 4	0 ± 0	160 ± 40
S/F/V	1.09×10^6	0 ± 0	9 ± 3	0 ± 0	99 ± 27
F/V/T	0.74×10^6	0 ± 0	44 ± 4	0 ± 0	326 ± 30
S/V/T	1.35×10^6	0 ± 0	70 ± 12	0 ± 0	945 ± 162
S/F/T/V	1.57×10^6	0 ± 0	104 ± 18	0 ± 0	1633 ± 188
S/F/T/V/3	2.14×10^6	5 ± 3	76 ± 12	2 ± 2	1766 ± 363

[a] Human ES cells (H1) were cocultured on irradiated OP9 stroma in ES induction medium and combinations of up to six different cytokine were compared. After 18 days, the cocultures were harvested and subjected to clonogenic (CFC) assays. The CFCs were scored after 2 weeks and data expressed as average $\pm$ SD.

[b] S, stem cell factor; F, Flt-3 ligand; V, VEGF; T, Tpo; and 3, IL-3. Dose of each cytokine is 10 ng/ml.

Tpo, Flt3L, GM-CSF, Epo, or VEGF derived from the Ad5 E1a-, partially E1b-, and partially E3-deficient vectors with an expression cassette in the E1a region containing human cytokine cDNA driven by the cytomegalovirus major immediate/early promoter/enhancer (Feugier *et al.*, 2005; Gammaitoni *et al.*, 2004). When stromal cells reach confluence, they can be transfected with 15 to 30 multiplicities of infection of the adenovector in serum-free medium (X-vivo, Biowhittaker, Walkerville, MA) for 12 h. Following transfection, the supernatant is removed and replaced with hESC differentiation medium and hESC. Sustained levels of ~100 ng/ml of cytokine as determined by an ELISA assay are produced through 4 to 5 weeks of culture.

Hematopoietic Differentiation of hESC Using Stromal-Conditioned Medium

The ability to generate hematopoietic cells from hESC in the absence of stromal-cell contamination presents certain advantages from both a practical and a clinically relevant standpoint. Irradiated (46 Gy) OP9 cells are cultured with IMDM supplemented with 20% heat-inactivated FBS, 0.1 mM β-mercaptoethanol, 3 mM L-glutamine (Invitrogen), 5 μM hydrocortisone, 10 U/ml of penicillin (Invitrogen), and 10 μg/ml of streptomycin. After a 12-h incubation at 37°, culture supernatants are collected and passed through Acrodisc syringe filters (0.2 μm, Pall Corp., Ann Arbor, MI). Conditioned medium (OP9-CM) can be harvested over 10 days. All the supernatants are stored at 4° and used within 1 week. For hematopoietic differentiation with OP9-CM, the centers of undifferentiated hESC colonies are harvested using a hand-made fine capillary. Approximately 20 colonies are transferred into 10-cm dishes with 5 ml of OP9-CM supplemented with 20 ng/ml of rhVEGF, 20 ng/ml of rhBMP-4, 20 ng/ml of SCF, 10 ng/ml of rhFlt3L, 20 ng/ml of rhIL-3, and 10 ng/ml of rhIL-6. Medium changes are performed every 3 to 4 days. When needed, single cell suspension is prepared by treatment of the hESC/OP9-CM cocultures with 0.05% trypsin/0.5 mM EDTA (Invitrogen) or "cell dissociation medium" (Accutase, Innovative Cell Technologies) for 15 min at 37°. Cells are washed twice with PBS + 5% FBS, filtered through a 100-μm cell strainer (BD Biosciences), counted, and used for clonogenic and flow cytometric assays and gene expression analysis.

Embryoid Body Formation and Hematopoietic Differentiation

Hanging Drop EB Cultures

This method is based on the murine system of Kyba *et al.* (2003). Fifteen-centimeter nontissue culture-treated dishes will support ~300 drops with each drop containing ~100 ES cells in 10 μl of differentiation medium. Cells are dispensed by micropipetting or using an eight-well multichannel pipettor. Dishes are inverted and incubated for 2 days at 37° in 5% CO$_2$. Single EBs form per drop and are collected by flushing the dish with PBS, transferring to 15-ml tubes, and allowing sedimentation by gravity for 3 min. Medium is aspirated and EBs are suspended in 10 ml fresh differentiation medium, transferred to 10-cm bacterial-grade dishes, and cultured under slow swirling conditions on a rotating shaker (50 rpm). Fresh medium is added every 2 days after removal of half the spent medium. It is important throughout these steps that EBs are prevented from attaching to the dish. At different stages of development, EBs can be dissociated by the addition of 0.5 ml of 0.25% trypsin and incubated 2 min at 37° followed by the addition of 5 ml IMDM + 10% FCS and passaged repeatedly through a 5-ml pipette until dissociated.

Aggregation Methods for EB Formation

Undifferentiated hESCs at confluence in six-well plates are treated with collagenase IV, scraped off their Matrigel attachment, and transferred to six-well low-attachment plates to allow for EB formation by overnight incubation in differentiation medium consisting of 80% knockout (KO) DMEM (GIBCO) supplemented with 20% nonheat-inactivated FBS (HyClone, Logan, UT), 1% nonessential amino acids, 1 mM L-glutamine, and 0.1 mM β-mercaptoethanol. After 12 h all cells and medium are transferred to a 15-ml tube and EBs and allowed to sediment by gravity for 3 min. This step is repeated to remove cell debris and ~30 EBs are then cultured on nonadherent 10-cm tissue culture plates (Corning) with 80% KO-DMEM (GIBCO), 20% FBS (Hyclone), 1% nonessential amino acids, 1 mM L-glutamine, and 0.1 mM β-mercaptoethanol (Sigma) in the presence of SCF (20 ng/ml), Flt3L (20 ng/ml), Tpo (20 ng/ml), VEGF (10 ng/ml, R&D Systems), activin A (10 ng/ml, R&D Systems), and BMP-4 (10 ng/ml, R&D Systems). Media and cytokines are replaced every 5 days. Chadwick *et al.* (2003) have used a comparable system, but have supplemented it with a somewhat different, and sometimes higher, concentration of cytokines (300 ng/ml SCF [Amgen], 300 ng/ml Flt3L [R&D Systems], 10 ng/ml IL-3 [R&D Systems], 10 ng/ml IL-6 [R&D Systems], 50 ng/ml G-CSF [Amgen], and 50 ng/ml BMP-4 [R&D Systems]).

Spin Technique for EB Generation

Ng *et al.* (2005) observed that only a subset of pieces of undifferentiated hESC colonies containing 500 to 1000 cells regularly formed blood cells, suggesting that a minimum number of hESC is required to generate EBs that differentiate into the mesodermal lineage. They developed a modified technique for hEB formation in which hESC are trypsinized into single cell suspension, washed in PBS, and resuspended in serum-free medium consisting of a 1:1 ratio of IMDM without phenol red and Ham's F12 (GIBCO, Invitrogen), 5 mg/ml bovine serum albumin (BSA; Sigma), 1:100 synthetic lipids (Sigma), 1:100 insulin/transferrin/selenium (ITS-X, GIBCO), 2 mM glutamine, 5% protein-free hybridoma mix (GIBCO), and 50 μg/ml ascorbic acid. Cells (300 to 10,000) are seeded in 100 μl in the serum-free medium supplemented with growth factors [(10 ng/ml BMP-4, 5 ng/ml hVEGF, 20 ng/ml SCF, 5 ng/ml Flt3L, 5 ng/ml IL-6, 5 ng/ml hIGF-II (R&D Systems)] in each well of 96-well round-bottomed, low attachment plates (Nunc, Roskilde, Denmark), centrifuged at 1500 rpm for 4 min at to aggregate the cells. After 10 to 12 days, EBs are transferred to 96-well flat-bottomed tissue culture plates precoated with gelatin in serum-free medium plus growth factors (VEGF, SCF, Flt3L, IL-3, Tpo, Epo) and allowed to differentiate further.

Erythroid Differentiation

In the murine ESC system, Carotta *et al.* (2004) were able to generate $>10^{11}$ erythroid cells from an input of 20,000 CCE mESC in 10 weeks. Methyl cellulose culture-generated EBs are harvested at 6 to 9 days and expand in Stem-Pro serum-free medium (StemPro34 plus nutrient supplementation; GIBCO), Epo (2 U/ml), SCF (100 ng/ml, R&D Systems),

10^{-6} dexamethasone (Sigma), and 40 ng/ml IGF-1 (Promega, Madison WI). Cell density is maintained between 2.5 and 4 $\times$ 10^6/ml. At days 1 and 3, cell aggregates and dead cells are removed by using a 70-μm cell strainer and Ficoll purification. To achieve differentiation, cells are cultured in StemPro serum-free medium with 10 U Epo, insulin (10 ng/ml Actrapid HM, Novo Nordisk), and a 3 $\times$ 10^6 M glucocorticoid receptor antagonist ZK112.993 and 1 mg/ml human transferrin (Sigma). The cells undergo three to four "differentiation divisions," reduced cell size, accumulated hemoglobin and formed enucleated erythrocytes within 72 h. Chang et al. (2006) used a modification of this method for the generation of human erythroid cells. CD45+ hematopoiesis peaked at the late day 14 EB differentiation stage, although low levels of CD45–ve erythroid differentiation were seen earlier. By morphology, hES-derived erythroid cells were of definitive type, but at both the RNA and the protein levels, cells coexpressed high levels of embryonic and fetal (γ) globins with little or no adult (β) globin observed. This was not altered by the presence or absence of FBS, VEGF, Flt3L, or coculture on OP9 and was not culture time dependent. Thus, coexpression of both embryonic and fetal globins by definitive erythroid cells did not faithfully mimic either YS or fetal liver ontogeny. In human yolk sac, primitive erythroid cells remain mostly nucleated and synthesized mainly embryonic globins (ε, ζ, α). Fetal cells have a macrocytic morphology and synthesize >80% adult globins ($\alpha2\gamma2$). Adult cells synthesize >90% adult globins ($\alpha2\beta2$). There are some discrepancies regarding the kinetics, morphology, and globin pattern of erythroid cells generated from human ES in various published reports. Qui et al. (2005) assigned ESC-derived erythroid cells exclusively to the primitive erythroid lineage, whereas Zambidis et al. (2005) supported transition from primitive to definitive phenotype because of increased levels of β-globin. Two studies showed exclusive or predominant expression of β-globin later in culture (Cerdan et al., 2004; Kaufman et al., 2001). It should be noted that erythroid differentiation has been reported predominantly with the H1 hES cell line (WiCell Research Inst. NIH Code WA01). However, robust erythropoiesis has been reported with hES2, hES3, and hES4 lines (Ng et al., 2005) and with our own studies with Miz-4. It did not occur or only at low levels in four other ES lines (hSF6-NIH Code UC06, UCSF, BG01, BG02, BG03-NIH Code BG01, BG02, BG03, BresaGen Inc., Masons, GA) (Chang et al., 2006).

Megakaryocyte Differentiation

CD34+ cells derived from hEBs or ESC stromal coculture are seeded onto 24-well plates (10,000 cells/well) in QBSF-60 serum-free medium

(Quality Biological, Inc., 2 ml/well), with the following human cytokines: 50 ng/ml SCF, 50 ng/ml Flt3L, 5 and 0 ng/ml Tpo for the first week. For megakaryocytic differentiation, 50 ng/ml SCF, 50 ng/ml IL-3, and 100 ng/ml Tpo are used in the second and third weeks. Only floating cells are harvested for cell phenotype analysis. The phenotype analysis by flow cytometry and cytospin are performed after the third week for CD41a+CD45+ megakaryocytes and developing platelets.

Neutrophil Differentiation

Lieber *et al.* (2004) developed a three-step liquid culture differentiation strategy enabling reliable and abundant production of neutrophils at high purity from murine ESC. Day 8 EBs are trypsinized for 5 min at room temperature and disaggregated into a cell suspension. The cells are washed in 20 ml IMDM containing 10% FBS, centrifuged, and resuspended in secondary differentiation medium and plated onto semiconfluent OP9 cells. The secondary differentiation mix contains 10% pretested heat-inactivated FBS (Summit Biotech), 10% horse serum (Biocell Laboratories, Rancho Dominguez, CA), 5% protein-free hybridoma medium (GIBCO BRL), 25 ng/ml oncostatin M (OSM), 10 ng/ml bFGF, 5 ng/ml IL-6, 5 ng/ml IL-11, and 1 ng/ml rLIF (R&D Systems) in 74% by volume IMDM containing 100 U/ml penicillin, 100 μg/ml streptomycin, and 1.5×10^{-4} M MTG. After 24 h, adherent cells associated with the monolayers are trypsinized and replated in the same medium onto new semiconfluent OP9 monolayers along with the cells in suspension to reduce monocyte/macrophage and fibroblast-like contaminants. After 3 days in the secondary differentiation mix, cells are transferred onto a semiconfluent OP9 monolayer at a concentration of approximately 4×10^5 cells/ml into a tertiary neutrophil differentiation mix containing 10% platelet-depleted serum (Animal Technologies, Tyler, TX), 2 mM L-glutamine, 88% by volume IMDM, 100 U/ml penicillin, 100 μg/ml streptomycin, 1.5×10^{-4} M MTG, 60 ng/ml G-CSF (Amgen), 3 ng/ml GM-CSF, and 5 ng/ml IL-6 (R&D Systems). After 4 to 20 days, the cells are harvested for assays.

Macrophage Differentiation

CD34+ cells ($\sim$2.5 to 4.0×10^5/ml) isolated from either hEBs or hESC stromal cocultures are plated in methyl cellulose culture (Methocult semi-solid medium; Stem Cell Technologies, Vancouver, BC) to generate myelo-monocytic colonies. At day 14 colonies are harvested by the addition of 5 ml DMEM containing 10% FBS and 10 ng/ml each of GM-CSF and M-CSF. Cells ($\sim$10^6) are placed in a 35-mm well and allowed to adhere for 48 h. At 2 and 4 days postharvest, medium is replaced with fresh complete

DMEM supplemented with 10 ng/ml GM-CSF and M-CSF. By 4 to 5 days, cells develop into mature macrophages that may be used for subsequent phenotypic and functional characterization (Anderson *et al.*, 2006). As assessed by FACS analysis, hES-CD34 cell-derived macrophages display characteristic cell surface markers CD14, CD4, CCR5, CXCR4, and HLA-DR, suggesting a normal phenotype. Tests evaluating phagocytosis, upregulation of the costimulatory molecule B7.1, and cytokine secretion in response to LPS stimulation showed that these macrophages are also functionally normal.

Lymphoid Differentiation from Human Embryonic Stem Cells

Fully grown hES colonies are cultured on irradiated OP9 stroma expressing the Delta ligand, DLL1, with α-MEM supplemented with 20% FBS (HyClone), 100 μM MTG, and 5 ng/ml Flt3L for 7 to 8 days. Floating cells from the hESC-OP9-DL1 coculture are harvested, and CD34+ cells are isolated using a "direct CD34 progenitor cell isolation kit" (Miltenyi Biotech Inc.) as recommended by the manufacturer. For T-cell progenitor (CD5+ and CD7+) differentiation, CD34+ cells are cultured for a further 3 weeks on irradiated OP9-DLL1 with 5 ng/ml Flt3L and 5 ng/ml IL-7 (R&D Systems). For B-cell (CD19+CD45+) differentiation, CD34+ cells from differentiation EBs are cultured on MS-5 stroma for >4 weeks with α-MEM supplemented with 10% FBS (HyClone), 100 μM MTG, and 10 ng/ml of each SCF and G-CSF. For NK cell (CD56+CD45+) differentiation, CD34+CD45 + cells from differentiated EBs are isolated and cultured on AFT-024 stroma with α-MEM supplemented with 10% FBS (HyClone), 100 μM MTG, 20 ng/ml SCF, 20 ng/ml IL-7, 10 ng/ml Flt3L, and 10 ng/ml IL-15 for >4 weeks. For dendritic cell differentiation, EB-derived CD34+ cells are seeded onto 24-well plates (10,000 cells/well) in QBSF-60 serum-free medium (Quality Biological, Inc., 2 ml/well) with the following human cytokines: 50 ng/ml SCF, 50 ng/ml Flt3L, and 50 ng/ml Tpo for the first week, replaced with 50 ng/ml SCF, 50 ng/ml Flt3L, 50 ng/ml Tpo, 10 ng/ml GM-CSF, and 20 ng/ml TNF-α for the second and subsequent weeks. Dendritic differentiation is evaluated by FACS expression of MHC class II, CD83, CD80, CD86, and CD40 with loss of the macrophage marker CD14.

In Vivo *Transplantation of EB-Derived CD34+ Cell into Immunodeficient Mice*

EB-derived CD34+ cell are harvested at days 10 to 12 as described earlier and transplanted into sublethally irradiated 8- to 10-week-old NOD/SCID β2m-/-mice using an intrafemoral bone marrow transplantation technique (IBMT) and into nonirradiated newborn mice by an intrahepatic transplant method (IHT) or by facial vein injection (FVI). Cell doses ranged from 50,000–100,000 for IHT and 100,000–500,000 for IBMT. In order to enhance engraftment, human bone marrow stromal cells (HS27 and/or HS5, 250,000 cells per mouse) (Roecklein and Torok-Storb, 1995) can be co-injected with hES-derived CD34+ cells into the femur. For IMBT, 8- to 10-week-old mice are exposed to 250 rad and on the following day, CD34+ cells (with or without human stromal cells) are suspended in PBS (20 μl) and loaded into a syringe ($\frac{1}{2}$ cc 281/2 gauge, U-100 insulin syringe; Beckton-Dickinson). Mice are anesthetized with a mixture of ketamine (0.8 μl/g of a 100-mg/ml stock solution) and xylazine (0.2 μl/g of a 20-mg/ml stock solution). The region from the groin to the knee joint is shaved with a razor, the knee is flexed to 90°, and a 28-gauge needle is inserted into the joint surface of the tibia through the patellar tendon and into the bone marrow cavity. For IHT, on the morning of birth, pups are transplanted without irradiation. CD34+ cells are suspended in PBS (50 μl) and loaded into a $\frac{1}{2}$ cc 281/2 gauge, U-100 insulin syringe The pup is held so that its body is pinned between the thumb and index finger of the left hand. The middle finger is placed on the pup's abdominal part so that the body can be tilted to one side to expose the liver, which is readily visualized through the skin. The needle is inserted into the liver from one side and the cells are injected toward the other side. After insertion and injection, the needle is withdrawn slowly to minimize seepage. For FVI, on the morning of birth, pups are exposed to 50 rad and returned to their mother's cage. The next day (~18 h), CD34+ cells suspended with PBS (40 μl) are loaded into a syringe (30-gauge ultra-fine insulin syringe, Beckton-Dickinson). The pup is held so that its body is pinned between the thumb and middle finger of the left hand. The index finger is placed on the pup's chin so that the head can be tilted to the side to expose the face and neck vessels. The needle is inserted into the facial vein and the cells are injected toward the heart. In the case of IHT and FVI, transplantation is done no later than 36 h after birth. All animal experiments require approval by institutional animal care and veterinary services.

Analysis of NOD/SCID Mouse Hematopoietic Engraftment

To prepare mouse bone marrow for flow cytometric analysis, BM cells are washed in PBS containing 3% BSA. The presence of human cells in BM of the transplanted immunodeficient mice is determined by flow cyto-metry using PE- or APC-conjugated antibody against human CD19, CD33, CD34, CD45, and glycophorin A (BD Pharmingen) (Fig. 5). In parallel, Southern blot and polymerase chain reaction (PCR) analyses can be per-formed to detect human DNA in mouse bone marrow. For extraction of genomic DNA, the DNeasy kit (Qiagen) is used according to the manufac-turer's recommendations. PCR for the human chromosome 17-specific PCR

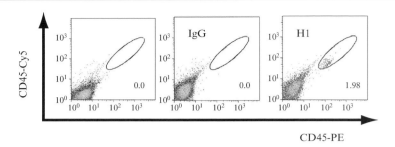

CD45-PE

FIG. 5. FACS analysis showing human hematopoietic (CD45+) bone marrow engraftment of NOD/SCID $\beta2-/-$ recipients 6 weeks after receiving hESC-derived hematopoietic cells by the intrahepatic route.

is performed using A forward primer 5'-ACACTCTTTTTGCAGGATC-TA-3' and backward primer 5'-AGCAATGTGAAACTCTGGGA-3' to amplify an 1171-bp sequence (40 cycles, 94° for 1 min, 60° for 1 min, and 72° for 1 min, followed by a final extension of 10 min at 72°). For Southern blot, 5 μg DNA is digested with an *Eco*RI restriction enzyme at 37° overnight, separated on an agarose gel, transferred to Hybond-N+ nylon membrane (Amersham Biosciences), and hybridized with a DIG-labeled (Roche Molecular System) human chromosome-specific probe (Bhatia *et al.*, 1998).

Bioluminescence Imaging

Stable transduction of transplantable cells with the GFP/luciferase fusion gene has provided an efficient quantitative measure of cell burden and location in human tumor xenografts transplant models (Wu *et al.*, 2005). Stable transduction of hESC with this GFP/luciferase fusion gene permits imaging of ESC-derived hematopoietic cells in NOD/SCID mice.

Production of a GFP/Luciferase-Expressing Lentivirus

A GFP/luciferase fusion gene, driven by the EF1 promoter, is cloned into the backbone of FUGW (kindly provided by Dr. David Baltimore) after deletion of a GFP gene controlled by the ubiquitin promoter (FUEGL). 293T cells are maintained in Dulbecco's modified Eagle's medium supplemented with 10% FBS, 100 units/ml penicillin, and 100 μg/ml streptomycin. The 293T cells are plated in 100-mm tissue culture dishes at least 12 h before transduction. The cell density should be 20 to 30% confluent when seeding and will be ~40% confluent for transfection. The culture medium is replaced with 10 ml of fresh medium 2 h before transfection. Prepare 2 ml of a calcium phosphate/DNA mixture suspension, which contains 1 ml of 2× HBS (0.05 *M* HEPES, 0.28 *M* NaCl, 1.5 m*M*

Na$_2$HPO$_4$, pH 7.12), 150 μl of 2 M CaCl2, 20 μg lentivirus vector, 10 μg pVSVG, 15 μg pΔ8.9, and distilled water (up to 2 ml) for each 100-mm plate. Allow the suspension to sit at room temperature for at least 10 min. Mix the precipitate well by vortexing and add 2 ml of calcium phosphate/DNA suspension to a 100-mm plate containing cells with a dropwise manner. Return the plates to the incubator and leave the precipitation for overnight (about 18 h). Replace fresh medium and check GFP expression the next morning. After 48 or 72 h of transfection, the virus is harvested and filtered with a 0.2-μm syringe filter. Filtered medium is concentrated with 100,000 MWCO centrifugal filter devices (Millipore) at 2000 rpm for 25 min and concentrated ~50 fold. (We harvest ~200 μl of concentrated virus suspension from 100 ml of nonconcentrated virus-containing medium.)

Transduction of Human Embryonic Stem Cells

hESC are cultured in a 24-well plate with low density (less than 10 colonies in a single well of 24-well plates) for 2 days. Concentrated viral supernatant (100 μl to 1 ml of Serum Replacement medium per well) is introduced with hESC and 4 μg/ml polybrene (Sigma) for 12 h. After 12 h, medium is replaced with fresh Serum Replacement medium and cultured for 4 to5 days. Undifferentiated hESC are isolated with a 1-ml micropipette and transferred to new fresh mitotically inactivated mouse embryonic fibroblast. After 4 to 5 days, GFP-expressing human embryonic stem cell colonies are checked manually and reisolated for further passage. GFP-expressing colonies are maintained for more than 2 months to isolate homogeneous colonies with uniform, stable GFP expression before further experiment. To confirm the luciferase activity *in vitro*, lysates of GFP-expressing colonies are analyzed by a Lumat LB9507 luminometer (EG&G Berthold) to measure the luciferase activity of luciferase reporter using the dual-luciferase reporter assay system (Promega) according to the manufacturer's recommendations.

In Vivo *Luciferase Imaging*

NOD/SCID, NOD/SCID-β2M($-/-$), and NOD/SCID-γ2($-/-$) mice are transplanted with GFP/luciferase-transduced ESC-derived hematopoietic cells sorted for CD34+ and GFP expression. At intervals (e.g., 3 and 5 weeks) animals are subject to whole body bioimaging. Luciferin (Xenogen), the substrate for firefly luciferase, is dissolved in phosphate-buffered saline at a concentration of 15.4 mg/ml and filtered through a 0.22-μm-pore-size filter before use. Mice are injected with 200 μl of luciferin (3 mg) and immediately anesthetized in an oxygen-rich induction chamber with 2% isoflurane (Baxter Healthcare, IL). The mice are maintained for

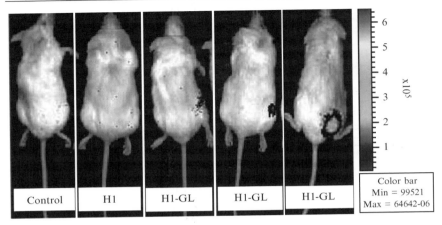

FIG. 6. Bioluminescent images at week 5 of NOD/SCID $\beta2-/-$ mice injected intrafemorally with PBS (negative control), nontransduced hESC-derived hematopoietic cells (H1), and H1-GFP/luciferase-derived hematopoietic cells (H1-GL).

at least 10 min so that there is adequate dissemination of the injected substrate. Anesthesia is maintained during the entire imaging process using a nose cone isoflurane/oxygen delivery device in the specimen chamber. Images are collected with 10- to 20-s integration times depending on the intensity of the bioluminescent signal. Data acquisition and analysis are performed using the LivingImage (Xenogen) software with the IgorPro image analysis package (WaveMetrics) (Fig. 6).

References

Anderson, J. S., Bandi, S., Kaufman, D. S., and Akkina, R. (2006). Derivation of normal macrophages from human embryonic stem (hES) cells for applications in HIV gene therapy. *Retrovirology* **3**, 24–30.

Berthier, R., Prandini, M. H., Schweitzer, A., Thevenon, D., Martin-Sisteron, H., and Uzan, G. (1997). The MS-5 murine stromal cell line and hematopoietic growth factors synergize to support the megakaryocytic differentiation of embryonic stem cells. *Exp. Hematol.* **25**, 481–490.

Bhatia, M. (2005). Derivation of the hematopoietic stem cell compartment from human embryonic stem cell lines. *Ann. N. Y. Acad. Sci.* **1044**, 24–28.

Bhatia, M., Bonnet, D., Murdoch, B., Gan, O. I., and Dick, J. E. (1998). A newly discovered class of human hematopoietic cells with SCID-repopulating activity. *Nature Med.* **4**, 1038–1045.

Bowles, K. M., Vallier, L., Smith, J. R., Alexander, M. R., and Pedersen, R. A. (2006). HOXB4 overexpression promotes hematopoietic development by human embryonic stem cells. *Stem Cells* **24**(5), 1359–1369.

Burt, R. K., Verda, L., Kim, D. A., Oyama, Y., Luo, K., and Link, C. (2004). Embryonic stem cells as an alternate marrow donor source: Engraftment without graft-versus-host disease. *J. Exp. Med.* **199**, 895–904.

Carotta, S., Pilat, S., Mairhofer, A., Schmidt, U., Dolznig, H., Steinlein, P., and Beug, H. (2004). Directed differentiation and mass cultivation of pure erythroid progenitors from mouse embryonic stem cells. *Blood* **104,** 1873–1880.

Cerdan, C., Rouleau, A., and Bhatia, M. (2004). VEGF-A165 augments erythropoietic development from human embryonic stem cells. *Blood* **103,** 2504–2512.

Chadwick, K., Wang, L., Li, L., Menendez, P., Murdoch, B., Rouleau, A., and Bhatia, M. (2003). Cytokines and BMP-4 promote hematopoietic differentiation of human embryonic stem cells. *Blood* **102,** 906–915.

Chang, K.-H., Nelson, A. M., Cao, H., Wang, L., Nakamoto, B., Ware, C. B., and Papayannopoulou, T. (2006). Definitive-like erythroid cells derived from human embryonic stem cells co-express high levels of embryonic and fetal globins with little or no adult globin. *Blood* **108**(5), 1515–123.

Cho, S. K., Webber, T. D., Carlyle, J. R., Nakano, T., Lewis, S. M., and Zuniga-Pflucker, J. C. (1999). Functional characterization of B lymphocytes generated *in vitro* from embryonic stem cells. *Proc. Natl. Acad. Sci. USA* **96,** 9797–9802.

Choi, K., Chung, Y. S., and Zhang, W. J. (2005). Hematopoietic and endothelial development of mouse embryonic stem cells in culture. *Methods Mol. Med.* **105,** 359–368.

Choi, K., Kennedy, M., Kazarov, A., Papadimitriou, J. C., and Keller, G. (1998). A common precursor for hematopoietic and endothelial cells. *Development* **125,** 725–732.

Collins, L. S., and Dorshkind, K. (1987). A stromal cell line from myeloid long-term bone marrow cultures can support myelopoiesis and B lymphopoiesis. *J. Immunol.* **138,** 1082–1087.

Cowan, C., Akutsu, H., and Melton, D. (2006). Human embryonic stem cells. *Methods Enzymol.* **418** (this volume).

Daley, G. Q. (2003). From embryos to embryoid bodies, generating blood from embryonic stem cells. *Ann. N. Y. Acad. Sci.* **996,** 122–131.

Dang, S. M., Gerecht-Nir, S., Chen, J., Itskovitz-Eldor, J., and Zandstra, P. W. (2004). Controlled, scalable embryonic stem cell differentiation culture. *Stem Cells* **22,** 275–282.

Dang, S. M., Kyba, M., Perlingeiro, R., Daley, G. Q., and Zandstra, P. W. (2002). Efficiency of embryoid body formation and hematopoietic development from embryonic stem cells in different culture systems. *Biotechnol. Bioeng.* **78,** 442–453.

Davidson, A. J., Ernst, P., Wang, Y., Dekens, M. P., Kingsley, P. D., Palis, J., Korsmeyer, S. J., Daley, G. Q., and Zon, L. I. (2003). cdx4 mutants fail to specify blood progenitors and can be rescued by multiple hox genes. *Nature* **425,** 300–306.

de Pooter, R. F., Cho, S. K., Carlyle, J. R., and Zuniga-Pflucker, J. C. (2003). *In vitro* generation of T lymphocytes from embryonic stem cell-derived prehematopoietic progenitors. *Blood* **102,** 1649–1653.

de Pooter, R. F., Cho, S. K., and Zuniga-Pflucker, J. C. (2005). *In vitro* generation of lymphocytes from embryonic stem cells. *Methods Mol. Biol.* **290,** 135–147.

Ema, M., Faloon, P., Zhang, W. J., Hirashima, M., Reid, T., Stanford, W. L., Orkin, S., Choi, K., and Rossant, J. (2003). Combinatorial effects of Flk1 and Tal1 on vascular and hematopoietic development in the mouse. *Genes Dev.* **17,** 380–393.

Fairchild, P. J., Brook, F. A., Gardner, R. L., Graca, L., Strong, V., Tone, Y., Tone, M., Nolan, K. F., and Waldmann, H. (2000). Directed differentiation of dendritic cells from mouse embryonic stem cells. *Curr. Biol.* **10,** 1515–1518.

Fairchild, P. J., Nolan, K. F., and Waldmann, H. (2003). Probing dendritic cell function by guiding the differentiation of embryonic stem cells. *Methods Enzymol.* **365,** 169–186.

Faloon, P., Arentson, E., Kazarov, A., Deng, C. X., Porcher, C., Orkin, S., and Choi, K. (2000). Basic fibroblast growth factor positively regulates hematopoietic development. *Development* **127,** 1931–1941.

Feugier, P., Li, N., Jo, D. Y., Shieh, J. H., MacKenzie, K. L., Lesesve, J. F., Latger-Cannard, V., Bensoussan, D., Crystal, R. G., Rafii, S., Stoltz, J. F., and Moore, M. A. (2005). Osteopetrotic mouse stroma with thrombopoietin, c-kit ligand, and flk-2 ligand supports long-term mobilized CD34+ hematopoiesis *in vitro*. *Stem Cells Dev.* **14**, 505–516.

Fujimoto, T. T., Kohata, S., Suzuki, H., Miyazaki, H., and Fujimura, K. (2003). Production of functional platelets by differentiated embryonic stem (ES) cells *in vitro*. *Blood* **102**, 4044–4051.

Fukuma, D., Matsuyoshi, H., Hirata, S., Kurisaki, A., Motomura, Y., Yoshitake, Y., Shinohara, M., Nishimura, Y., and Senju, S. (2005). Cancer prevention with semi-allogeneic ES cell-derived dendritic cells. *Biochem. Biophys. Res. Commun.* **335**, 5–13.

Gammaitoni, L., Weisel, K. C., Gunetti, M., Wu, K. D., Bruno, S., Pinelli, S., Bonati, A., Aglietta, M., Moore, M. A., and Piacibello, W. (2004). Elevated telomerase activity and minimal telomere loss in cord blood long-term cultures with extensive stem cell replication. *Blood* **103**, 4440–4448.

Hamaguchi, I., Morisada, T., Azuma, M., Murakami, K., Kuramitsu, M., Mizukami, T., Ohbo, K., Yamaguchi, K., Oike, Y., Dumont, D. J., and Suda, T. (2006). Loss of Tie2 receptor compromises embryonic stem cell-derived endothelial but not hematopoietic cell survival. *Blood* **107**, 1207–1213.

Helgason, C. D., Sauvageau, G., Lawrence, H. J., Largman, C., and Humphries, R. K. (1996). Overexpression of HOXB4 enhances the hematopoietic potential of embryonic stem cells differentiated *in vitro*. *Blood* **87**, 2740–2749.

Hematti, P., Obrtlikova, P., and Kaufman, D. S. (2005). Nonhuman primate embryonic stem cells as a preclinical model for hematopoietic and vascular repair. *Exp. Hematol.* **33**, 980–986.

Hidaka, M., Stanford, W. L., and Bernstein, A. (1999). Conditional requirement for the Flk-1 receptor in the *in vitro* generation of early hematopoietic cells. *Proc. Natl. Acad. Sci. USA* **96**, 7370–7375.

Hirata, S., Senju, S., Matsuyoshi, H., Fukuma, D., Uemura, Y., and Nishimura, Y. (2005). Prevention of experimental autoimmune encephalomyelitis by transfer of embryonic stem cell-derived dendritic cells expressing myelin oligodendrocyte glycoprotein peptide along with TRAIL or programmed death-1 ligand. *J. Immunol.* **174**, 1888–1897.

Hole, N., Graham, G. J., Menzel, U., and Ansell, J. D. (1996). A limited temporal window for the derivation of multilineage repopulating hematopoietic progenitors during embryonal stem cell differentiation *in vitro*. *Blood* **88**, 1266–1276.

Honig, G. R., Li, F., Lu, S. J., and Vida, L. (2004). Hematopoietic differentiation of rhesus monkey embryonic stem cells. *Blood Cells Mol. Dis.* **32**, 5–10.

Itoh, K., Tezuka, H., Sakoda, H., Konno, M., Nagata, K., Uchiyama, T., Uchino, H., and Mori, K. J. (1989). Reproducible establishment of hemopoietic supportive stromal cell lines from murine bone marrow. *Exp. Hematol.* **17**, 145–153.

Jo, D. Y., Rafii, S., Hamada, T., and Moore, M. A. (2000). Chemotaxis of primitive hematopoietic cells in response to stromal cell-derived factor-1. *J. Clin. Invest.* **105**, 101–111.

Kaufman, D. S., Hanson, E. T., Lewis, R. L., Auerbach, R., and Thomson, J. A. (2001). Hematopoietic colony-forming cells derived from human embryonic stem cells. *Proc. Natl. Acad. Sci. USA* **98**, 10716–10721.

Keller, G. (2005). Embryonic stem cell differentiation: Emergence of a new era in biology and medicine. *Genes Dev.* **19**, 1129–1155.

Keller, G., Kennedy, M., Papayannopoulou, T., and Wiles, M. V. (1993). Hematopoietic commitment during embryonic stem cell differentiation in culture. *Mol. Cell. Biol.* **13**, 473–486.

Kennedy, M., Firpo, M., Choi, K., Wall, C., Robertson, S., Kabrun, N., and Keller, G. (1997). A common precursor for primitive erythropoiesis and definitive haematopoiesis. *Nature* **386**, 488–493.

Kim, S. J., Kim, B. S., Ryu, S. W., Yoo, J. H., Oh, J. H., Song, C. H., Kim, S. H., Choi, D. S., Seo, J. H., Choi, C. W., Shin, S. W., Kim, Y. H., and Kim, J. S. (2005). Hematopoietic differentiation of embryoid bodies derived from the human embryonic stem cell line SNUhES3 in co-culture with human bone marrow stromal cells. *Yonsei Med. J.* **46**, 693–699.

Kim, S. J., Yoo, J. H., Kim, B. S., Oh, J. H., Song, C. H., Shin, H. J., Kim, S. H., Choi, C. W., and Kim, J. S. (2006). Mesenchymal stem cells derived from chorionic plate may promote hematopoietic differentiation of a human embryonic stem cell line, SNUhES3. *Acta Haematol.* **116** (in press).

Kitajima, K., Tanaka, M., Zheng, J., Sakai-Ogawa, E., and Nakano, T. (2003). *In vitro* differentiation of mouse embryonic stem cells to hematopoietic cells on an OP9 stromal cell monolayer. *Methods Enzymol.* **365**, 72–83.

Kodama, H., Nose, M., Niida, S., and Nishikawa, S. (1994). Involvement of the c-kit receptor in the adhesion of hematopoietic stem cells to stromal cells. *Exp. Hematol.* **22**, 979–984.

Kyba, M., Perlingeiro, R. C., and Daley, G. Q. (2002). HoxB4 confers definitive lymphoid-myeloid engraftment potential on embryonic stem cell and yolk sac hematopoietic progenitors. *Cell* **109**, 29–37.

Kyba, M., Perlingeiro, R. C. R., and Daley, G. Q. (2003). Development of hematopoietic repopulating cells from embryonic stem cells. *Methods Enzymol.* **365**, 114–129.

Lacud, G., Kouskoff, V., Trumble, A., Schwantz, S., and Keller, G. (2004). Haploinsufficiency of Runx1 results in the acceleration of mesodermal development and hemangioblast specification upon *in vitro* differentiation of ES cells. *Blood* **103**, 886–889.

Lengerke, C., and Daley, G. Q. (2005). Patterning definitive hematopoietic stem cells from embryonic stem cells. *Exp. Hematol.* **33**, 971–979.

Lensch, M. W., and Daley, G. Q. (2006). Scientific and clinical opportunities for modeling blood disorders with embryonic stem cells. *Blood* **107**, 2605–2612.

Lerou, P. H., and Daley, G. Q. (2005). Therapeutic potential of embryonic stem cells. *Blood Rev.* **19**, 321–331.

Li, F., Lu, S., Vida, L., Thomson, J. A., and Honig, G. R. (2001). Bone morphogenetic protein 4 induces efficient hematopoietic differentiation of rhesus monkey embryonic stem cells *in vitro*. *Blood* **98**, 335–342.

Lian, R. H., Maeda, M., Lohwasser, S., Delcommenne, M., Nakano, T., Vance, R. E., Raulet, D. H., and Takei, F. (2002). Orderly and nonstochastic acquisition of CD94/NKG2 receptors by developing NK cells derived from embryonic stem cells *in vitro*. *J. Immunol.* **168**, 4980–4987.

Lieber, J. G., Webb, S., Suratt, B. T., Young, S. K., Johnson, G. L., Keller, G. M., and Worthen, G. S. (2004). The *in vitro* production and characterization of neutrophils from embryonic stem cells. *Blood* **103**, 852–859.

Liu, B., Sun, Y., Jiang, F., Zhang, S., Wu, Y., Lan, Y., Yang, X., and Mao, N. (2003). Disruption of Smad5 gene leads to enhanced proliferation of high-proliferative potential precursors during embryonic hematopoiesis. *Blood* **101**, 124–133.

Liu, H., and Roy, K. (2005). Biomimetic three-dimensional cultures significantly increase hematopoietic differentiation efficacy of embryonic stem cells. *Tissue Eng.* **11**, 319–330.

Lu, S. J., Li, F., Vida, L., and Honig, G. R. (2002). Comparative gene expression in hematopoietic progenitor cells derived from embryonic stem cells. *Exp. Hematol.* **30**, 58–66.

Lu, S. J., Li, F., Vida, L., and Honig, G. R. (2004). CD34+CD38– hematopoietic precursors derived from human embryonic stem cells exhibit an embryonic gene expression pattern. *Blood* **103,** 4134–4141.

Martin, C. H., and Kaufman, D. S. (2005). Synergistic use of adult and embryonic stem cells to study human hematopoiesis. *Curr. Opin. Biotechnol.* **16,** 510–515.

Matsuyoshi, H., Senju, S., Hirata, S., Yoshitake, Y., Uemura, Y., and Nishimura, Y. (2004). Enhanced priming of antigen-specific CTLs *in vivo* by embryonic stem cell-derived dendritic cells expressing chemokine along with antigenic protein: Application to antitumor vaccination. *J. Immunol.* **172,** 776–786.

McGrath, K. E., and Palis, J. (1997). Expression of homeobox genes, including an insulin promoting factor, in the murine yolk sac at the time of hematopoietic initiation. *Mol. Reprod. Dev.* **48,** 145–153.

Menendez, P., Wang, L., Chadwick, K., Li, L., and Bhatia, M. (2004). Retroviral transduction of hematopoietic cells differentiated from human embryonic stem cell-derived CD45(neg) PFV hemogenic precursors. *Mol. Ther.* **10,** 1109–1120.

Miyagi, T., Takeno, M., Nagafuchi, H., Takahashi, M., and Suzuki, N. (2002). Flk1+ cells derived from mouse embryonic stem cells reconstitute hematopoiesis *in vivo* in SCID mice. *Exp. Hematol.* **30,** 1444–1453.

Moore, K. A., Ema, H., and Lemischka, I. R. (1997). *In vitro* maintenance of highly purified, transplantable hematopoietic stem cells. *Blood* **89,** 4337–4347.

Muller, A. M., and Dzierzak, E. A. (1993). ES cells have only a limited lymphopoietic potential after adoptive transfer into mouse recipients. *Development* **118,** 1343–1351.

Nakano, T., Era, T., Takahashi, T., Kodama, H., and Honjo, T. (1997). Development of erythroid cells from mouse embryonic stem cells in culture: Potential use for erythroid transcription factor study. *Leukemia* **11**(Suppl. 3), 496–500.

Nakano, T., Kodama, H., and Honjo, T. (1994). Generation of lymphohematopoietic cells from embryonic stem cells in culture. *Science* **265,** 1098–1101.

Nakano, T., Kodama, H., and Honjo, T. (1996). *In vitro* development of hematopoietic system from mouse embryonic stem cells, a new approach for embryonic hematopoiesis. *Int. J. Hematol.* **65,** 1–8.

Nakayama, N., Lee, J., and Chiu, L. (2000). Vascular endothelial growth factor synergistically enhances bone morphogenetic protein-4-dependent lymphohematopoietic cell generation from embryonic stem cells *in vitro*. *Blood* **95,** 2275–2283.

Narayan, A. D., Chase, J. L., Lewis, R. L., Tian, X., Kaufman, D. S., Thomson, J. A., and Zanjani, E. D. (2006). Human embryonic stem cell-derived hematopoietic cells are capable of engrafting primary as well as secondary fetal sheep recipients. *Blood* **107,** 2180–2183.

Ng, E. S., Davis, R. P., Azzola, L., Stanley, E. G., and Elefanty, A. G. (2005). Forced aggregation of defined numbers of human embryonic stem cells into embryoid bodies fosters robust, reproducible hematopoietic differentiation. *Blood* **106,** 1601–1603.

Nishikawa, S. I., Nishikawa, S., Hirashima, M., Matsuyoshi, N., and Kodama, H. (1998). Progressive lineage analysis by cell sorting and culture identifies FLK1+VE-cadherin+ cells at a diverging point of endothelial and hemopoietic lineages. *Development* **125,** 1747–1757.

Olsen, A. L., Stachura, D. L., and Weiss, M. J. (2006). Designer blood: Creating hematopoietic lineages from embryonic stem cells. *Blood* **107,** 1265–1275.

Otani, T., Inoue, T., Tsuji-Takayama, K., Ijiri, Y., Nakamura, S., Motoda, R., and Orita, K. (2005). Progenitor analysis of primitive erythropoiesis generated from *in vitro* culture of embryonic stem cells. *Exp. Hematol.* **33,** 632–640.

Park, C., Afrikanova, I., Chung, Y. S., Zhang, W. J., Arentson, E., Fong Gh, G., Rosendahl, A., and Choi, K. (2004). A hierarchical order of factors in the generation of FLK1- and

SCL-expressing hematopoietic and endothelial progenitors from embryonic stem cells. *Development* **131,** 2749–2762.

Pilat, S., Carotta, S., Schiedlmeier, B., Kamino, K., Mairhofer, A., Will, E., Modlich, U., Steinlein, P., Ostertag, W., Baum, C., Beug, H., and Klump, H. (2005). HOXB4 enforces equivalent fates of ES-cell-derived and adult hematopoietic cells. *Proc. Natl. Acad. Sci. USA* **102,** 12101–12106.

Pineault, N., Helgason, C. D., Lawrence, H. J., and Humphries, R. K. (2002). Differential expression of Hox, Meis1, and Pbx1 genes in primitive cells throughout murine hematopoietic ontogeny. *Exp. Hematol.* **30,** 49–57.

Potocnik, A. J., Kohler, H., and Eichmann, K. (1997). Hemato-lymphoid *in vivo* reconstitution potential of subpopulations derived from *in vitro* differentiated embryonic stem cells. *Proc. Natl. Acad. Sci. USA* **94,** 10295–10300.

Priddle, H., Jones, D. R., Burridge, P. W., and Patient, R. (2006). Hematopoiesis from human embryonic stem cells: Overcoming the immune barrier in stem cell therapies. *Stem Cells* **24,** 815–824.

Qiu, C., Hanson, E., Olivier, E., Inada, M., Kaufman, D. S., Gupta, S., and Bouhassira, E. E. (2005). Differentiation of human embryonic stem cells into hematopoietic cells by coculture with human fetal liver cells recapitulates the globin switch that occurs early in development. *Exp. Hematol.* **33,** 1450–1458.

Robertson, S. M., Kennedy, M., Shannon, J. M., and Keller, G. (2000). A transitional stage in the commitment of mesoderm to hematopoiesis requiring the transcription factor SCL/tal-1. *Development* **127,** 2447–2459.

Roecklein, B. A., and Torok-Storb, B. (1995). Functionally distinct human marrow stromal cell lines immortalized by transduction with the human papilloma virus E6/E7 genes. *Blood* **85,** 997–1005.

Sasaki, K., Nagao, Y., Kitano, Y., Hasegawa, H., Shibata, H., Takatoku, M., Hayashi, S., Ozawa, K., and Hanazono, Y. (2005). Hematopoietic microchimerism in sheep after *in utero* transplantation of cultured cynomolgus embryonic stem cells. *Transplantation* **79,** 32–37.

Schiedlmeier, B., Klump, H., Will, E., Arman-Kalcek, G., Li, Z., Wang, Z., Rimek, A., Friel, J., Baum, C., and Ostertag, W. (2003). High-level ectopic HOXB4 expression confers a profound *in vivo* competitive growth advantage on human cord blood CD34+ cells, but impairs lymphomyeloid differentiation. *Blood* **101,** 1759–1768.

Schmitt, T. M., de Pooter, R. F., Gronski, M. A., Cho, S. K., Ohashi, P. S., and Zuniga-Pflucker, J. C. (2004). Induction of T cell development and establishment of T cell competence from embryonic stem cells differentiated *in vitro*. *Nature Immunol.* **5,** 410–417.

Senju, S., Hirata, S., Matsuyoshi, H., Masuda, M., Uemura, Y., Araki, K., Yamamura, K., and Nishimura, Y. (2003). Generation and genetic modification of dendritic cells derived from mouse embryonic stem cells. *Blood* **101,** 3501–3508.

Shalaby, F., Ho, J., Stanford, W. L., Fischer, K. D., Schuh, A. C., Schwartz, L., Bernstein, A., and Rossant, J. (1997). A requirement for Flk1 in primitive and definitive hematopoiesis and vasculogenesis. *Cell* **89,** 981–990.

Tian, X., and Kaufman, D. S. (2005). Hematopoietic development of human embryonic stem cells in culture. *Methods Mol. Med.* **105,** 425–436.

Tian, X., Morris, J. K., Linehan, J. L., and Kaufman, D. S. (2004). Cytokine requirements differ for stroma and embryoid body-mediated hematopoiesis from human embryonic stem cells. *Exp. Hematol.* **32,** 1000–1009.

Umeda, K., Heike, T., Yoshimoto, M., Shiota, M., Suemori, H., Luo, H. Y., Chui, D. H., Torii, R., Shibuya, M., Nakatsuji, N., and Nakahata, T. (2004). Development of primitive and definitive hematopoiesis from nonhuman primate embryonic stem cells *in vitro*. *Development* **131,** 1869–1879.

Vodyanik, M. A., Bork, J. A., Thomson, J. A., and Slukvin, J. A., II (2005). Human embryonic stem cell-derived CD34+ cells: Efficient production in the coculture with OP9 stromal cells and analysis of lymphohematopoietic potential. *Blood* **105,** 617–626.

Wang, J., Zhao, H. P., Lin, G., Xie, C. Q., Nie, D. S., Wang, Q. R., and Lu, G. X. (2005a). *In vitro* hematopoietic differentiation of human embryonic stem cells induced by co-culture with human bone marrow stromal cells and low dose cytokines. *Cell Biol. Int.* **29,** 654–661.

Wang, L., Li, L., Shojaei, F., Levac, K., Cerdan, C., Menendez, P., Martin, T., Rouleau, A., and Bhatia, M. (2004). Endothelial and hematopoietic cell fate of human embryonic stem cells originates from primitive endothelium with hemangioblastic properties. *Immunity* **21,** 31–41.

Wang, L., Menendez, P., Cerdan, C., and Bhatia, M. (2005b). Hematopoietic development from human embryonic stem cell lines. *Exp. Hematol.* **33,** 987–996.

Wang, L., Menendez, P., Shojaei, F., Li, L., Mazurier, F., Dick, J. E., Cerdan, C., Levac, K., and Bhatia, M. (2005c). Generation of hematopoietic repopulating cells from human embryonic stem cells independent of ectopic HOXB4 expression. *J. Exp. Med.* **201,** 1603–1614.

Wang, Y., Yates, F., Naveiras, O., Ernst, P., and Daley, G. Q. (2005d). Embryonic stem cell-derived hematopoietic stem cells. *Proc. Natl. Acad. Sci. USA* **102,** 19081–19086.

Wang, Z., Skokowa, J., Pramono, A., Ballmaier, M., and Welte, K. (2005e). Thrombopoietin regulates differentiation of rhesus monkey embryonic stem cells to hematopoietic cells. *Ann. N.Y. Acad. Sci.* **1044,** 29–40.

Weiss, M. J., Keller, G., and Orkin, S. H. (1994). Novel insights into erythroid development revealed through *in vitro* differentiation of GATA-1 embryonic stem cells. *Genes Dev.* **8,** 1184–1197.

Wiles, M. V., and Johansson, B. M. (1997). Analysis of factors controlling primary germ layer formation and early hematopoiesis using embryonic stem cell *in vitro* differentiation. *Leukemia* **11**(Suppl. 3), 454–456.

Wiles, M. V., and Keller, G. (1991). Multiple hematopoietic lineages develop from embryonic stem (ES) cells in culture. *Development* **111,** 259–267.

Woll, P. S., Martin, C. H., Miller, J. S., and Kaufman, D. S. (2005). Human embryonic stem cell-derived NK cells acquire functional receptors and cytolytic activity. *J. Immunol.* **175,** 5095–5103.

Wu, K.-D., Cho, Y. S., Katz, J., Ponomarev, V., Chen-Kiang, S., Danishefsky, S. J., and Moore, M. A. S. (2005). Investigation of anti-tumor effects of synthetic epothilone analogs in human myeloma models *in vitro* and *in vivo*. *Proc. Natl. Acad. Sci. USA* **102,** 10640–10645.

Yu, J., Vodyanik, M. A., He, P., Slukvin, P., II, and Thomson, J. A. (2006). Human embryonic stem cells reprogram myeloid precursors following cell-cell fusion. *Stem Cells* **24,** 168–176.

Zambidis, E. T., Peault, B., Park, T. S., Bunz, F., and Civin, C. I. (2005). Hematopoietic differentiation of human embryonic stem cells progresses through sequential hematoen-dothelial, primitive, and definitive stages resembling human yolk sac development. *Blood* **106,** 860–870.

Zhan, X., Dravid, G., Ye, Z., Hammond, H., Shamblott, M., Gearhart, J., and Cheng, L. (2004). Functional antigen-presenting leucocytes derived from human embryonic stem cells *in vitro*. *Lancet* **364,** 163–171.

Zou, G. M., Chan, R. J., Shelley, W. C., and Yoder, M. C. (2006). Reduction of Shp-2 expression by small interfering RNA reduces murine embryonic stem cell-derived *in vitro* hematopoietic differentiation. *Stem Cells* **24,** 587–594.

[14] Hematopoietic Cells from Primate Embryonic Stem Cells

By FEI LI, SHI-JIANG LU, and GEORGE R. HONIG

Abstract

Embryonic stem (ES) cells, derived from early stage embryos, are pluripotent precursors of all of the tissues and organs of the body. ES cells from the mouse have been shown to undergo differentiation *in vitro* to form a variety of different cell types, including the differentiated progeny of hematopoietic precursors. These hematopoietic cells, however, exhibit numerous differences from those of human cells, and it has become increasingly clear that mouse ES cell differentiation has significant limitations as a model of human developmental biology. The more recent isolation and characterization of nonhuman primate ES cell lines have made available an experimental model with characteristics considerably more close to human biology. We have developed experimental conditions that promote efficient differentiation of these cells to produce progeny cells with considerable similarity to hematopoietic precursors harvested from bone marrow of adult animals.

Introduction

With the widespread use of human embryonic stem (ES) cells in studies directed toward transplantation for clinical purposes, the strong need for a suitable primate model becomes ever more apparent. Stable ES cell lines have been developed for several nonhuman primate species, including rhesus monkey (Thomson *et al.,* 1995), common marmoset (Thomson *et al.,* 1996), and cynomolgus monkey (Suemori *et al.,* 2001). These cell lines have provided new opportunities for the study of the emergence of hematopoietic stem cells and their development, with implications closely related to human biology and medical science.

In Vitro *Hematopoiesis from Nonhuman Primate ES Cells*

The differentiation of nonhuman primate ES cells *in vitro* to form hematopoietic precursors has been investigated extensively (Hiroyama *et al.,* 2006; Honig *et al.,* 2004; Li *et al.,* 2001; Umeda *et al.,* 2004, 2006; Zhang *et al.,* 2006). These studies have established the ability of nonhuman primate ES cells to

METHODS IN ENZYMOLOGY, VOL. 418
0076-6879/06 $35.00
DOI: 10.1016/S0076-6879(06)18014-3

differentiate into hematopoietic progenitor cells with progeny characteristic of primitive and definitive erythroid, myeloid, and lymphoid lineages. Nevertheless, efforts to achieve multilineage reconstitution of NOD/SCID mouse bone marrow utilizing these hematopoietic precursors have not been successful; it must be recognized, however, that hematopoietic precursors harvested from rhesus monkey bone marrow also do not engraft in NOD/SCID animals (Li *et al.*, unpublished data). Intrauterine fetal liver injection of cynomolgus monkey ES cell-derived hematopoietic precursors in sheep and in fetal cynomolgus monkeys has also met with only limited success (Sasaki *et al.*, 2005; Shibata *et al.*, 2006), suggesting that the present culture conditions are not favorable for generating functional stem cells.

Embryonic Mesoderm Regulators and Hematopoietic Onset of Rhesus Monkey ES Cells

Bone morphogenetic proteins (BMP) have been shown to be critical to the effective directed hematopoietic differentiation of rhesus monkey ES cells (Li *et al.*, 2001). BMPs are known to play a pivotal role in the patterning of embryonic ventral mesoderm and in the specification of precursors for hematopoietic and endothelial cells (Huber *et al.*, 1998; Miyanaga *et al.*, 1999; Winnier *et al.*, 1995). BMP-4 was also found to be required for mesoderm and hemoglobin induction during the hematopoietic differentiation of mouse ES cells (Johansson *et al.*, 1995; Wiles and Johansson, 1999). It has been demonstrated that BMP-4 stimulates the early expression of a group of hematopoiesis-associated genes and exerts a potent and direct effect on the hematopoietic differentiation of rhesus monkey and human ES cells *in vitro* (Chadwick *et al.*, 2003; Li *et al.*, 2001; Lu *et al.*, 2002a,b, 2004). In subsequent studies we observed an additive effect on rhesus monkey ES cell differentiation from the combined activity of BMP-2, BMP-4, and BMP-7 (Honig *et al.*, 2004). The achievement of efficient hematopoietic differentiation of ES cells *in vitro* suggests BMPs as the key elements permissive to the recapitulation of the hematopoiesis during embryonic development.

Although the transplantation studies by other investigators involving BMP-4-induced cynomolgus monkey hematopoietic precursors resulted in only limited engraftment (Sasaki *et al.*, 2005; Shibata *et al.* 2006), the *in vitro* culture-based approach nevertheless offers great promise for identifying other as yet unknown factors and elements crucial to the goal of developing functional stem cells for bone marrow reconstitution. In light of the considerable biological similarity of nonhuman primate ES cells to human ES cells, the nonhuman primate could ultimately serve as an outstanding animal model for preclinical transplantation studies.

Protocol for Propagation of Rhesus Monkey ES Cell Lines

Overview

ES cell lines have been developed from several nonhuman primate species. Rhesus monkey (*Macaca mulatta*) ES cell lines, the first of these, were developed at the Wisconsin Regional Primate Center (Thomson *et al.*, 1995), later followed by the isolation of ES cell lines from common marmoset (*Callithrix jacchus* or new world monkey) (Thomson *et al.*, 1996) and cynomolgus monkey (*Macaca fascicularis*) (Suemori *et al.*, 2001). Rhesus monkey ES cell lines have been made available to the general research community through the Wisconsin Regional Primate Center and its affiliate, the WiCell Research Institute. Although it is not difficult to propagate rhesus monkey ES cell lines in an established cell culture laboratory, the expandability and growth rates may vary considerably among different rhesus monkey cell lines. Of eight rhesus ES cell lines available from the Wisconsin Regional Primate Center (R278.5, R366.4, R367, R394.3, R420, R456, R460, and R475), R366.4 and R420 have been found to grow well and yield reproducible hematopoietic differentiation results in our laboratory (Fig. 1). Both of these lines have been studied extensively for their potential for hematopoietic development.

Reagents for Propagation and Storage of Undifferentiated ES Cells

1. ES cell growth medium: Dulbecco's modified Eagle medium (DMEM with high glucose) (GIBCO, Grand Island, NY) supplemented with 15% fetal bovine serum (FBS) (Hyclone, Logan, UT), 1 mM glutamine, $1 \times 10^{-4}\,M$ β-mercaptoethanol, and 2% MEM amino acids solution (all from GIBCO)
2. 0.05% trypsin/EDTA solution (GIBCO)

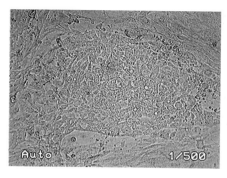

FIG. 1. Undifferentiated rhesus monkey ES cell colony (from cell line R366.4).

3. 0.1% gelatin solution: type A gelatin (3000 bloom) (Sigma, St. Louis, MO) prepared in distilled and deionized water

4. Cryopreservation medium: fetal bovine serum (Hyclone) supplemented with 10% dimethyl sulfoxide (DMSO) (Fisher Scientific, Fair Lawn, NJ)

5. Dulbecco's phosphate-buffered saline (D-PBS) (without calcium and magnesium chlorides) (GIBCO)

Preparation of Mouse Embryonic Fibroblasts (MEF)

1. Euthanize day 13 to 14 (postcoitus) pregnant CF-1 mice using Institutional Animal Care and Use Committee (IACUC)-approved procedures.

2. Dissect and remove embryos from the uterine horns of the pregnant mice under sterile conditions.

3. Separate and discard internal organs (i.e., liver and heart) and head as much as possible from the embryo proper. Rinse the blood off the remainder of the embryo tissue with D-PBS solution and immerse the washed embryo tissue in 0.05% trypsin/EDTA solution (GIBCO) and mince with iris scissors in 100-mm culture dishes.

4. Digest the minced tissues with 10 ml trypsin/EDTA solution (from 10 embryos) for 15 min at 37°. Thoroughly pipette the cells once every 5 min while incubating. Following digestion, add 20 ml ES growth medium to the cell suspension and continue to pipette 10 to 20 times.

5. Remove aggregates of the undigested tissues using a 70-μm cell strainer (Becton Dickinson Labware, Franklin Lakes, NJ). Wash the cell suspension in ES growth medium twice, followed by incubating in a T75 flask at 1 embryo/T75 in 20 ml ES growth medium.

6. Harvest the cells when the cultures reach confluence, using 0.05% trypsin/EDTA. Replate the cells at a 1:3 dilution ratio as second passage. When they reach confluence, harvest the MEF cells using trypsin solution and wash the detached cells once. Freeze the MEF cells at one-eighth of a confluence harvest of cells/vial in 0.5 ml of 10% DMSO fetal bovine serum at –70° overnight. Transfer the frozen vials the following day to a liquid nitrogen freezer. The stored MEF cells can be kept in liquid nitrogen for more than 9 years without losing any proliferation potential. The thawed cells can be propagated for up to three more passages without losing their activity in supporting ES cell growth.

Propagation of Undifferentiated Rhesus Monkey ES Cells

1. Preparation of MEF layers. Thaw one vial of frozen MEF cells in a 37° water bath and wash the cells once in ES cell growth medium by

centrifugation at $250g$ for 6 min. Initiate the culture by seeding one T75 culture flask with 15 ml ES cell growth medium. Grow the cells at 37° in 5% CO_2 for 72 to 96 h. Coat one T25 tissue culture flask by immersing the culture surface with 0.1% gelatin solution for 16 to 24 h prior to seeding. Two hours before seeding the irradiated MEF cells, remove the gelatin solution and leave the treated flask in a cell culture hood to dry. At day 3 or day 4 of MEF culture, harvest the MEF cells (at 90% confluence) using 0.05% trypsin/EDTA solution (5 min at 37°). Wash and resuspend the detached cells in 10 ml of ES growth medium. Irradiate the MEF cells with a γ ray source at 3000 to 5000 cGy (or, alternatively, treat adherent cells for 5 to 6 h with 10 μg/ml mitomycin C in the culture flask before harvesting cells for washing and plating). Wash and count the irradiated or mitomycin C-treated MEF cells. Seed the cells in one gelatin-coated T25 flask at 0.0375×10^6 cells/cm^2 growth area and incubate at 37° overnight to allow attachment. The treated cells can also be frozen in liquid nitrogen in FBS containing 10% DMSO. Mitotically blocked frozen MEF cells can be thawed at a later time to use as feeders to allow more flexibility for the research schedule.

2. Seeding of rhesus monkey ES cells. Thaw one vial of frozen rhesus monkey ES cells (one-fifth of a T25 culture from a previous passage) in a 37° water bath and wash the cells once with ES cell growth medium. Seed the ES cells in one T25 flask containing a preformed mitotically blocked MEF layer in 7 ml of ES cell growth medium and incubate at 37° in 5% CO_2. At day 6 and day 7, the ES cells will be ready for differentiation studies. Unused ES cells are cryopreserved at one-fifth of the culture harvest/vial in fetal bovine serum containing 10% DMSO.

Induction of Hematopoietic Differentiation of Rhesus Monkey ES Cells

Overview

Two major culture systems have been developed to induce hematopoietic differentiation of ES cells. One involves coculture with bone marrow stromal feeder. The other involves the generation of ES cell aggregates without the presence of feeder cells to allow embryoid body (EB) formation. In work with rhesus and human ES cells, we have found that the stromal coculture system offers greater efficiency in terms of the yield of either CD34+ precursors or hematopoietic colonies in the primary differentiation culture. The stromal coculture allows ES cells to adhere and develop differentiated colonies with visible hematopoietic clusters. Because the adhered ES cells are not in aggregated form, it may allow consistent exposure of the testing agents in the culture. Although the majority of published work on hematopoietic differentiation of nonhuman primate ES cells adopted the stromal coculture method, Zhang *et al.* (2006) described a novel approach to achieve robust hematopoietic differentiation by growing an adherent monolayer of cynomolgus monkey ES cells without stromal feeders or EB formation.

Materials and Reagents for Rhesus ES Cell Differentiation Culture

1. Hematopoietic differentiation medium: Iscove's modified Dulbecco's medium (IMDM) (GIBCO) supplemented with 7.5% FBS (Hyclone), 7.5% horse serum, 1 mM glutamine, 5×10^{-5} M β-mercaptoethanol (all from GIBCO), and 5×10^{-6} M hydrocortisone (Sigma)
2. S-17 mouse bone marrow stromal cell growth medium: IMDM (GIBCO) supplemented with 15% FBS (Hyclone), 1 mM glutamine, and 5×10^{-5} M β-mercaptoethanol (all from GIBCO)
3. The S-17 mouse bone marrow stromal cell line (Collins *et al.*, 1987)
4. Mesoderm regulatory factors: bone morphogenetic proteins (BMP-2, 4, 7) (R&D Systems, Minneapolis, MN)
5. Hematopoietic growth factors (R&D Systems): recombinant human SCF, interleukin (IL)-3, granulocyte-colony stimulating factor (G-CSF), vascular endothelial growth factor (VEGF), IL-6, flt3 ligand, and erythropoietin

Methods

1. Preparation of S17 stromal cell layers. Thaw one vial of frozen S17 cells, containing one-fifth of a confluent culture from one T75 flask, and wash once in S17 growth medium. Seed the cells into one T75 flask with 15 ml of S17 growth medium. When they reach confluence, trypsinize and wash the cells once in S17 stromal growth medium. Replate the cells into six-well tissue culture plates at 0.025×10^6 cells/well in 2 ml of S17 growth medium. Allow the cell layers to grow for 72 h to reach partial confluence; they are then ready to be used for rhesus monkey ES cell differentiation cultures. Alternatively, irradiate S17 cells at 3000 rad and plate at 80 to 90% confluence.

2. Differentiation cultures of rhesus ES cells. Trypsinize and wash rhesus monkey ES cells from day 6 to 7 expansion cultures in ES cell growth medium. Determine the ES cell concentration by cytometer counting. Seed a total of 5000 ES cells in 4 ml ES cell differentiation medium into one well of a six-well culture plate with preformed subconfluent S17 stromal cells. Add BMPs at various testing doses at the initiation of the differentiation (day 0). Add a combination of recombinant human hematopoietic cytokines at day 6 of the differentiation cultures. The doses for each of the growth factors are 20 ng/ml of SCF, IL-3, G-CSF, and VEGF; 10 ng/ml of IL-6 and flt3 ligand; and 2 units/ml of erythropoietin. No fresh BMPs are added at day 6.

3. Characterization of differentiated ES colonies and primary hemato-poietic clusters. At days 13 to 14 of differentiation culture, clusters of small and round hematopoietic cells (=20 cells/cluster) inside the large crater-like differentiated ES cell colonies emerge (Fig. 2). Some of the clusters are encircled by endothelial-like cell structures, which are reminiscent of embryonic blood islands. The clusters initially adhere to the S17 cells presenting typical cobblestone area morphology. Some clusters will detach and migrate when the culture continues. *In situ* immunofluorescence staining demonstrates that some of the blast cells within the clusters are CD34$^+$. Subcultures of these CD34$^+$ cells with S17 stromal cells give rise to secondary and tertiary cobblestone area forming colonies. Progeny cells from the secondary cultures exhibit myeloid and erythroid characteristics when analyzed by gene expression and cytospin examination (Li *et al.*, 2001). The hematopoietic blast cells can be rinsed gently from the cultures. Remove large aggregates of differentiated ES cell colonies from the cell suspension after passing them through a 40-μm cell strainer. The collected cells can then be used for further biological and molecular analyses.

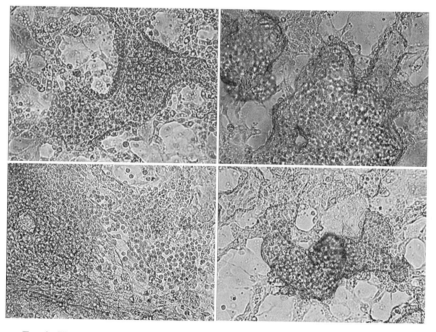

Fig. 2. Hematopoietic clusters from differentiated rhesus monkey ES cell colonies.

Technical Notes

1. The growth of S17 mouse bone marrow stromal cells usually stops once they reach confluence, judging by stable medium pH (pink or red) in the culture. It has been our observation that, in a few cases, the medium turned acidic (orange or yellow) after the cells reach confluence in the preexpansion culture. In such circumstances the effect of supporting hematopoietic differentiation of rhesus ES cells by these cells will be compromised greatly. Using irradiated S17 cells may offer more consistency.

2. Undifferentiated rhesus monkey ES cells seen under the cytometer have large blast cell-like morphology (about three to four times the size of a lymphocyte). They show homogeneous low light reflection, smooth or bleb-like edges, and most of the ES cells have discernible nuclei. Using these criteria in counting ES cells provides a good approximation of the starting number of ES cells and makes the comparison of differentiated hematopoietic clusters more accurate

3. After day 6 of cytokine feeding during the differentiation culture, there is no need to replenish cells with fresh medium as well as BMPs and cytokines. Because some differentiated hematopoietic clusters may migrate or detach from differentiated ES colonies, greater care should be exercised in handling the culture plates to prevent disaggregation of the clusters. Cluster counting should be conducted in a consistent and timely manner to ensure the accuracy of data.

References

Chadwick, K., Wang, L., Li, L., Menendez, P., Murdoch, B., Rouleau, A., and Bhatia, M. (2003). Cytokines and BMP-4 promote hematopoietic differentiation of human embryonic stem cells. *Blood* **102,** 906–915.

Collins, L. S., and Dorshkind, K. (1987). A stromal cell line from myeloid long-term bone marrow cultures can support myelopoiesis and B lymphopoiesis. *J. Immunol.* **138,** 1082–1087.

Hiroyama, T., Miharada, K., Aoki, N., Fujioka, T., Sudo, K., Danjo, I., Nagasawa, T., and Nakamura, Y. (2006). Long-lasting *in vitro* hematopoiesis derived from primate embryonic stem cells. *Exp. Hematol.* **34,** 760–769.

Honig, G. R., Li, F., Lu, S. J., and Vida, L. (2004). Hematopoietic differentiation of rhesus monkey embryonic stem cells. *Blood Cell Mol. Dis.* **32,** 5–10.

Huber, T. L., Zhou, Y., Mead, P. E., and Zon, L. I. (1998). Cooperative effects of growth factors involved in the induction of hematopoietic mesoderm. *Blood* **92,** 4128–4137.

Johansson, B. M., and Wiles, M. W. (1995). Evidence for involvement of activin A and bone morphogenetic protein 4 in mammalian mesoderm and hematopoietic development. *Mol. Cell Biol.* **15,** 141–151.

Li, F., Lu, S., Vida, L., Thomson, J. A., and Honig, G. R. (2001). Bone morphogenetic protein 4 induces efficient hematopoietic differentiation of rhesus monkey embryonic stem cells *in vitro. Blood* **98,** 335–342.

Lu, S. J., Li, F., Vida, L., and Honig, G. R. (2002a). Comparative gene expression in hematopoietic progenitor cells derived from embryonic stem cells. *Exp. Hematol.* **30,** 58–66.

Lu, S. J., Li, F., Vida, L., and Honig, G. R. (2004). CD34$^+$CD38$^-$ hematopoietic precursors derived from human embryonic stem cells exhibit an embryonic gene expression pattern. *Blood* **103,** 4134–4141.

Lu, S. J., Quan, C., Li, F., Vida, L., and Honig, G. R. (2002b). Hematopoietic progenitor cells derived from embryonic stem cells: Analysis of gene expression. *Stem Cells* **94,** 428–437.

Miyanaga, Y., Shiurba, R., and Asashima, M. (1999). Blood cell induction in xenopus animal cap explants: Effects of fibroblast growth factor, bone morphogenetic proteins, and activin. *Dev. Genes Evol.* **209,** 69–76.

Sasaki, K., Nagao, Y., Kitano, Y., Hasegawa, H., Shibata, H., Takatoku, M., Hayashi, S., Ozawa, K., and Hanazono, Y. (2005). Hematopoietic microchimerism in sheep after *in utero* transplantation of cultured cynomolgus embryonic stem cells. *Transplantation* **79,** 32–37.

Shibata, H., Ageyama, N., Tanaka, Y., Kishi, Y., Kyoko Sasaki, K., Nakamura, S., Muramatsu, S., Hayashi, S., Kitano, Y., Terao, K., and Hanazonoa, Y. (2006). Improved safety of hematopoietic transplantation with monkey ES cells in the allogeneic setting. *Stem Cells* **24,** 1450–1457.

Suemori, H., Tada, T., Torii, R., Hosoi, Y., Kobayashi, K., Imahie, H., Kondo, Y., Iritani, A., and Nakatsuji, N. (2001). Establishment of embryonic stem cell lines from cynomolgus monkey blastocysts produced by IVF or ICSI. *Dev. Dyn.* **222,** 273–279.

Thomson, J. A., Kalishman, J., Golos, T. G., Durning, M., Harris, C. P., Becker, R. A., and Hearn, J. P. (1995). Isolation of a primate embryonic stem cell line. *Proc. Natl. Acad. Sci. USA* **92,** 7844–7848.

Thomson, J. A., Kalishman, J., Golos, T. G., Durning, M., Harris, C. P., and Hearn, J. P. (1996). Pluripotent cell lines derived from common maroset (callithrix jacchus) blastocysts. *Biol. Reprod.* **55,** 254–259.

Umeda, K., Heike, T., Yoshimoto, M., Shinoda, G., Shiota, M., Suemori, H., Luo, H. Y., Chui, D. H. K., Torii, R., Shibuya, M., Nakatsuji, N., and Nakahata, T. (2006). Identification and characterization of hemoangiogenic progenitors during cynomolgus monkey embryonic stem cell differentiation. *Stem Cells* **24,** 1348–1358.

Umeda, K., Heike, T., Yoshimoto, M., Shiota, M., Suemori, H., Luo, H. Y., Chui, D. H. K., Torii, R., Shibuya, M., Nakatsuji, N., and Nakahata, T. (2004). Development of primitive and definitive hematopoiesis from nonhuman primate embryonic stem cells *in vitro*. *Development* **131,** 1869–1879.

Wiles, M. V., and Johansson, B. M. (1999). Embryonic stem cell development in a chemically defined medium. *Exp. Cell Res.* **247,** 241–248.

Winnier, G., Blessing, M., Labosky, P. A., and Hogan, B. L. (1995). Bone morphogenetic protein-4 is required for mesoderm formation and patterning in the mouse. *Genes Dev.* **9,** 2105–2116.

Zhang, H., Saeki, K., Kimura, A., Saeki, K., Nakahara, M., Doshi, M., Kondo, Y., Nakano, T., and You, A. (2006). Efficient and repetitive production of hematopoietic and endothelial cells from feeder-free monolayer culture system of primate embryonic stem cells. *Biol. Reprod.* **74,** 295–306.

[15] Vascular Cells

By ILANA GOLDBERG-COHEN, GILAD BECK,
ANNA ZISKIND, and JOSEPH ITSKOVITZ-ELDOR

Abstract

Embryonic stem (ES) cells are cells derived from the inner cell mass of a blastocyst stage embryo. These self-renewing multipotent cells are able to differentiate to the three embryonic germ layers, the endoderm, ectoderm, and mesoderm, and are thus able to produce virtually all cell types. The ES cell capacity to generate various cell types has been studied extensively, and exploitation of ES cell characteristics allowed the production of several differentiated cell types of multiple tissues. Moreover, the process of ES cell differentiation provides a unique opportunity to observe early embryonic developmental events that are unattainable in the embryo itself. This chapter addresses the *in vitro* differentiation procedure of endothelial and vascular smooth muscle cells from human ES cells, with reference to similar studies performed in mouse and nonhuman primate ES cells, and provides several tools for the detailed characterization of differentiated cells.

Derivation and Characterization of Embryonic Stem (ES) Cells

Embryonic stem cell research began with the initial derivation of ES cells from the mouse embryo in the year 1981 (Evans and Kaufman, 1981; Martin, 1981), and with the well-established development and propagation of mouse ES cells, attainment of ES cells from rhesus monkeys (Thomson *et al.*, 1995) and eventually from human sources (Thomson *et al.*, 1998) was soon to follow.

Embryonic stem cells are cells derived from the inner cell mass (ICM) of an embryo in the blastocyst stage and comprise a population of cells that hold the potential to both self-renew and differentiate to multiple cell types.

Embryonic stem cells are unique due to their ability to proliferate in culture for prolonged periods of time while maintaining a uniform undifferentiated phenotype and a normal karyotype (Thomson *et al.*, 1998). Unlike mouse ES cells that are cultured in the absence of any feeder layer (Smith *et al.*, 1988), human ES cells, following separation, require plating on a feeder layer that is composed of mouse embryonic fibroblasts (MEFs) for their continuous culture (Amit *et al.*, 2000). However, great importance is attributed to identifying suitable conditions that will allow human ES cells to

METHODS IN ENZYMOLOGY, VOL. 418
0076-6879/06 $35.00
DOI: 10.1016/S0076-6879(06)18015-5

be cultured either on a fibroblast layer of a human origin (Amit *et al.*, 2003) or independently of any feeder layer whatsoever (Ludwig *et al.*, 2006) so that implementation of ES cells in regenerative medicine is facilitated.

Differentiation of Embryonic Stem Cells

A fundamental feature of ES cells that serves to motivate the extensive research into their nature is their ability to undergo differentiation and generate cells of various lineages. The developmental potential that ES cells hold is easily demonstrated when the cells are injected intramuscularly or subcutaneously into immunodeficient mice. Following injection, a benign tumor termed teratoma is formed that contains elaborate structures composed of differentiated cells that are derivatives of the three embryonic germ layers: the endoderm, ectoderm, and mesoderm (Thomson *et al.*, 1998). Several techniques have been developed to promote differentiation of ES cells *in vitro*. First, ES cells can be induced to differentiate when cultured in suspension (Itskovitz-Eldor *et al.*, 2000; Keller, 1995). Under these culture conditions aggregates of ES cells form spherical moieties termed embryoid bodies (EBs) that contain cells typical of the three embryonic germ layers. Differentiation of ES cells in this three-dimensional organization is advantageous due to the cell–cell interactions that are facilitated within these structures and may promote specific developmental events. However, the three-dimensional structure may hinder the ability to determine any signal transduction pathways that influence the developmental programs that take place. A second strategy used to induce ES cell differentiation *in vitro* is coculture of the undifferentiated ES cells with a differentiation-inducing stromal feeder layer such as the commonly utilized stromal feeder layer OP9 (Nakano *et al.*, 1994). The benefit/weakness in the employment of stromal feeder layers is the various factors secreted by the cocultured stromal cells. These factors can effectively promote the differentiation of the desired cell lineage but can also yield a differentiating population of undesired cells. Furthermore, suitable protocols must be available for successful separation of the differentiated cell population from the cocultured stromal cells. The third approach for *in vitro*-induced differentiation of ES cells is culture of the undifferentiated cells on various extracellular matrix proteins so that no interference of foreign cells, supportive or other, can influence the course of differentiation (Gerecht-Nir *et al.*, 2003; Nishikawa *et al.*, 1998). This method, however, requires careful selection of the chosen extracellular matrix substrate, as the nature of this specific protein will determine the course of differentiation. This chapter demonstrates the development of endothelial and smooth muscle cells from ES cells cultured on type IV collagen, which serves as the preferred extracellular matrix protein for the development of vascular cells.

Development of the Blood Vascular System

During the initial stages of formation, the metabolic necessities of the developing embryo, including delivery of oxygen and nutrients and disposal of metabolic waste products, are answered by diffusion. This, however, occurs as long as the embryo is small in dimension; as soon as it increases in mass the diffused supply can no longer reach the inner core of the embryo. Thus, an efficient transport mechanism must form to facilitate the imminent developmental events and appears in the form of the blood vascular system.

Three different pathways enable the formation of blood vessels during embryonic development: vasculogenesis, angiogenesis, and arteriogenesis.

Vasculogenesis is the process that encompasses all the developmental events that result in the *de novo* formation of a primary vascular network (Risau, 1997). It initiates in the extraembryonic yolk sac where a common precursor to both hematopoietic and vascular systems, termed the hemangioblast, forms aggregates known as blood islands. Cells in the heart of these blood islands will differentiate into cells of the hematopoietic system, whereas cells in the periphery will migrate to distinct sites, differentiate to endothelial cells, and eventually assemble a capillary plexus (Risau, 1997). The subsequent development of mature blood vessels occurs through sprouting of the preexisting primitive vascular structures in the process of angiogenesis. Endothelial cells of the newly formed primary vessels differentiate further and migrate to assemble into mature vasculature (Risau, 1997). The angiogenic events can be crudely separated into three stages. First, the vascular endothelial growth factor (VEGF) induces vasodilatation, which is accompanied by increased vessel permeability (Carmeliet, 2000). This is followed by loosening of cell–cell contacts between adjacent endothelial cells and their supporting cells and matrix and migration of differentiating endothelium (Coussens *et al.*, 1999) and finally the migrating endothelial cells assemble to form mature vessels and become quiescent. Unlike vasculogenesis, which is mostly limited to embryonic developmental events, angiogenesis is detected quite frequently in the adult life as well. Physiological angiogenesis can be encountered during various events, such as the female reproductive cycle and wound healing, and a number of pathologies, including tumorigenesis, also demonstrate angiogenic occurrences (Folkman and Shing, 1992).

The third and least understood mechanism of blood vessel formation is arteriogenesis, where either maturation of preexisting collaterals or *de novo* formation of mature blood vessels results in the appearance of new arteries with fully developed tunica media (Buschmann and Schaper, 2000).

Vascular Developmental Potential of Embryonic Stem Cells

One significant feature of ES cell study is the ability to follow early developmental events, including the earliest occurrences of lineage commitment and specification. These initial developmental stages are difficult, if not altogether impossible, to attain when studying the intact human embryo. Advantages of the ES cell differentiation model in the study of early development were well demonstrated with investigation of the hematopoietic lineage differentiation of the mouse embryo. In this study, mouse ES cell-derived EBs were shown to generate cells of the hematopoietic lineage with developmental processes that mimic the hematopoietic lineage development of the mouse embryo itself (Keller et al., 1993; Palis et al., 1999). The contribution of human ES cells to early blood vessel appearance was determined in a study carried out in our laboratory with the evaluation of vasculature formation in 4- to 8-week old human embryos as compared to human ES cell-derived teratomas (Gerecht-Nir et al., 2004). To this end, detection of well-recognized endothelial and vascular smooth muscle cell (vSMC) markers was available by means of reverse transcription-polymerase chain reaction (RT-PCR) assays and utilization of immunohistochemical staining. In both the embryos and the teratomas investigated, CD34, CD31, von Willebrand factor (vWF), and VEGF receptor 2 (also called Flk-1) were used to label endothelial cells in the developing vasculature, and smooth muscle actin (SMA) was utilized for the detection of vascular smooth muscle cells present. This study also ascertained that roughly 7% of the small blood vessels detected in human ES cell-derived teratomas were of human origin by means of HLA staining. The potential of human ES cells to produce cells of the vascular lineage was investigated further with utilization of the microarray analysis. A broad screening of genes expressed in ES cell-derived EBs was facilitated with particular attention attributed to prominent genes of the vascular lineage (Gerecht-Nir et al., 2005), as is described later.

The ability of human ES cells differentiating spontaneously in the form of EBs to generate endothelial-like cells organized in vessel-like structures was demonstrated previously (Levenberg et al., 2002). This study ascertained by RT-PCR analysis the increased expression of various recognized endothelial cell markers, including CD31, CD34, and VE cadherin, during the course of EB differentiation. Furthermore, all EBs examined demonstrated the presence of endothelial-like cells by CD31 labeling. These endothelial-like cells first appear in clusters and shape into vessel-like structures with increased capillary dimensions as EB differentiation proceeds.

The microarray analysis performed in our laboratory enabled one to compare the pattern of gene expression of developing 1- to 4-week-old EBs as well as undifferentiated human ES cells (Gerecht-Nir *et al.*, 2005). Genes detected in this broad screening were divided into two main clusters, one of which contained genes that were significantly downregulated with continuous development, whereas the second cluster was composed of genes significantly upregulated during the course of differentiation. Detailed examination of the genes upregulated during EB development revealed a significant increase in expression of genes that participate in the development of the vascular lineage. These include the vascular adhesion molecules PECAM1 (CD31) and VCAM1, the vascular specific receptors VEGF and ANG1, and several transcription factors linked to developmental events of blood cells and endothelium, including TAL1 and LMO1 (Table I). Confirmation of microarray data was available with quantitative PCR analysis of the developing EBs that similarly identified the reported increase in the unique pattern of gene expression associated with ongoing differentiation. Thus, it is now evident that human ES cells hold a potential to both develop into vascular cells by initiating the appropriate developmental signal transduction cascades as well as aggregate and sprout to form vessel-like structures. These intrinsic features are now implemented in the induction of ES cell differentiation to create cells with a vascular fate *in vitro*.

Vascular Differentiation of ES Cells *In Vitro*

Mesodermal Specification in Two-Dimensional Culture on Type IV Collagen-Coated Dishes

Studies of vascular lineage formation via pathways of vasculogenesis or angiogenesis facilitated the identification of various molecules of significance in vasculature initial formation and development. These include VEGF (Ferrara and Henzel, 1989; Leung *et al.*, 1989) and VEGF receptors 1 (Matthews *et al.*, 1991) and 2 (Shibuya *et al.*, 1990) (flt-1 and flk-1, respectively), VE cadherin (Breier *et al.*, 1996; Lampugnani *et al.*, 1992), and others. However, more profound understanding of the function of key regulators in vascular-forming events and of the mechanisms that govern blood vessel development is still lacking. Two *in vitro* models are frequently employed in the study of blood vessel differentiation, namely, spontaneously differentiating EBs and embryonic mesodermal cell culture (Palis *et al.*, 1995; Risau *et al.*, 1988; Wang *et al.*, 1992). In these models, however, the desired vascular development is accompanied by simultaneous differentiation of additional cell populations that impede the ability to distinguish specifically the developmental occurrences that the vascular

TABLE I
GENES OF THE VASCULAR LINEAGE UPREGULATED WITH EB DEVELOPMENT

	Symbol	Title
v-SMC		
1	MYH11	Myosin, heavy polypeptide 11, smooth muscle
2	LMOD1	Leiomodin 1
3	PDGFB	Platelet-derived growth factor
4	PDGFRB	PDGF receptor,
5	TGFB3	Transforming growth factor
6	TGFBR2	TGF receptor II
7	TGFBR3	TGF receptor III
ECs		
1	PECAM1	CD31 antigen
2	VCAM1	Vascular cell adhesion molecule 1
3	PCDH12	Protocadherin 12
4	CDH5	VE-cadherin
5	VEGF	Vascular endothelial growth factor
6	VEGFC	Vascular endothelial growth factor C
7	FIGF	VEGF D
8	EPAS1	Endothelial PAS domain protein 1
9	FLT1	Vascular endothelial growth factor
10	FLT4	fms-related tyrosine kinase 4
11	ANGPT1	Angiopoietin 1
12	ANGPT2	Angiopoietin 2
13	GATA2	GATA-binding protein 2
14	GATA3	GATA-binding protein 3
Hematopoietic/ECs		
1	CD34	CD34 antigen
2	TAL1	T-cell acute lymphocytic leukemia 1
3	BMI1	B lymphoma insertion region
4	LMO2	LIM domain only 2
5	TIE	Tyrosine kinase
6	GATA1	GATA-binding protein 1
7	RUNX1	Runt-related transcription factor 1
8	216966 at	CD41(Gpllb)
9	PTPRC	CD45

cells undergo. Successful observation and study of such molecular mechanisms that endothelial cells experience are available in a system that allows the culture of isolated endothelial cells *in vitro*. As mentioned previously, a well-recognized method for induction of ES cell differentiation is culture of the undifferentiated ES cells on matrices that originate from proteins of the extracellular matrix (Gerecht-Nir *et al.*, 2003; Nishikawa *et al.*, 1998). Prudent selection of the differentiation-inducing matrix will determine the course of ES cell differentiation as the matrix properties direct, and thus may generate the desired cell populations. Nishikawa *et al.* (1998) studied the potential of murine ES cells cultured on various matrices to differentiate to mesodermal cells without utilization of spontaneous differentiation in the form of EBs. The matrices investigated for their ability to induce mesodermal differentiation were gelatin, fibronectin, type I collagen, and type IV collagen. Undifferentiated ES cells were placed on culture dishes covered with the various matrices with no supplementation of exogenous factors and were analyzed by flow cytometry for mesodermal characteristic features. These mesodermal features include the absence of E cadherin expression, a marker demonstrated previously to be significantly down-regulated during mesodermal differentiation (Burdsal *et al.*, 1993), and the appearance of the mesodermal marker Flk1 (Mercola *et al.*, 1990); that is, ES cells were cultured on the matrix-coated dishes and assayed for the differentiation of E cadherin-negative Flk1-positive cells. The assay determined that culture on type IV collagen proved most efficient for attainment of mesodermal-like cells, although some degree of mesodermal differentiation was demonstrated when cells were cultured in the presence of other extracellular matrix proteins as well (Nishikawa *et al.*, 1998).

Embryonic Stem Cell-Derived Endothelial Cells Cultured on Type IV Collagen

The successful cultivation of mesodermal-like cells from ES cells cultured on type IV collagen-coated plates promoted extensive research to determine the culture conditions that would facilitate the attainment of endothelial and mural cells from ES cells *in vitro*. As mentioned previously, ES cells cultured on type IV collagen acquire expression of Flk1 (Nishikawa *et al.*, 1998), a VEGF receptor that acts as an early marker for the lateral plate mesoderm (Palis *et al.*, 1995). Furthermore, Flk1 has been distinguished as an early marker for differentiating blood and vascular cells (Kabrun *et al.*, 1997; Shalaby *et al.*, 1995). The ability of Flk1-positive cells, isolated by flow cytometry from ES cells cultured on type IV collagen to develop and form endothelial and vascular smooth muscle cells *in vitro*, was demonstrated by Yamashita *et al.* (2000). In this study, Yamashita and colleagues (2000) were able to selectively direct the differentiation of mouse ES cells to either endothelial or vascular smooth muscle cells by reculture of the sorted cells on type IV collagen-coated six-well dishes upon addition of exogenous growth factors. Results indicate that, in the absence of growth factors, cultured Flk1-positive cells, which lack expression of endothelial or smooth muscle cell-specific markers, favor differentiation to smooth muscle actin (SMA)-expressing mural cells. However, the selective addition of

50 ng/ml of VEGF, a growth factor with powerful mitogenic effects on both differentiating and mature endothelial cells (Leung *et al.*, 1989), to serum-free recultured Flk1-positive cells transforms the differentiation altogether to the endothelial cell lineage. Characterization of these endothelial-like cells revealed that the cells express the prominent endothelial markers VE cadherin, CD34, and endoglin. The cells also appear to uptake acetylated LDL readily, which is another typical feature of endothelial cells. Similarly, the differentiation potential of serum-free recultured Flk1-positive cells was evaluated upon the addition of platelet-derived growth factor-BB (PDGF-BB). The developmental significance of PDGF-BB to vascular formation was reported previously when embryos lacking the PDGF-B gene, as well as the receptor for PDGF-B, demonstrated impaired recruitment of pericytes and vascular smooth muscle cells to sites of neovascularization (Hellstrom *et al.*, 1999). The absent secretion of PDGF-B by the activated endothelial cells reduced both proliferation of the vascular smooth muscle cells and their recruitment to the sprouting vascular network. The exogenous supplementation of PDGF-B to recultured Flk1-positive cells enabled the differentiation of vascular smooth muscle-like cells, as was evident by their immunostaining with the recognized smooth muscle cell marker SMA. The SMA-positive cells also took a spindle-like form similar to that of native vascular smooth muscle cells. While the endothelial-like cells generated with the addition of VEGF exogenously retained expression of flk1, the PDGF-B-induced vascular smooth muscle cells lost flk1 expression with the course of differentiation and thus flk1 continuous expression is greatly dependent on the constant presence of VEGF. The dual characteristic of the sorted flk1-positive cells that allows them to differentiate to endothelial as well as smooth muscle cells was further ascertained in both *in vitro* and *in vivo* models. *In vitro*, the sorted flk1-positive cells placed in type I collagen gel in the presence of VEGF aggregate and migrate to form tube-like structures composed of cells immunostained with the endothelial cell-specific marker PECAM1 as well as with the smooth muscle cell marker SMA. *In vivo* experiments that consisted of injection of flk1-positive sorted cells into chick embryos confirmed the ability of these cells to contribute to the generation of both endothelial and vascular smooth muscle cells that form the developing embryonic vasculature.

Vascular Cell Generation from Human Embryonic Stem Cells

The gene array analysis of human ES cells discussed earlier clearly demonstrated that it is well within the power of human ES cells to differentiate to vascular cells, thus contributing to the development of the blood vascular lineage both *in vitro* and *in vivo*. These observations are further substantiated in sectioned ES cell-derived EBs. The EBs, sectioned serially, evidently exemplify that following 15 days of spontaneous differentiation endothelial cells, tracked by immunostaining with the endothelial cell marker CD34, display vessel-like formation within various regions of the EB sectioned (Gerecht-Nir *et al.*, 2003). Furthermore, the endothelial net assumes a three-dimensional organization, apparently imitating the early developmental appearance of the vascular network. Elongated smooth muscle cells, labeled with SMA for identification, were also detected surrounding voids within the sectioned EBs. Although endothelial and smooth muscle cells do spontaneously differentiate form human ES cells as a unique cell population within the EBs, the variability of the EBs and the complex purification of vascular cells from the mixed population of differentiating cells of the EB require a more directed approach for the generation of endothelial and of smooth muscle cells from human ES cells.

Therefore, a method for the directed differentiation of these vascular cells from undifferentiated ES cells has been devised in our laboratory that consists of dissociation of the undifferentiated ES cells and subsequent reculture on type IV collagen-coated six-well dishes (Gerecht-Nir *et al.*, 2003).

Dissociation of Human ES Cells

A standard protocol for the removal of human ES cells from their supporting feeder layer either for continuous culture or for the purpose of differentiation is the addition of 0.12 to 0.2% of type IV collagenase (Worthington) to the ES cells and subsequent culture for 30 min to 3 h so that it is evident that the ES cell colonies are dissociated from their feeder layer. However, this removal of intact or even fragmented ES cell colonies and not dissociation to single cells, which is advantageous for continuous passaging of undifferentiated ES cells, does not allow the ES cells to initiate differentiation to the mesodermal lineage when cells are moved for continuous culture on type IV collagen-coated plates. To avoid ES cell reculture as colonies, cells are dissociated to single cell suspension not by collagenase, but rather with EDTA splitting medium (0.5 mM EDTA [Promega], 0.1 mM β-mercaptoethanol [Sigma], 1% defined fetal bovine serum [FBS] [HyClone] in phosphate-buffered saline [PBS] not containing Ca and Mg [GIBCO-BRL]) added for a period of 30 min to 2 h until dissociation of the ES cells is clearly visible. Following separation of ES cells to a single cell suspension, cells are collected with cold Ca^{+2}, Mg^{+2} free PBS (GIBCO-BRL), pelleted (1200 rpm, 5 min, 4°), and resuspended in αMEM medium ([GIBCO-BRL] supplemented with 10% defined FBS [HyClone], 0.2% β-mercaptoethanol [GIBCO-BRL], and 0.2% ribonucleosides and deoxyribonucleosides [Biological Industries]) for pending reculture. Resuspended cells are then cultured on type IV collagen-coated six-well plates (Becton Dickinson). Coating the six-well plates with type IV collagen is facilitated by the placement of 1 ml of 5 to 10 μg/ml type IV collagen (Cultrex) diluted in cold, twice-distilled water to the culture dishes 1 to 2 h prior to reculture. Alternatively, culture dishes can be coated and stored at 4° up to 24 h before plating. Collagen must be removed before cells are placed for reculture.

Cell Seeding Concentration

Upon seeding the resuspended cells at a cell concentration of 1×10^4 cells/cm^2, as described by the Nishikawa group in their study with mouse ES cells (Yamashita *et al.*, 2000), it was evident that significant cell death took place.

Reculture at a 10-fold seeding concentration of 1 to 1.5×10^5 cells/cm^2 resulted in the increased appearance of undifferentiated ES cell colonies encircled by differentiated cells with a mesodermal phenotype.

However, when resuspended cells are seeded at a concentration of 5 to 7×10^4 cells/cm^2 and allowed to differentiate on the collagen-coated six-well dishes for a period of 6 days, cells resume a somewhat uniform appearance and specifically differentiate to two discrete, phenotypically identifiable cell populations. These two distinct cell populations are composed of a population of smaller cells with evident bulky nuclei and of a population of larger flattened cells with a fiber-like arrangement. BrdU labeling identified the smaller cells as a proliferating cell population, whereas the larger flattened cells were recognized as a nonproliferating cell population. Thus, for continuous differentiation of the proliferating cells alone, all cells cultured for a period of 6 days on type IV collagen-coated six-well dishes are passed through a 40-μm strainer (Falcon), as described later, that retains the larger, nonproliferating cells, permitting the smaller proliferating cells to pass through the mesh and therefore reculture on collagen-coated six-well plates for further differentiation (Fig. 1).

For the purpose of this specific separation of the proliferating cell population from the mixed cell populations that appear following culture of undifferentiated ES cells for 6 days on collagen-coated six-well dishes, cells are dissociated from the collagen-coated plates by culture in the presence of EDTA splitting medium for a period of 30 min to 2 h, until cells detach from the collagenous coating. For complete detachment of the cells, mechanical scraping is implemented after which EDTA splitting medium is neutralized with 5 volumes of culture medium and cells are pelleted (1200 rpm, 5 min, 4°) and resuspended in 5 ml of culture medium. The resuspended cells are then passed through a 40-μm mesh, and cells that

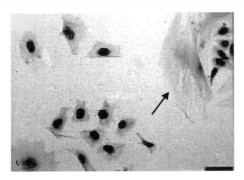

FIG. 1. Cells labeled with BrdU for determination of their proliferative state following a 6-day culture on type IV collagen. Culture on collagen resulted in differentiation of the human ES cells to two discrete cell populations: small proliferating cells that are labeled efficiently with BrdU and larger flattened cells that are not labeled with BrdU (marked by an arrow). From Gerecht-Nir et al. (2003).

pass through are recultured on type IV collagen-coated six-well plates at a seeding concentration of 2.5×10^4 cells/cm^2.

Lineage-Specific Differentiation

Constant culture of the undifferentiated ES cells on type IV collagen-coated six-well dishes followed by specific cell partition after a 6-day culture and reculture of the segregated cells on collagen-coated plates yield a semidifferentiated mesodermal cell population that shows increased expression of unique markers that define endothelial progenitor cells both by immunostaining and by RT-PCR analysis. These markers, which include PECAM-1, CD34, CD133, Gata-2, and Tie2, are either absent or only poorly expressed in undifferentiated ES cells and were demonstrated previously to be upregulated in 6-day-old spontaneously differentiating EBs (Levenberg et al., 2002).

Specific markers that identify differentiation to cells of the smooth muscle cell lineage are not detected at this point. To encourage differentiation to specific lineages, the mesodermal-like cells obtained when ES cells are cultured on collagen-coated plates are subjected to the induction of endothelial and vascular smooth muscle cell-specific growth factors. The growth factors are added to the culture medium following filtration of the cells cultured on collagen for a period of 6 days. For development of endothelial-like cells, the culture medium is supplemented with human VEGF. The VEGF added is the splice variant 165 (R&D Systems, Inc.), which is considered the predominant VEGF isoform (Keyt et al., 1996) at a concentration of 50 ng/ml.

The reculture of filtrated cells in the presence of VEGF promotes organization of differentiating cells to ring-like structures (Fig. 2). These differentiating cells are able to rapidly uptake acetylated LDL in a similar manner to recognized endothelial cells and present expression of the endothelial markers vWF and PECAM-1. Furthermore, when filtrated cells are cultured on Matrigel (Becton Dickinson) or collagen I gel (Roche) in the presence of 50 ng/ml of VEGF, cells penetrate to form three-dimensional structures *in vitro* that resemble tube-like structures (Fig. 3). Immunostaining of the sprouting cells for the presence of CD34 enables the identification of the cells as endothelial-like cells.

To encourage production of vascular smooth muscle-like cells from the differentiating mesodermal cells cultured on collagen, the filtrated cells culture medium is supplemented with 10 ng/ml of human PDGF-BB (R&D Systems, Inc.) that, as described previously, was identified as a factor crucial for the proliferation and recruitment of vascular smooth muscle cells to sites of neovascularization. The majority of cells cultured for 10 to 12 days in the presence of PDGF-BB express specific smooth muscle cell-specific markers that include SMA, calponin, and smooth muscle-myosin heavy chain.

Although ES cells in general and human ES cells in particular have been demonstrated previously to have the capacity to create cells of the vascular lineage, mainly through the spontaneous differentiation of EBs, the notion of directed differentiation has been quite appealing. Either for

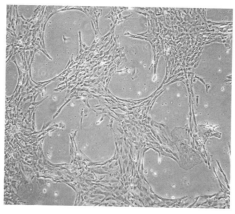

FIG. 2. Following filtration of human ES cells cultured on type IV collagen, separated cells were recultured on type IV collagen in the presence of VEGF. Recultured cells are organized in ring-like structures on the collagen plating. (See color insert.)

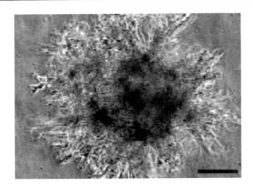

Fig. 3. Cells cultured on type IV collagen were seeded on Matrigel in the presence of VEGF to determine their ability to form vessel-like structures in a three-dimensional gel. A cell cluster is shown sprouting outward from the seeded cells to form a network structure within the gel. Bar: 100 μm. From Gerecht-Nir *et al.* (2003).

pure research or for practice, cells induced to differentiate by the guidance of specific growth factors hold immense potential.

This model for the directed differentiation of human ES cells to cells of the vascular lineage, specifically to endothelial and vascular smooth muscle cells, provides a powerful tool for subsequent implementation. It allows the identification of events that transpire at the various stages of vascular differentiation and, at the same time, provides the means to acquire a significant number of vascular cells so that their implementation in regenerative medicine can be advanced to practice.

References

Amit, M., Carpenter, M. K., Inokuma, M. S., Chiu, C. P., Harris, C. P., Waknitz, M. A., Itskovitz-Eldor, J., and Thomson, J. A. (2000). Clonally derived human embryonic stem cell lines maintain pluripotency and proliferative potential for prolonged periods of culture. *Dev. Biol.* **227,** 271–278.

Amit, M., Margulets, V., Segev, H., Shariki, K., Laevsky, I., Coleman, R., and Itskovitz-Eldor, J. (2003). Human feeder layers for human embryonic stem cells. *Biol. Reprod.* **68,** 2150–2156.

Breier, G., Breviario, F., Caveda, L., Berthier, R., Schnurch, H., Gotsch, U., Vestweber, D., Risau, W., and Dejana, E. (1996). Molecular cloning and expression of murine vascular endothelial cadherin in early stage development of cardiovascular system. *Blood* **87,** 630–641.

Burdsal, C. A., Damsky, C. H., Pedersen, C. H., and R. A., C. H. (1993). The role of E-cadherin and integrins in mesoderm differentiation and migration at the mammalian primitive streak. *Development* **118,** 829–844.

Buschmann, I., and Schaper, W. (2000). The pathophysiology of the collateral circulation (arteriogenesis). *J. Pathol.* **190,** 338–342.

Carmeliet, P. (2000). Mechanisms of angiogenesis and arteriogenesis. *Nature Med.* **6,** 389–395.

Coussens, L. M., Raymond, W. W., Bergers, G., Laig-Webster, M., Behrendtsen, O., Werb, Z., Caughey, G. H., and Hanahan, D. (1999). Inflammatory mast cells up-regulate angiogenesis during squamous epithelial carcinogenesis. *Genes Dev.* **13,** 1382–1397.

Evans, M. J., and Kaufman, M. H. (1981). Establishment in culture of pluripotential cells from mouse embryos. *Nature* **292,** 154–156.

Folkman, J., and Shing, Y. (1992). Angiogenesis. *J. Biol. Chem.* **267,** 10931–10934.

Gerecht-Nir, S., Dazard, J. E., Golan-Mashiach, M., Osenberg, S., Botvinnik, A., Amariglio, N., Domany, E., Rechavi, G., Givol, D., and Itskovitz-Eldor, J. (2005). Vascular gene expression and phenotypic correlation during differentiation of human embryonic stem cells. *Dev. Dyn.* **232,** 487–497.

Gerecht-Nir, S., Osenberg, S., Nevo, O., Ziskind, A., Coleman, R., and Itskovitz-Eldor, J. (2004). Vascular development in early human embryos and in teratomas derived from human embryonic stem cells. *Biol. Reprod.* **71,** 2029–2036.

Gerecht-Nir, S., Ziskind, A., Cohen, S., and Itskovitz-Eldor, J. (2003). Human embryonic stem cells as an *in vitro* model for human vascular development and the induction of vascular differentiation. *Lab. Invest.* **83,** 1811–1820.

Hellstrom, M., Kalen, M., Lindahl, P., Abramsson, A., and Betsholtz, C. (1999). Role of PDGF-B and PDGFR-beta in recruitment of vascular smooth muscle cells and pericytes during embryonic blood vessel formation in the mouse. *Development* **126,** 3047–3055.

Itskovitz-Eldor, J., Schuldiner, M., Karsenti, D., Eden, A., Yanuka, O., Amit, M., Soreq, H., and Benvenisty, N. (2000). Differentiation of human embryonic stem cells into embryoid bodies comprising the three embryonic germ layers. *Mol. Med.* **6,** 88–95.

Kabrun, N., Buhring, H. J., Choi, K., Ullrich, A., Risau, W., and Keller, G. (1997). Flk-1 expression defines a population of early embryonic hematopoietic precursors. *Development* **124,** 2039–2048.

Keller, G., Kennedy, M., Papayannopoulou, T., and Wiles, M. V. (1993). Hematopoietic commitment during embryonic stem-cell differentiation in culture. *Mol. Cell. Biol.* **13,** 473–486.

Keller, G. M. (1995). *In-vitro* differentiation of embryonic stem cells. *Curr. Opin. Cell Biol.* **7,** 862–869.

Keyt, B. A., Berleau, L. T., Nguyen, H. V., Chen, H., Heinsohn, H., Vandlen, R., and Ferrara, N. (1996). The carboxyl-terminal domain (111–165) of vascular endothelial growth factor is critical for its mitogenic potency. *J. Biol. Chem.* **271,** 7788–7795.

Lampugnani, M. G., Resnati, M., Raiteri, M., Pigott, R., Pisacane, A., Houen, G., Ruco, L. P., and Dejana, E. (1992). A novel endothelial-specific membrane-protein is a marker of cell–cell contacts. *J. Cell Biol.* **118,** 1511–1522.

Leung, D. W., Cachianes, G., Kuang, W. J., Goeddel, D. V., and Ferrara, N. (1989). Vascular endothelial growth-factor is a secreted angiogenic mitogen. *Science* **246,** 1306–1309.

Levenberg, S., Golub, J. S., Amit, M., Itskovitz-Eldor, J., and Langer, R. (2002). Endothelial cells derived from human embryonic stem cells. *Proc. Natl. Acad. Sci. USA* **99,** 4391–4396.

Ludwig, T. E., Levenstein, M. E., Jones, J. M., Berggren, W. T., Mitchen, E. R., Frane, J. L., Crandall, L. J., Daigh, C. A., Conard, K. R., Piekarczyk, M. S., Llanas, R. A., and Thomson, J. A. (2006). Derivation of human embryonic stem cells in defined conditions. *Nat. Biotechnol.* **24,** 185–187.

Martin, G. R. (1981). Isolation of a pluripotent cell line from early mouse embryos cultured in medium conditioned by teratocarcinoma stem cells. *Proc. Natl. Acad. Sci. USA* **78,** 7634–7638.

Matthews, W., Jordan, C. T., Gavin, M., Jenkins, N. A., Copeland, N. G., and Lemischka, I. R. (1991). A receptor tyrosine kinase cdna isolated from a population of enriched primitive hematopoietic-cells and exhibiting close genetic-linkage to C-kit. *Proc. Natl. Acad. Sci. USA* **88,** 9026–9030.

Mercola, M., Wang, C. Y., Kelly, J., Brownlee, C., Jacksongrusby, L., Stiles, C., and Bowenpope, D. (1990). Selective expression of Pdgf-A and its receptor during early mouse embryogenesis. *Dev. Biol.* **138,** 114–122.

Nakano, T., Kodama, H., and Honjo, T. (1994). Generation of lymphohematopoietic cells from embryonic stem cells in culture. *Science* **265,** 1098–1101.

Nishikawa, S., Nishikawa, S., Hirashima, M., Matsuyoshi, N., and Kodama, H. (1998). Progressive lineage analysis by cell sorting and culture identifies FLK1(+)VE-cadherin(+) cells at a diverging point of endothelial and hemopoietic lineages. *Development* **125,** 1747–1757.

Palis, J., Mcgrath, K. E., and Kingsley, P. D. (1995). Initiation of hematopoiesis and vasculogenesis in murine yolk-sac explants. *Blood* **86,** 156–163.

Palis, J., Robertson, S., Kennedy, M., Wall, C., and Keller, G. (1999). Development of erythroid and myeloid progenitors in the yolk sac and embryo proper of the mouse. *Development* **126,** 5073–5084.

Risau, W. (1997). Mechanisms of angiogenesis. *Nature* **386,** 671–674.

Risau, W., Sariola, H., Zerwes, H. G., Sasse, J., Ekblom, P., Kemler, R., and Doetschman, T. (1988). Vasculogenesis and angiogenesis in embryonic-stem-cell-derived embryoid bodies. *Development* **102,** 471–478.

Shalaby, F., Rossant, J., Yamaguchi, T. P., Gertsenstein, M., Wu, X. F., Breitman, M. L., and Schuh, A. C. (1995). Failure of blood-island formation and vasculogenesis in Flk-1-deficient mice. *Nature* **376,** 62–66.

Shibuya, M., Yamaguchi, S., Yamane, A., Ikeda, T., Tojo, A., Matsushime, H., and Sato, M. (1990). Nucleotide-sequence and expression of a novel human receptor-type tyrosine kinase gene (Flt) closely related to the Fms family. *Oncogene* **5,** 519–524.

Smith, A. G., Heath, J. K., Donaldson, D. D., Wong, G. G., Moreau, J., Stahl, M., and Rogers, D. (1988). Inhibition of pluripotential embryonic stem-cell differentiation by purified polypeptides. *Nature* **336,** 688–690.

Thomson, J. A., Itskovitz-Eldor, J., Shapiro, S. S., Waknitz, M. A., Swiergiel, J. J., Marshall, V. S., and Jones, J. M. (1998). Embryonic stem cell lines derived from human blastocysts. *Science* **282,** 1145–1147.

Thomson, J. A., Kalishman, J., Golos, T. G., Durning, M., Harris, C. P., Becker, R. A., and Hearn, J. P. (1995). Isolation of a primate embryonic stem-cell line. *Proc. Natl. Acad. Sci. USA* **92,** 7844–7848.

Wang, R., Clark, R., and Bautch, V. L. (1992). Embryonic stem cell-derived cystic embryoid bodies form vascular channels: An *in vitro* model of blood-vessel development. *Development* **114,** 303–316.

Yamashita, J., Itoh, H., Hirashima, M., Ogawa, M., Nishikawa, S., Yurugi, T., Naito, M., Nakao, K., and Nishikawa, S. (2000). Flk1-positive cells derived from embryonic stem cells serve as vascular progenitors. *Nature* **408,** 92–96.

[16]   Cardiomyocytes

By Xiangzhong Yang, Xi-Min Guo, Chang-Yong Wang,
and X. Cindy Tian

Abstract

Derivation of cardiomyocytes from embryonic stem cells would be a boon for treatment of the many millions of people worldwide who suffer significant cardiac tissue damage in a myocardial infarction. Such cells could be used for transplantation, either as loose cells, as organized pieces of cardiac tissue, or even as pieces of organs. Eventual derivation of human embryonic stem cells via somatic cell nuclear cloning would provide cells that not only may replace damaged cardiac tissue, but also would replace tissue without fear that the patient's immune system will reject the implant. Embryonic stem cells can differentiate spontaneously into cardiomyocytes. *In vitro* differentiation of embryonic stem cells normally requires an initial aggregation step to form structures called embryoid bodies that differentiate into a wide variety of specialized cell types, including cardiomyocytes. This chapter discusses methods of encouraging embryoid body formation, causing pluripotent stem cells to develop into cardiomyocytes, and expanding the numbers of cardiomyocytes so that the cells may achieve functionality in transplantation, all in the mouse model system. Such methods may be adaptable and/or modifiable to produce cardiomyocytes from human embryonic stem cells.

Introduction

Coronary heart disease is responsible for more deaths worldwide in people 60 years of age and older than any other cause. Stroke, also a product of cardiovascular disease, ranks second as a cause of death in this age group (MacKay and Mensah, 2004). In 2002, India, China, and the Russian Federation had the greatest numbers of deaths from coronary heart disease in the world. In the United States, where cardiovascular disease is the number one cause of death in both men and women, there were more than 910,000 deaths due to cardiovascular disease in 2003 (American Heart Association, 2005). Furthermore, in 2003, nearly seven million people had in-patient cardiovascular surgery in the United States and more than seven million (or 3.5% of the population) had suffered a myocardial infarction at some time.

METHODS IN ENZYMOLOGY, VOL. 418   0076-6879/06 $35.00
           DOI: 10.1016/S0076-6879(06)18016-7

Perhaps as many as a billion or more myocytes are damaged or lost as a result of myocardial infarction (Laflamme and Murry, 2005). The end result of myocardial damage is heart failure, which, according to Laflamme and Murry (2005), results in the greatest number of hospitalizations of people over 65 years of age in the United States. People who have suffered from myocardial infarction would be a prime group for stem cell cardiomyocyte transplantation to repair muscle damage.

The most promising cell sources for regenerative medicine are embryonic and adult stem cells. The capacity of murine embryonic stem (ES) cells to differentiate into cardiomyocytes has been demonstrated by a number of research groups (Doetschman et al., 1985; Wobus et al., 1991). The subsequent demonstration of spontaneous differentiation of human ES cells into the three germ layers (Schuldiner et al., 2000), including the development of functional cardiomyocytes (Kehat et al., 2001), has opened the avenue to a human cell source for cell-based therapies, including cardiac tissue engineering. Unlimited differentiation capacity and indefinite propagation represent the strongest advantages for use of ES cells.

Somatic cell nuclear transfer, a technique that has been used successfully to clone cattle, rabbits, sheep, pigs, mice, and other animals, were it to work successfully in humans, would be a prime technique for production of human ES cells (hES cells). The goal of production of such nuclear transfer embryonic stem cells would be to use autologous transplantation to avoid immune rejection. It is, perhaps, the strongest rationale for therapeutic cloning. To date, such cells have not been produced, but they remain a clear goal of researchers in this field.

The functional cardiomyocytes derived from ES cells can be used in cell transplantation for restoration of heart function by replacement of diseased myocardium (Reinlib and Field, 2000) or the design of artificial cardiac muscle constructs *in vitro* for later implantation *in vivo* (Zimmermann and Eschenhagen, 2003).

This chapter describes protocols for generation, differentiation and enrichment of cardiomyocytes.

Embryoid Body (EB) Generation

The most robust method for generating the most differentiated cell types is through the EB system, where ES cells differentiate spontaneously as tissue-like spheroids in suspension culture. EB differentiation has been shown to recapitulate aspects of early embryogenesis, including the formation of a complex, three-dimensional architecture wherein cell–cell and cell–matrix interactions are thought to support the development of the three embryonic germ layers and their derivatives (Itskovitz-Eldor et al., 2000; Keller, 1995).

Presently, all human and most mouse ES cell lines require aggregation of multiple ES cells to initiate EB formation efficiently (Dang *et al.*, 2002; Itskovitz-Eldor *et al.*, 2000). Standard methods of EB formation include hanging drop, liquid suspension, and methyl cellulose culture. These culture systems maintain a balance between allowing ES cell aggregation necessary for EB formation and preventing EB agglomeration for efficient cell growth and differentiation (Dang *et al.*, 2002). However, these culture systems are limited in their production capacity and are not easily amenable to process-control strategies.

In the mouse system, the ES cells spontaneously form three-dimensional aggregates and differentiate after withdrawal of leukemia inhibitory factor (LIF) and transfer onto a nonadherent surface (Maltsev *et al.*, 1994) (Fig. 1). These three-dimensional aggregates recapitulate early embryological development in the mouse and allow derivatives of the three germ layers to form *in vitro*. In hES cells, spontaneous differentiation toward the ectoderm, endoderm, and mesoderm has also been reported when cultivated either as EBs or as a monolayer at high cell density. EB formation in ES cells is normally achieved by dissociating colonies into single cells and promoting agglomeration by seeding at high cell densities in nonadherent Petri dishes. Another way to form EBs is to suspend cells in small droplets hanging from the underside of a culture plate, often referred to as the hanging drop method. For hES cells, EB formation is promoted by detaching small clumps of hES colonies by enzymatic (collagenase/dispase) or chemical dissociation (EDTA) and keeping them in suspension in nonadherent culture dishes.

Hanging Drop Method

Mouse ES Cell Dissociation

1. Warm trypsin to 37° in a water bath.
2. Discard the culture medium for undifferentiated medium, wash with 1× phosphate-buffered saline (PBS) twice.
3. Add the following amount of trypsin (0.25%:0.04% EDTA=1:1)
   a. 0.5 ml/well of four-well plate
   b. 1.0 ml/well of six-well plate
4. Incubate at 37° with 5% CO_2 for about 1 min. The cells are ready when the edges of the colony are rounded up and curled away from the mouse embryonic fibroblasts (MEFs) on the plate.
5. Add 1 to 2 ml of fetal calf serum (FCS) to block the trypsin.
6. Scrape and wash the colonies off the plate with a pipette.

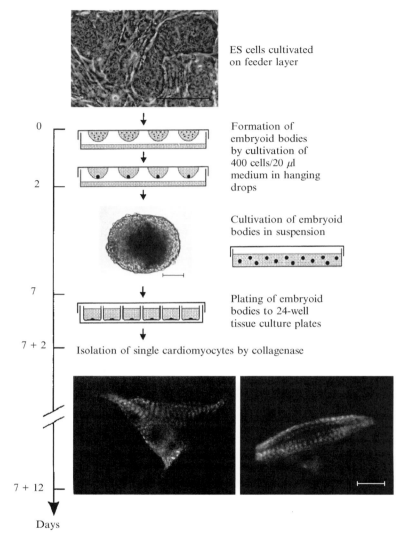

FIG. 1. Schematic presentation of the *in vitro* differentiation protocol of embryonic stem (ES) (D3) cells into cardiomyocytes. The cross-striated structure of sarcomeres of single cardiomyocytes at the terminal differentiation stage was visualized by indirect immunofluorescence with monoclonal antibodies against cardiac-specific α-cardiac myosin heavy chain and troponin TI-1 (bottom, left, and right). Bars: 100 μm (top and middle) and 10 μm (bottom). From Maltsev *et al.* (1994) with permission.

7. Transfer the cell suspension to a 10-ml conical tube.
8. Break up the colonies by pipetting up and down against the bottom of the tube until there appears to be a fine suspension of cells, that is, without clumps of cells remaining.
9. Spin the cells at 1400 rpm for 5 min.
10. Aspirate the supernatant off and resuspend the cells with 10 ml of differentiation medium.
11. Transfer the cell suspension into a T75 flask precoated with 0.1% gelatin and incubate at 37° with 5% CO_2 for an hour to allow for the adherence of fibroblasts onto the surface of the flask.
12. Collect the medium containing cells that remain unattached.
13. Spin at 1000 rpm for 5 min.
14. Aspirate off the wash medium.
15. Add another 1 ml of differentiation medium and resuspend the cells by repetitive pipetting until there appears to be a fine suspension of cells.
16. Count the cells using a hemocytometer.

Hanging Drop Culture

1. Dilute the cell suspension to 400 to 500 cells/30 μl (1.5 to 2 × 10^4/ml) by differentiation medium.
2. Invert a 100-mm tissue culture plate cover and put it on a "pipetting guide." The "pipetting guide" is prepared by drawing cross lines on a piece of paper with an interline distance of 8 mm; the crosses are where the cell drops are located.
3. Pipette 30 μl cell suspension onto the inner surface of the plate cover corresponding to each *cross*.
4. Add 10 ml PBS into the plate.
5. *Gently* turn the plate cover to enlarge the attaching surface for the droplets.
6. *Quickly* invert the plate cover and cover the plate *gently*.
7. Carefully transfer the plate into the incubator.
8. Culture for 2 or 3 days.

Suspension Culture

1. Aspirate the suspended droplets and transfer them into a 100-mm bacteriological plate. Each plate receives approximately 100 droplets.
2. Add 10 ml culture medium into each plate.
3. Shake the plate gently.
4. Incubate at 37° with 5% CO_2 for 4 to 5 days.

Human ES Cell Suspension Culture

1. Let human ES cells grow until the colonies are large and the cells are pretty piled up—about the time when you would normally split the colonies or even a day past that.

2. Treat cells with 0.2 to 0.5 mg/ml dispase. Use the lowest possible concentration of dispase, but it tends to vary a bit.

3. Wait until the colonies completely detach from the plate. Do not blow colonies off with a pipette. This should take about 20 to 30 min. If nothing is happening by that point, add more dispase.

4. Once the colonies come up, gently transfer them to a 15-ml conical tube with a 10-ml pipette. Do not break up the colonies.

5. The cells should sink to the bottom of the tube after a minute or two without any spinning. Aspirate off the medium and wash once in hES medium. If you are in a hurry and need to spin the colonies down, 1 min at 500 rpm is enough.

6. Transfer cells to a flask containing ES medium without basic fibroblast growth factor (bFGF). Put all of the EBs from one six-well plate into a T80 flask with about 25 ml medium.

7. The cells will round up into actual embryoid bodies after about 12 to 24 h. They should then be fed every day by exchanging half the medium with fresh medium. The EBs should not attach; if they do, tap the flask gently to dislodge the EBs.

Scalable Production of EBs in a Bioreactor

All of the current protocols for hEB generation list aggregation of hESCs as a prerequisite for initiating EB formation. At later stages, however, agglomeration of the EBs may have negative effects on cell proliferation and differentiation, as was shown in the mouse system (Dang *et al.*, 2002). When formed in static cultures (mostly in flasks), agglomerated large EBs revealed extensive cell death and, eventually, large necrotic centers due to mass transport limitations. To maintain balance between these two processes and to achieve control of the extent of EB agglomeration, several methods have been developed for mouse ES cells. These methods include hanging drops and methyl cellulose cultures (Dang *et al.*, 2002). Although efficient, to some extent, in preventing the agglomeration of EBs, the complex nature of these systems makes upscaling them a rather difficult task. However, a much simpler process in spinner flasks resulted in the formation of large cell clumps within a few days, indicative of significant cell aggregations in the cultures (Wartenberg *et al.*, 2001). Attempts to increase the stirring rate to avoid agglomeration may result in massive hydrodynamic damage to the cells due to extensive mixing in the vessels (Chisti, 2001). Gerecht-Nir *et al.* (2004) introduced rotating cell culture systems (RCCS), developed by NASA, as milder bioreactors for hEB formation and differentiation. In RCCS, the operating principles are: (1) whole-body rotation around a horizontal axis, which is characterized by extremely low fluid shear stress, and (2) oxygenation by active or passive diffusion to the exclusion of all but dissolved gases from the reactor chamber, yielding a vessel devoid of gas bubbles and gas/fluid interfaces (Lelkes and Unsworth, 2002). The resulting flow pattern in the RCCS is laminar with mild mixing, as the vessel rotation is slow. The settling of the cell clusters, which is associated with oscillations and tumbling, generates fluid mixing. The outcome is a very low-shear environment. Another advantage of the RCCS is that they are designed geometrically so that the membrane area to volume of medium ratio is high, thus enabling efficient gas exchange. Gerecht-Nir and associates (2004) indicated that cultivation of hESCs in the slow-turning lateral vessel (STLV) system, a type of RCCS with oxygenator membrane in the center (Fig. 2), yielded nearly fourfold more hEB particles compared to static, conventional Petri dishes. These EBs were intact in shape and had less necrosis in the center (Fig. 3). Under the dynamic cultivation in STLV, differentiation of hEBs progressed in a normal course (Gerecht-Nir *et al.*, 2004).

1. Prepare confluent six-well plates (60 cm^2) of undifferentiated hESCs.
2. Disperse the cells into small clumps (3 to 20 cells) using 0.5 mM EDTA supplemented with 1% FBS (HyClone).
3. Prepare STLV bioreactor (Synthecon, Inc., Houston, TX) according to the instruction manual.
   a. Using an Allen wrench, unscrew and remove center bolt at the top of the vessel. Gently twist outer wall while holding top end cap to disassemble vessel. Repeat procedure for rear end cap. Remove O-rings from each end cap.
   b. Place all pieces of vessel in a 4-liter beaker filled with a warm solution of mild detergent designed for tissue culture labware. Soak for 1 h.
   c. Scrub plastic parts (except oxygenator core) with a soft bristle brush as necessary to remove any residues.
   d. Very gently clean oxygenator membrane with the tip of your finger using latex laboratory gloves.
   Note: harsh scrubbing will damage membrane material. Do not use a brush to cleanse the membrane.
   e. Rinse vessel parts with a continuous flow of ultrapure water for 15 to 20 min.

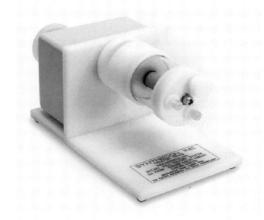

FIG. 2. Illustration of STLV system. From Synthecon, Inc. with permission.

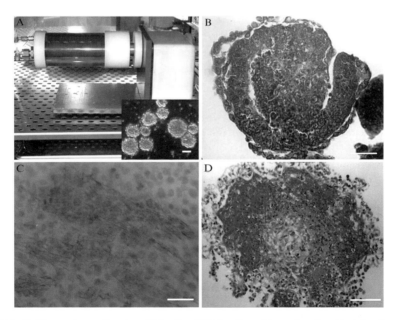

FIG. 3. Comparison of embryoid bodies (EBs) cultured in a slow-turning lateral vessel (STLV) and suspended in flasks. EBs cultured in a 250-ml STLV (A) for 5 days are intact and well differentiated, without necrosis in the center (B, H&E); these EBs demonstrated productive differentiation into cardiomyocytes (C, cardiac troponin-T [cTnT] immunohisto-chemistry); EBs cultured in suspension for 5 days are smaller in size and have significant necrosis in the middle (D, H&E). Scale bars: 50 μm (A), 100 μm (B, D), or 10 μm. (Guo *et al.*, 2006).

f. Soak vessel parts in fresh, ultrapure water overnight.
g. Remove vessel parts from water and place on absorbent pads to dry.
h. Assemble vessel.
i. Fill unit with 70% ethanol and allow it to soak for 24 h.
j. Sterilize as described.

Autoclave Method

i. Empty vessel of 70% ethanol.
ii. Remove and dispose of plastic valves and their caps. Remove fill port cap and autoclave it separate from the vessel. Cover all three ports with aluminum foil.
iii. Loosen center screw one turn.
iv. Wrap vessel and caps and autoclave for 30 min at 105° to 110°. It is not necessary to slow vent the autoclave.
v. Remove from autoclave; cool to room temperature.

4. Seed hESCs into the STLV at initial cell concentrations ranging from 1×10^4 to 1×10^5 cells per 1 ml medium. The medium may be KO-DMEM, supplemented with 20% FBS, penicillin–streptomycin, 1 mM L-glutamine, 0.1 mM β-mercaptoethanol, and 1% nonessential amino acid stock.

Manipulation of the STLV bioreactor should follow the following instructions.

a. Transfer the vessel to a sterile hood. Remove the end caps and place them on sterile alcohol pads or sterile Petri plates.
b. Aspirate medium from the unit through a $\frac{1}{4}$-in. port.
c. Fill the vessel to 50% of total volume with growth medium minus serum. Allow space to load cells. (Serum addition at this time increases foaming and leads to difficulty in removing the air bubbles later.)
d. Count cells to be used or mince primary tissue (ten 1-mm pieces per 5 ml of medium).
e. Dilute the cells into separate containers of medium to yield desired final concentration (2 to 3×10^5/ml has been used by some authors).
f. Add appropriate amount of washed, prepared microcarrier beads (5 mg/ml) to diluted cells.
g. With a 10-ml pipette, load cell/bead/medium solution through the $\frac{1}{4}$-in. port.

    h. Add the appropriate amount of serum and top off the vessel with medium.

    i. Wipe the port with an alcohol pad. Replace the cap and tighten. Close the syringe port valves.

    j. Fill a 10-ml sterile syringe with growth medium. Wipe one syringe port with an alcohol pad and attach syringe.

    k. Wipe the other syringe port with an alcohol pad and attach an empty 3- or 5-ml sterile syringe. Open valves of both syringe ports.

    l. Gently invert vessel and tap on sides to expel air bubbles from under the ports. Maneuver air bubbles under the empty syringe. With both valves open, press on the syringe gently to replace air bubbles with medium.

    m. Discard the small syringe. Wipe the port with an alcohol pad and replace with the cap or another syringe.

    n. Leave the large, medium-filled syringe on the unit with the valve open, as the volume of the medium in the vessel may vary slightly with temperature.

    o. Attach the vessel to the rotator base in a humidified CO_2 incubator. Check that the unit is level.

    p. Turn power on and adjust to an initial rotation speed of 10 rpm.

Note: cells and cell aggregates should rotate with the vessel and not settle within the vessel, nor should they collide with the cylinder wall or oxygenator core of the vessel. When the speed is adjusted properly, the cells and cell aggregates will orbit within the vessel. The rotation speed will need to be increased to compensate for the increased sedimentation rates of anchorage-dependent cells as the aggregate particles increase in size.

5. Set the bioreactor to rotate at a speed at which the suspended cell aggregates remain close to a stationary point within the reactor vessel; 10 rpm is recommended as the initial speed.

6. Change the medium every 3 days as described next.

    a. Turn off power and immediately remove the vessel from the base and take it to a sterile environment (biological hood).

    b. Stand the vessel vertically on its base (valves up) and let the cell/bead aggregates settle to the bottom.

    c. Close any valves that may be open.

    d. Remove and discard any syringes that may be attached. Wipe ports with sterile alcohol pads.

    e. Remove the $\frac{1}{4}$-in. port cap and any Luer lock caps that may be attached and place on sterile alcohol pads.

    f. Aspirate medium through the $\frac{1}{4}$-in. port. Usually, one-fourth to one-half of the conditioned medium is left in the vessel. Aspirate droplets from syringe ports.

    g. Fill the vessel with medium using a Luer lock syringe or a sterile pipette. Flow the medium down the wall of the vessel. *(Do not disturb the cell aggregate particles.)*

    h. Fill a 10-ml sterile syringe with growth medium. Wipe one syringe port with an alcohol pad and attach syringe.

    i. Wipe the other syringe port with an alcohol pad and attach an empty 3- or 5-ml sterile syringe.

    j. Gently invert vessel and tap on sides to expel air bubbles from under the ports. Maneuver air bubbles under the empty syringe. With both valves open, press on the syringe gently to replace air bubbles with medium.

    k. When all bubbles are removed, close the syringe valves and discard the small syringe. Wipe the port with an alcohol pad and replace the cap or leave in a small syringe for later sampling.

    l. Leave the large, medium-filled syringe on the unit with the valve open, as the volume of the medium in the vessel may vary slightly with temperature change.

    m. Attach the vessel to the rotator base and replace them in the humidified CO_2 incubator.

    n. Turn on the power and adjust the speed as necessary.

7. Take samples at different time intervals, if necessary, to monitor the EBs growing in the vessel, following the instructions described next.

    a. If a sampling syringe is not in place, stop the vessel rotation. Remove the syringe port cap and place it in a sterile Petri dish. Attach a sterile, empty 1.5- or 10-ml syringe to the valve. Both syringe port valves should be open.

    b. Turn on the power to allow the cell/bead aggregates to be distributed evenly (2 min).

    c. Push medium into the vessel with the medium-containing 10-ml syringe that is still attached from inoculation. A slight pull on the smaller, sampling syringe may also be necessary. This procedure provides a homogeneous, representative sample; however, it may take some practice, as the vessel is still rotating.

    d. When the desired sample has been drawn (usually 1 to 5 ml), turn off the power. Close the valve on the sampling syringe port and remove the sampling syringe.

    e. Attach another sampling syringe or replace the port cap.

    f. Turn on the power and adjust speed if necessary.

g. If any bubbles are visible, turn off the power and utilize the bubble removal procedures provided earlier.

Note: tissue particles too large to be drawn into a syringe can, in some cases, be removed with forceps through the fill port. Extreme care should be taken to avoid damaging the oxygenator membrane with the forceps.

Cardiomyocyte Differentiation

Embryonic stem cells can differentiate spontaneously into cardiomyocytes. *In vitro* differentiation of ES cells normally (except for neurogenesis) requires an initial aggregation step to form structures, termed embryoid bodies (EBs), which differentiate into a wide variety of specialized cell types, including cardiomyocytes. A number of parameters specifically influence the differentiation potency of ES cells to form cardiomyocytes in culture: (1) the starting number of cells in the EB, (2) media, FBS, growth factors, and additives, (3) ES cell lines, and (4) the time of EB plating (Wobus et al., 2002). Cardiomyocytes are located between an epithelial layer and a basal layer of mesenchymal cells within the developing EB (Hescheler et al., 1997). Cardiomyocytes are readily identifiable because, within 1 to 4 days after plating, they contract spontaneously. With continued differentiation, the number of spontaneously beating foci increases and all the EBs may contain localized beating cells. The rate of contraction within each beating area increases rapidly with differentiation, followed by a decrease in average beating rate with maturation. Depending on the number of cells in the initial aggregation step, the change in beating rate and the presence of spontaneous contractions continue from several days to more than 1 month. Fully differentiated cardiomyocytes often stop contracting, but can be maintained in culture for many weeks. Thus, developmental changes of cardiomyocytes may be correlated with the length of time in culture and can be divided readily into three stages of differentiation: early (pacemaker-like or primary myocardial-like cells), intermediate, and terminal (atrial-, ventricular-, nodal-, His-, and Purkinje-like cells) (Hescheler et al., 1997).

Similar to mouse cells, hES cells differentiate when they are removed from feeder layers and grown in suspension. EBs of hES cells are heterogeneous and can express markers specific to neuronal, hematopoietic, and cardiac origin (Itskovitz-Eldor et al., 2000; Schuldiner et al., 2000).

Several chemicals have proven helpful for the enhancement of cardiogenic differentiation of mouse or human ES cells. They are retinoic acid (Wobus et al., 1997; Xu et al., 2002), 0.5 to 1.5% dimethylsulfoxide (DMSO) (Ventura and Maioli, 2000), and 5-aza-dC (Xu et al., 2002).

1. Transfer the EBs into a 100-mm tissue culture plate. For each plate, seed 100 EBs.
2. Add 15 ml induction medium to each plate. Shake or pipette gently so as to distribute the EBs evenly.

The induction medium varies according to different protocols. It may be differentiation medium supplemented with ascorbic acid (Takahashi *et al.*, 2003), retinoic acid (Wobus *et al.*, 1997; Xu *et al.*, 2002), DMSO (Klug *et al.*, 1996), or 5-aza-dC (Xu *et al.*, 2002).

3. Make sure the EBs are distributed evenly across the entire plate.
4. Place plate gently into incubator. Make sure the EBs are not disturbed.
5. Let the EBs settle overnight in an incubator.
6. Change medium every 3 days.
7. Observe the cells every day under an inverted optical microscope to monitor the appearance of beating areas.
8. Continue culture for approximately 7 to 10 days or more, as needed.

Enrichment of Cardiomyocytes

To use hES cell-derived cardiomyocytes in therapeutic applications, it will be beneficial to produce a population of cells highly enriched for cardiomyocytes. Xu *et al.* (2002) first demonstrated the enrichment of hES cell-derived cardiomyocytes by Percoll gradient separation and proliferation capacity of the enriched cells. These cells express appropriate cardiomyocyte-associated proteins. A subset of them appears to be proliferative, as determined by BrdU incorporation or expression of Ki-67, suggesting that these cardiomyocytes represent an early stage of cells. This strategy has been further proved efficient for the enrichment of cardiomyocytes from mouse ES cells and neonatal rat ventricular cardiomyocytes (E *et al.*, 2006; Guo *et al.*, 2006) (Fig. 4 and Table I).

1. Wash the differentiated cultures containing beating cardiomyocytes three times with PBS or a low calcium solution (Maltsev *et al.*, 1993). The low calcium solution contains 120 mM NaCl, 5.4 mM KCl, 5 mM MgSO$_4$, 5 mM sodium pyruvate, 20 mM glucose, 20 mM taurine, and 10 mM HEPES at pH 6.9.
2. Discard the PBS. Add an appropriate amount of 1 mg/ml collagenase B to cover the cells. The 1 mg/ml collagenase B is prepared in the low calcium solution supplemented with 30 μM CaCl$_2$.
3. Incubate at 37° for 1 to 2 h.
4. Resuspend the cells in a high potassium solution. The high potassium solution contains 85 mM KCl, 30 mM K$_2$HPO$_4$, 5 mM

MgSO₄, 1 mM EGTA, 2 mM Na₂ATP, 5 mM sodium pyruvate, 5 mM creatine, 20 mM taurine, and 20 mM glucose at pH 7.2.

5. Incubate at 37° for 15 min for more complete dissociation.
6. Gently pipette to achieve a uniform cell suspension.
7. Transfer the cell suspension into a 10-ml conical tube. Spin at 1200 rpm for 5 min.
8. Aspirate the supernatant. Add 3 ml high glucose DMEM containing 20% FBS and resuspend the cells by gentle pipetting.
9. Prepare a Percoll gradient as described.
   a. Mix Percoll with 8.5% NaCl (9:1) to reach a physiological osmotic equilibrium.

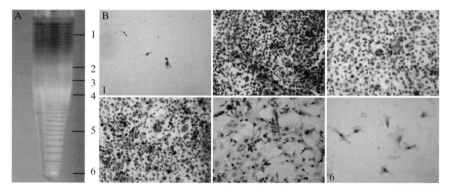

FIG. 4. Characterization of Percoll-enriched cardiomyocytes from mESCs. (A) Six fractions of mES cell-derived cardiomyocytes after Percoll enrichment. (B) Characterization of cells in each fraction by anticardiac troponin T (cTnT) staining after plating for 3 days in culture. cTnT-positive cells were mainly in fractions 4 and 5 (E *et al.*, 2006).

TABLE I
ENRICHMENT OF mES CELL-DERIVED CARDIOMYOCYTES BY PERCOLL GRADIENT[a]

Fraction	Cells collected	Beating cells[b]	%cTnT-positive cells (day 3)
Input cells	$1\sim2 \times 10^8$	++	$18 \pm 3\%$
II	1.77×10^7	+	$3 \pm 2\%$
III	3.25×10^6	+	$6 \pm 3\%$
IV	5.65×10^6	+++	$40 \pm 2\%$
V	2.60×10^6	++++	$88.7 \pm 4\%$

[a] mES cell-derived cardiomyocytes differentiated for 14 days were enriched by Percoll gradient separation (see text). After separation, each layer was collected, and cells were counted and replated. Cultures were maintained for 3 days before evaluation of cTnT immunoreactivity. From E *et al.* (2006).
[b] Amount of beating cells: ++++ > +++ > ++ > +.

    b. Dilute the Percoll–8.5% NaCl solution with 8.5% NaCl to a final Percoll concentration of 40.5 and 58.5%, which correspond to a physical density of 1.065 and 1.069 g/ml, respectively.

10. Add 3 ml of 58.5% Percoll to the bottom of a 10-ml conical tube and then gently add 3 ml of 40.5% Percoll on top of the 58.5% Percoll.
11. Add 3 ml of cell suspension onto the top of the Percoll solution by using a pipette leaning against the inner wall of the tube. Be sure to do it *very gently* so as not to disturb the Percoll layers.
12. Centrifuge at 1500 rpm for 30 min.
13. After centrifugation, two layers of cells will be observed: one on top of the Percoll (fraction I) and a layer of cells at the interface of the two layers of Percoll (fraction III). Cells can also be found in the 40.5% Percoll layer (fraction II) and the 58.5% Percoll layer (fraction IV).

Generally, 20 to 40% of cells in fraction III and 50 to 70% of cells in fraction IV express cardiac-specific troponin I (cTnI), a subunit of the troponin complex that provides a calcium-sensitive molecular switch for the regulation of striated muscle contraction (Bhavsar *et al.*, 1996).

14. Carefully aspirate the cells in fraction III and cells in fraction IV.
15. Wash the cell fractions twice with PBS.

Conclusion

These methods have worked successfully in producing cardiomyocytes from mouse ES cells. Cardiomyocytes derived from mESCs have been successfully introduced into the process of engineering cardiac muscle. Nevertheless, in their review of differentiation of ES cells from mouse and human to produce cardiomyocytes, Wei *et al.* (2005) noted the numerous problems in deriving human cells and differentiating them into cardiomyocytes. They point out that some studies show chromosomal abnormalities in hES cells as a result of enzymatic dissociation methods. They also indicate that differentiation of hES cells to cardiomyocytes is slower and less efficient than comparative differentiation using mouse ES cells. However, Laflamme and Murry (2005) noted, "In contrast to the limited proliferative capacity of mouse ES cell-derived cardiomyocytes, human ES cell-derived cardiomyocytes show sustained cell cycle activity both *in vitro* and after *in vivo* transplantation into the nude rat heart." Nevertheless, Wei *et al.* (2005) suggested that the numerous differences that have been noted between human and murine ES cells may be attributed to the longer gestation time allowed for heart development in the human embryo versus the mouse embryo. They outline methods of inducing cardiomyocyte development from cells derived from EBs through the addition of growth factors and cytokines into the culture medium.

Mummery *et al.* (2002) used a coculture method to derive cardiomyocytes from hES cells. She and her co-workers wrote a fairly extensive review of derivation and use of cardiomyocytes (van Laake *et al.*, 2005).

These are all beyond the scope of the current chapter, which is focused on techniques. Of course, techniques may change rapidly, as new methods for growing cells are developed, and in concert with tests both in large animal models and in humans.

References

American Heart Association (2005). "Heart Disease and Stroke Statistics—2006 Update." American Heart Association, Dallas, TX.

Bhavsar, P. K., Brand, N. J., Yacoub, M. H., and Barton, P. J. (1996). Isolation and characterization of the human cardiac troponin I gene (TNNI3). *Genomics* **35**, 11–23.

Chisti, Y. (2001). Hydrodynamic damage to animal cells. *Crit. Rev. Biotechnol.* **21**, 67–110.

Dang, S. M., Kyba, M., Perlingeiro, R., Daley, G. Q., and Zandstra, P. W. (2002). Efficiency of embryoid body formation and hematopoietic development from embryonic stem cells in different culture systems. *Biotechnol. Bioeng.* **78**, 442–453.

Doetschman, T. C., Eistetter, H., Katz, M., Schmidt, W., and Kemler, R. (1985). The *in vitro* development of blastocyst-derived embryonic stem cell lines: Formation of visceral yolk sac, blood islands and myocardium. *J. Embryol. Exp. Morphol.* **87**, 27–45.

E, L.-L., Zhao, Y.-S., Guo, X.-M., Wang, C.-Y., Jiang, H., Li, J., Duan, C.-M., and Song, Y. (2006). Enrichment of cardiomyocytes derived from mouse embryonic stem cells. *J. Heart Lung Transplant.* **25**, 664–674.

Gerecht-Nir, S., Cohen, S., and Itskovitz-Eldor, J. (2004). Bioreactor cultivation enhances the efficiency of human embryoid body (hEB) formation and differentiation. *Biotechnol. Bioeng.* **86**, 493–502.

Guo, X.-M., Zhao, Y.-S., Wang, C.-Y., E, L.-L., Chang, H.-X., Zhang, X.-A., Duan, C.-M., Dong, L.-Z., Jiang, H., Li, J., Song, Y., and Yang, X. (2006). Creation of engineered cardiac tissue *in vitro* from mouse embryonic stem cells. *Circulation* **113**, 2229–2237.

Hescheler, J., Fleischmann, B. K., Lentini, S., Maltsev, V. A., Rohwedel, J., Wobus, A. M., and Addicks, K. (1997). Embryonic stem cells: A model to study structural and functional properties in cardiomyogenesis. *Cardiovasc. Res.* **36**, 149–162.

Itskovitz-Eldor, J., Schuldiner, M., Karsenti, D., Eden, A., Yanuka, O., Amit, M., Soreq, H., and Benvenisty, N. (2000). Differentiation of human embryonic stem cells into embryoid bodies compromising the three embryonic germ layers. *Mol. Med.* **6**, 88–95.

Kehat, I., Kenyagin-Karsenti, D., Snir, M., Segev, H., Amit, M., Gepstein, A., Livne, E., Binah, O., Itskovitz-Eldor, J., and Gepstein, L. (2001). Human embryonic stem cells can differentiate into myocytes with structural and functional properties of cardiomyocytes. *J. Clin. Invest.* **108**, 407–414.

Keller, G. M. (1995). *In vitro* differentiation of embryonic stem cells. *Curr. Opin. Cell Biol.* **7**, 862–869.

Klug, M. G., Soonpaa, M. H., Koh, G. Y., and Field, L. J. (1996). Genetically selected cardiomyocytes from differentiating embryonic stem cells form stable intracardiac grafts. *J. Clin. Invest.* **98**, 216–224.

Laflamme, M. A., and Murry, C. E. (2005). Regenerating the heart. *Nature Biotechnol.* **23**, 845–856.

Lelkes, P. I., and Unsworth, B. R. (2002). Neuroectodermal cell culture: Endocrine cells. *In* "Methods of Tissue Engineering" (A. Atala and R. P. Lanza, eds.), pp. 371–382. Academic Press, London.

MacKay, J., and Mensah, G. A. (2004)."The Atlas of Heart Disease and Stroke," pp. 48–49. World Health Organization, Zurich, Switzerland, and Centers for Disease Control and Prevention, Atlanta, GA.

Maltsev, V. A., Rohwedel, J., Hescheler, J., and Wobus, A. M. (1993). Embryonic stem cells differentiate *in vitro* into cardiomyocytes representing sinusnodal, atrial and ventricular cell types. *Mech. Dev.* **44**, 41–50.

Maltsev, V. A., Wobus, A. M., Rohwedel, J., Bader, M., and Hescheler, J. (1994). Cardiomyocytes differentiated *in vitro* from embryonic stem cells developmentally express cardiac-specific genes and ionic currents. *Circ. Res.* **75**, 233–244.

Mummery, C., Ward-van Oostwaard, D., Doevendans, P., Spijker, R., van den Brink, S., Hassink, R., van der Heyden, M., Opthof, T., Pera, M., de la Riviere, A. B., Passier, R., and Tertoolen, L. (2002). Differentiation of human embryonic stem cells to cardiomyocytes: Role of coculture with visceral endoderm-like cells. *Circulation* **107**, 2733–2740.

Reinlib, L., and Field, L. (2000). Cell transplantation as future therapy for cardiovascular disease? A workshop of the National Heart, Lung, and Blood Institute. *Circulation* **101**, E182–E187.

Schuldiner, M., Yanuka, O., Itskovitz-Eldor, J., Melton, D. A., and Benvenisty, N. (2000). Effects of eight growth factors on the differentiation of cells derived from human embryonic stem cells. *Proc. Natl. Acad. Sci. USA* **97**, 11307–11312.

Takahashi, T., Lord, B., Schulze, P. C., Fryer, R. M., Sarang, S. S., Gullans, S. R., and Lee, R. T. (2003). Ascorbic acid enhances differentiation of embryonic stem cells into cardiac myocytes. *Circulation* **107**, 1912–1916.

van Laake, L. W., van Hoof, D., and Mummery, C. L. (2005). Cardiomyocytes derived from stem cells. *Ann. Med.* **37**, 499–512.

Ventura, C., and Maioli, M. (2000). Opioid peptide gene expression primes cardiogenesis in embryonal pluripotent stem cells. *Circ. Res.* **87**, 189–194.

Wartenberg, M., Dönmez, F., Ling, F. C., Acker, H., Hescheler, J., and Sauer, H. (2001). Tumor-induced angiogenesis studied in confrontation cultures of multicellular tumor spheroids and embryoid bodies grown from pluripotent embryonic stem cells. *FASEB J.* **15**, 995–1005.

Wei, H., Juhasz, O., Li, J., Tarasova, Y. S., and Boheler, K. R. (2005). Embryonic stem cells and cardiomyocyte differentiation: Phenotypic and molecular analyses. *J. Cell. Mol. Med.* **9**, 804–817.

Wobus, A. M., Guan, K., Yang, H. T., and Boheler, K. R. (2002). Embryonic stem cells as a model to study cardiac, skeletal muscle, and vascular smooth muscle cell differentiation. *Methods Mol. Biol.* **185**, 127–156.

Wobus, A. M., Kaomei, G., Shan, J., Wellner, M. C., Rohwedel, J., Ji, G., Fleischmann, B., Katus, H. A., Hescheler, J., and Franz, W. M. (1997). Retinoic acid accelerates embryonic stem cell-derived cardiac differentiation and enhances development of ventricular cardiomyocytes. *J. Mol. Cell. Cardiol.* **29**, 1525–1539.

Wobus, A. M., Wallukat, G., and Hescheler, J. (1991). Pluripotent mouse embryonic stem cells are able to differentiate into cardiomyocytes expressing chronotropic responses to adrenergic and cholinergic agents and Ca^{2+} channel blockers. *Differentiation* **48**, 173–182.

Xu, C., Police, S., Rao, N., and Carpenter, M. K. (2002). Characterization and enrichment of cardiomyocytes derived from human embryonic stem cells. *Circ. Res.* **91**, 501–508.

Zimmermann, W. H., and Eschenhagen, T. (2003). Cardiac tissue engineering for replacement therapy. *Heart Failure Rev.* **8**, 259–269.

[17] Oocytes

By Karin Hübner, James Kehler, and Hans R. Schöler

Abstract

Embryonic stem cells (ESCs), derivatives of totipotential cells of early mammalian embryos, have proven to be one of the most powerful tools for studying developmental and stem cell biology. When injected into embryos, ESCs can contribute to tissues derived from all three germ layers and to the germ line. Prior studies have successfully shown that ESCs can recapitulate features of embryonic development by spontaneously forming somatic lineages in culture. More recent studies using differentiating monolayer cultures and embryoid bodies have shown that mouse ESCs can also form germ cells that are capable of undergoing meiosis and forming both male and female gametes. This chapter provides detailed instruction on how to differentiate ESCs in monolayer cultures to derive germ cells and oocyte-like structures and presents standard methodologies for detecting expression of key genetic pathways required for primordial germ cell (PGC) development and oogenesis *in vivo*. While the full potential of these ESC-derived germ cells and oocyte-like structures remains to be demonstrated, this assay provides a new approach to studying reproductive developmental biology *in vitro*.

Introduction

Mouse ESCs, derivatives of inner cell mass (ICM) cells, are capable of reentering the totipotency cycle both *in vivo* and *in vitro*. When ESCs are supported by a recipient embryo, they are capable of differentiating into all three germ layers (endoderm, mesoderm, and ectoderm) and germ cells (Nagy *et al.*, 1990; Robertson *et al.*, 1986; Tam and Rossant, 2003). Surprisingly, in the absence of an embryonic body plan to direct development, ESCs are capable of spontaneously recapitulating early embryonic stages culminating in the specification of the germ line in culture. Three studies have demonstrated that mouse ESCs can form early germ cells that are subsequently capable of forming female and male gametes, respectively, in both monolayer cultures and embryoid bodies (EBs) (Geijsen *et al.*, 2004; Hübner *et al.*, 2003; Toyooka *et al.*, 2003). More recently, early oocyte-like structures have also been detected in EBs grown in conditioned medium from neonatal testicular cell cultures. However, further maturation of these oocyte-like structures remains to be demonstrated

METHODS IN ENZYMOLOGY, VOL. 418   0076-6879/06 $35.00

(Lacham-Kaplan *et al.*, 2005). In monolayer cultures, rare oocyte-like structures can progress to form structures that exhibit the appropriate polarity, morphology, and gene expression of blastocysts, completing the totipotency cycle in culture (Hübner *et al.*, 2003). While the ability of these gametes and embryonic structures to support development to term still has to be shown, this *in vitro* assay provides a powerful approach to test the full developmental capacity of cell lines currently accepted as pluripotential.

The transcription factor Oct4 was one of the first genes identified as being required for the continuity of the germ line *in vivo,* and progress in understanding its function and regulation has provided an effective tool for studying ESC maintenance and differentiation *in vitro*. Oct4 has at least two different known functions: it is required for (1) the maintenance of totipotential cells of the early mouse embryo and their ESC derivatives (Nichols *et al.*, 1996; Pesce *et al.*, 1998) and (2) the survival of PGCs (Kehler *et al.*, 2004). Oct4 is differentially regulated in ESCs and PGCs. Both ESCs and PGCs require the presence of a distal element (DE) within the *Oct4* promoter (Yeom *et al.*, 1996; Yoshimizu *et al.*, 1999). Deletion within the proximal element (PE) of CR3, a conserved region identical in several mammalian orthologs (Nordhoff *et al.*, 2001), reduces the activity of an Oct4–GFP transgene within ESCs, but maintains strong expression within PGCs, thus facilitating the detection of germ cells in monolayer cultures of differentiating ESCs (Hübner *et al.*, 2003).

Before inducing the differentiation of ESCs, it is critical that ESCs are proliferating in a pluripotential state (reviewed by Nagy and Vintersten, 2006). In addition to *Oct4* expression, several additional genes have been implicated in the maintenance of pluripotency in mouse ESCs. A feeder layer of mouse embryonic fibroblasts (MEFs) that secrete several soluble factors, including leukemia inhibitory factor (LIF) into the culture media, and/or the addition of recombinant LIF is required to support mouse ESCs in an undifferentiated state (Smith *et al.*, 1988). Binding of LIF and other members of the interleukin-6 family of cytokines to gp-130/LIFRβ heterodimer receptors on the surface of mouse ESCs (Nakamura *et al.*, 1998; Stahl *et al.*, 1994) activates the Jak family kinase/Stat3 signaling pathway required for the maintenance of pluripotency (Matsuda *et al.*, 1999; Niwa *et al.*, 1998, 2000). Independent expression of *nanog*, a divergent homeodomain gene, is also required for the maintenance of pluripotency in ESCs (Chambers *et al.*, 2003; Mitsui *et al.*, 2003). In both monolayer and EB cultures, removal of MEFs and LIF that support pluripotency is required to initiate germ cell differentiation.

In the past, overlap in the expression of several cell surface receptors and transcription factors such as Oct4 has confounded the detection of germ cells from ESCs *in vitro*. Despite this limitation, the expression of some additional markers, such as tissue-nonspecific alkaline phosphatase (TNAP), Blimp1, stella, and fragilis, which indicate the onset of germ cell competence and commitment in the epiblast (Ohinata *et al.*, 2005; Saitou *et al.*, 2002), has been used with experimental finesse to detect PGC formation *in vitro* (Geijsen *et al.*, 2004). The subsequent expression and activation of the c-Kit tyrosine kinase receptor by stem cell factor (SCF) are required for the survival and proliferation of migratory stage PGCs *in vivo* (De Miguel *et al.*, 2002; Matsui *et al.*, 1991) and can also be used to detect and isolate germ cells forming *in vitro* (Hübner *et al.*, 2003). Likewise, post-migratory changes in PGCs, such as the onset of expression of the *murine vasa homolog* (*mvh*) gene (Toyooka *et al.*, 2000), can also be used to identify cells committed to the germ line in culture (Toyooka *et al.*, 2003).

While the detection of meiotic markers provides additional proof of the irreversible progression of ESC-derived cells into the germ line, and delayed expression of SRY in monolayer cultures provides a mechanism for the entry of these XY ES cells into the default pathway of female sexual differentiation, observed morphological changes provide the most compelling evidence of oogenesis occurring after 2 weeks of differentiation *in vitro*. In the developing female genital ridge, PGCs first express the mouse homologue of the yeast meiosis-specific gene *DMC1* by 13.5 days postcoitus (dpc), coincident with their entry and arrest in prophase I of meiosis (Menke *et al.*, 2003). Whereas in the male genital ridge the onset of *SRY* expression in pre-Sertoli cells between 11.5 and 12.5 dpc initiates testicular sex cord formation (Tilmann and Capel, 2002), PGCs in this environment undergo mitotic arrest by 13.5 dpc and become committed to a spermatogenic fate by 14.5 dpc (Albrecht and Eicher, 2001; Ohta *et al.*, 2004). In the absence of *SRY* expression in ectopic locations such as the adrenal gland or in primary cultures, XY PGCs enter the default, female pathway, adopting morphological changes consistent with oogonial differentiation and meiotic arrest (Adams and McLaren, 2002; Upadhyay and Zamboni, 1982). *SRY* expression has been detected in EBs as early as 5 days of differentiation (when germ cells were also present), presumably triggering the concurrent development of pre-Sertoli cells capable of supporting the subsequent formation of rare haploid male germ cells in culture (Geijsen *et al.*, 2004). In monolayer cultures, *SRY* expression by reverse transcription-polymerase chain reaction (RT-PCR) is not normally detected until after the onset of *DMC1* expression and the formation of follicle-like structures containing oocyte-like cells between 15 and 25 μm in diameter (Hübner *et al.*, 2003). Further staining of these early oocyte-like cells for the synaptonemal complex protein 3 (SCP3) by immunohisto-chemistry (IHC) may reveal a range in nuclear staining patterns that is suggestive of meiotic progression concurrent with oocyte growth (Dobson et al., 1994) that is similar to in vivo controls (Hübner et al., 2003).

While direct observation of monolayer cultures between 16 and 23 days of differentiation should reveal growing follicle-like structures, several ancillary assays can be used to infer whether key two-way signaling interactions between oocytes and granulosa cells critical for primordial follicular development and the resumption of meiosis in vivo are also occurring in vitro. During this time period, transient expression of the growth differentiation factor 9 (GDF9) that is normally produced by oocytes to stimulate granulosa cell proliferation (Findlay et al., 2002) can be detected by quantitative RT-PCR in differentiating cultures. In addition, activation of the estrogenic biosynthetic pathway can be detected indirectly by RT-PCR for an increase in key enzymes or directly by ELISA for an increase in estradiol that is normally formed by granulosa cells and is required for oocyte meiotic maturation. More importantly, careful observations during this time period should reveal the growth of primary and secondary follicle-like structures containing single, central oocyte-like cells (Hübner et al., 2003).

Between 23 and 25 days of differentiation, key morphological changes indicative of progression of the reductional division stages of meiosis I should become evident in these oocyte-like structures arising in follicle-like structures in culture. In the mouse and most mammalian species, a fully grown oocyte must first undergo nuclear, epigenetic, and cytoplasmic maturation and resume meiosis during estrus or in culture starting with the breakdown of the germinal vesicle and ending with the extrusion of first polar body (PB1) and formation of a second meiotic spindle (MII) (Eppig et al., 2003). Small oocyte-like structures surrounded by a zona pellucida (ZP)-like matrix reactive for the ZP2 protein can be detected in both primary and secondary cultures that resemble mature MII oocytes produced from cultured primary follicles. Further staining of these rare oocyte-like structures with the DNA-binding dye Hoechst 33342 may reveal nuclear material within a PB-like vesicle in the presumptive perivitelline space, parallel to an area of diffuse nuclear material within the cytoplasm, suggestive of a MII spindle (Kehler et al., 2005). While mouse oocytes normally remain arrested at MII in vivo until after ovulation and activation by fertilization, this process can be disturbed in vitro.

Between 42 and 45 days of differentiation, multicellular structures may arise spontaneously in primary and secondary monolayer cultures, and their embryonic-like organization can be supported by further IHC studies. Parthenogenetic activation provides a likely explanation for this occurrence, as oocytes in most mammalian species can progress to MII readily in vitro and in vivo due to various environmental stimuli and misexpression of key components of a meiotic cytostatic factor (Fan and Sun, 2004).

In the absence of spermatozoa in our primary and secondary cultures, multicellular (2 to 64 cell) embryo-like, morula, and blastocyst-like structures were detected floating in the supernatant and trapped within follicle-like structures during this time. Some resembled *in vivo* matured blastocysts with an inner cluster of cells expressing the endogenous Oct4 protein surrounded by an outer single cell layer expressing the trophectodermal marker TROMA as detected by IHC (Hübner *et al.*, 2003). While their karyotype and developmental potential remain to be determined, the generation of these blastocyst-like structures represents the completion of the totipotency cycle *in vitro* and demonstrates the utility of monolayer cultures for studying oogenesis and early embryonic development.

General Technical Comments

The majority of laboratories that have successfully demonstrated the capacity of ESCs to form germ cells *in vitro* have used EBs (Geijsen *et al.*, 2004; Lacham-Kaplan *et al.*, 2005; Toyooka *et al.*, 2003). This protocol relies upon differentiating ESCs in monolayer cultures to induce the concurrent formation of germ cells and, equally importantly, the ovarian stromal cells necessary to support their further development as oocytes. Establishment of this method was facilitated by the use of ESCs carrying an Oct4-GFP transgene that is expressed only in germ cells (Hübner *et al.*, 2003). While useful for their initial identification, this transgene is not a prerequisite for the successful derivation of germ cells from ESCs *in vitro*. Furthermore, the distinct size and behavior of female germ cells, in particular the maturing ones, make their identification possible even without a genetic marker.

The main concern of any differentiation procedure is yield and quality of the desired cell type. Yield and quality of oocytes in this differentiation protocol depend greatly on the quality of the initial ESC culture: the better the quality of the starting ESCs, the better the quantity of resulting oocytes. Partially differentiated ESCs will not produce sufficient follicle-like structures and subsequently will yield only a few to no oocytes. Another critical step in this protocol is the complete removal of MEF cells, before the initial plating of ESCs for differentiation. Even a small percentage of contaminating feeder cells has a detrimental effect on the initial germ cell differentiation and on the formation of follicle-like structures, even though their mechanism of action in this context is not known.

It is also noteworthy to mention that the cell populations developing within a plate are not synchronized and that all of the time points mentioned here are intended merely as a guideline rather than as fixed time points for certain key events during ESC differentiation.

Embryonic Stem Cell Culture

ESCs can be maintained indefinitely in culture as a pluripotential population of undifferentiated cells if handled carefully (Smith *et al.*, 2001). Maintenance of undifferentiated mouse ESCs is reliant on appropriate culture conditions and requires pretesting of serum and supporting MEF feeder cells, as well as close monitoring of size and morphology of the colonies. As the likelihood of developing a heterogeneous or prematurely differentiating ESC culture increases with time, start with early passage ESCs.

ESCs are plated as a single cell suspension on mitomycin C-treated or irradiated MEF cells on 0.1% gelatin-coated tissue culture plates in LIF-supplemented ESC medium or knockout/Serum-Replacement medium (KO/SR medium), respectively. Cultures are incubated at 37° in a humidified incubator at 5% CO_2. Cultures are trypsinized and expanded at a ratio of 1:5 on fresh feeder cells every 48 h, when the colony size reaches approximately 80 to 100 μm in diameter. The colonies should be solid with smooth edges and no apparent cell structure within (Fig. 1).

Medium and Reagents

ESC medium: DMEM containing 4.5 g/liter glucose (Invitrogen), 15% fetal calf serum (FCS) (Hyclone), 2 mM L-glutamine (Invitrogen), 100 μM nonessential amino acids (Invitrogen), 1 μM β-mercaptoethanol (Sigma), 50 μg/ml each penicillin and streptomycin (Invitrogen), and 1000 U/ml murine LIF (ESGRO TM; Chemicon)

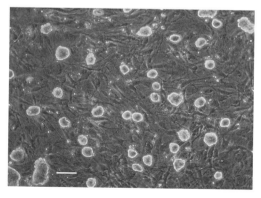

FIG. 1. Phase contrast image of undifferentiated ESCs growing on MEF feeder cells. Scale bar: 100 μm. Magnification 100×.

KO/SR medium: KO-DMEM (Invitrogen), 15% Serum Replacement
(Invitrogen), 2 mM L-glutamine (Invitrogen), 100 μM nonessen-
tial amino acids (Invitrogen), 1 μM β-mercaptoethanol (Sigma),
50 μg/ml each penicillin and streptomycin (Invitrogen), and
1000 U/ml murine LIF (ESGRO TM; Chemicon)
FCS: (Hyclone) heat inactivated for 30 min at 56°
Phosphate-buffered saline (PBS): 1× PBS, Ca^{2+} and Mg^{2+} free
(Invitrogen)
Gelatin: 0.1% in PBS (Invitrogen), 0.45 μm filtered
Trypsin: 0.25% trypsin/EDTA (Invitrogen)

MEF Removal from ESC Cultures

Trypsinize ESCs grown on MEF cells and collect the entire cell suspension in ESC medium in a 50-ml Falcon tube. Gently disaggregate cells by
pipetting the suspension repeatedly up and down. To remove feeders, one
utilizes their ability to reattach faster to a tissue culture plate than ESCs.
Replate the cell suspension on a gelatinized tissue culture (TC) plate and let
feeder cells attach for 15 to 30 min in an incubator. Good MEFs will attach to
the plate within 15 to 20 min, while ESCs should remain suspended in the
supernatant. Take the supernatant off and replate the cell suspension a
second time for another 15 to 20 min. The resulting secondary supernatant
should now contain about 98% ESCs and only about 2% MEFs.

However, the ability of MEF feeder batches to adhere rapidly varies and
decreases with increasing passage number. Therefore, it is advisable to also
use early passage (three to five) MEFs. In our hands, some feeder batches
did not attach well and required up to 45 min of replating. Careful monitoring is advisable! In this scenario, ESCs will also attach loosely to the already
attached feeder cell population, but they can be dislodged easily by tapping
the plate gently. Spin the supernatant down at 200g and resuspend the cell
pellet in an appropriate volume for cell counting. A manual, differential cell
count can be performed based on differences in size and morphology to
determine the total number and purity of the ESCs recovered.

ESC Differentiation and the Emergence of Germ Cells

Day 0. Plate MEF-free ESCs at a density of 1 to 2.5 × 10^4 cells/cm^2 in
gelatinized 6-cm tissue culture plates in ESC medium without LIF (differentiation medium). Please note that not all batches of fetal calf serum that
will support undifferentiated ESCs will also support germ cell differentiation. It is critical to test several batches of FCS from different sources, as
we have found that on average only 25% of lots of ESC-certified FCS will
support germ cell differentiation. In addition, the cell density is crucial for

the induction of germ cell formation. A lower cell number will not yield sufficient germ cell colonies, and a significantly higher cell density triggers the spontaneous formation of EBs that eventually develop into cyst-like structures. Undesired EB formation will also occur if ESCs are homogenized incompletely and not seeded as a single cell suspension. Plated cells adhere within 24 h and exhibit a morphology atypical of ESC cultures: they grow adherently as a monolayer and do not form the round to oblong colonies typical of undifferentiated ESCs.

Days 3 to 8. Replace medium on day 3, if cultures have grown confluent. Over the next few days the cell layer will overgrow, and substantial cell death takes place. The medium should then be changed every day and care should be taken to remove as many dead cells and debris as possible. Even though the cultures appear to die, germ cell formation is initiated during this same time.

Detection of Early Stage Germ Cells

Flow cytometry using Oct4-GFP ESCs or antibodies for specific cell surface markers provides an effective method for detecting germ cells as early as day 4 of differentiation. By day 7 of differentiation, most cultures contain on average 25% germ cells as suggested by Oct4-GFP expression, increasing to about 40% by day 8. The majority of these 7-day-old, *in vitro*-derived Oct4-GFP+ germ cells coexpress the c-Kit receptor at levels similar to *in vivo*-derived male, late-stage PGCs (Fig. 2).

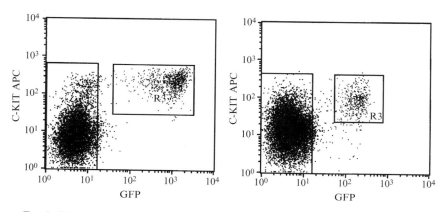

FIG. 2. Flow cytometric dot plots comparing coexpression of Oct4-GFP (*X* axis) and the c-Kit receptor (*Y* axis) in *in vivo*-derived PGCs (left R3 gate) from 13.5-dpc male gonadal ridges and *in vitro*-derived germ cells (right R3 gate) after 7 days of differentiation. Axes are marked in a logarithmic scale of arbitrary light intensity units.

Flow cytometric isolation based on c-Kit and Oct4-GFP expression into subpopulations for subsequent mRNA analyses of early germ cell markers suggests that monolayer cultures contain heterogeneous germ cells at different developmental stages by 1 week of differentiation (Hübner *et al.*, 2003).

Detection of Late Stage Germ Cells

Days 8 to 10. From this stage on the cultures usually contain only small amounts of dead cells, and the medium should be replaced every other day. Morphological changes within the culture should now be evident and a variety of different cell types can be observed. Early germ cells grow initially as flat-appearing patches and develop into large colonies of up to a couple of hundred cells with close cell–cell contacts (Fig. 3).

Around day 8 of differentiation cells within these colonies round up and physically separate from each other. In phase contrast at a 100× magnification these colonies can be identified easily as bright three-dimensional growing cell clusters that host round and light-refracting cells (Fig. 4). IHC for Oct4 and Vasa on colonies demonstrated an overall reduced Oct4 expression and a high percentage of Vasa+ cells (Hübner *et al.*, 2003), indicative of the development of postmigratory PGCs.

Formation of Follicle-Like Structures

Days 10 to 12. Small clusters of 5 to 20 viable cells containing both somatic and Vasa+ germ cells detach from the large colonies. These small clusters do not detach simultaneously, but around 60% of all detaching

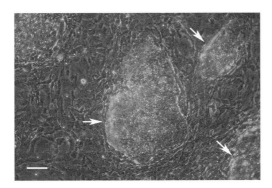

Fig. 3. Phase contrast image of an ESC differentiation culture at day 8 showing patches of early germ cells (arrows). Scale bar: 100 μm. Magnification 100×.

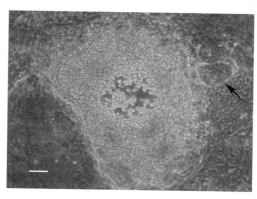

FIG. 4. Phase contrast image of a large germ cell colony at day 10. Note that individual cells start to detach from the center. The black arrow points to a small early germ cell cluster. Scale bar: 100 μm. Magnification 100×.

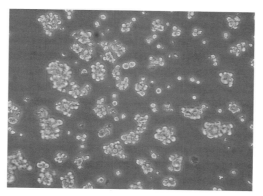

FIG. 5. Phase contrast image of detaching small cell clusters of an ESC differentiation culture at day 8. Scale bar: 50 μm. Magnification 100×.

cells are reproducibly observed around days 11 to 12. Collect and combine cell clusters from about five 6-cm dishes and spin cells down at 100g. Gently replate the entire cell population onto a 6-cm bacterial culture, Petri dish (Fig. 5) and aggregate overnight in suspension culture. The majority will form large aggregates within 16 h. Do not use tissue culture-treated dishes for this aggregation step, as this will promote premature readhesion.

Days 13 to 18. Transfer aggregates to *in vitro* maturation (IVM) medium and plate them in four-well plates at a density of around 5 to

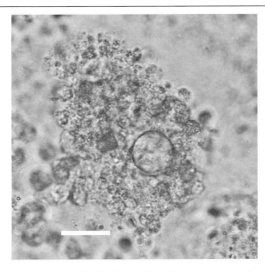

FIG. 6. Phase contrast image of a floating follicle-like structure 4 days after replating of aggregates. Scale bar: 30 μm. Magnification 400×.

10 aggregates/well. Five 6-cm plates should yield at least 200 aggregates. Within 4 days these aggregates organize into follicle-like structures with central oocyte-like cells of 15 to 25 μm in size surrounded by proliferating cells of around 10 μm (Fig. 6). Morphologically similar follicle-like structures also develop in the master plates. However, these structures are embedded in the primary cell layers and are more difficult to identify. Interestingly, a large number of the remaining germ cell colonies apparently get embedded in a jelly-like matrix and do not disintegrate. While somatic cells keep on proliferating, and the overall morphology of the culture changes toward tissue-like structures, germ cells within this matrix do not seem to differentiate or develop any further over the next weeks. Keep master plates in differentiation medium and change the medium every other day.

RT-PCR should reveal the onset of *StAR, Cyp17,* and *p450 aromatase* expression in primary and secondary aggregate cultures by day 16, concurrent with rising estradiol levels in the medium, which indicates the presence of functional ovarian stromal cells. Additional evidence for folliculogenesis is the transient expression of *GDF-9* in all cultures by day 16 and a significant increase of *GDF-9* mRNA levels in replated aggregate cultures.

Medium and Reagents

Differentiation medium: ESC medium (DMEM with FCS not KO medium with SR) without LIF

IVM medium: MEMα (without ribonucleosides and deoxyribonucleosides, Invitrogen), supplemented with 3 mg/ml bovine serum albumin (BSA) (Serologicals Proteins, Inc.), 0.23 mM pyruvic acid, 5 mg/ml transferrin, 5 ng/ml selenium, 10 mg/ml insulin (ITS; Invitrogen), 1 ng/ml epidermal growth factor (EGF) (Chemicon), and 1 U/ml of gonadotropin pregnant mare serum gonadotropin (Chemicon)

Oocyte Growth

Days 19 to 30. Follicle-like structures in the aggregate cultures maintain their three-dimensional organization for several days but eventually attach to the dish, when their size exceeds approximately 250 μm. The majority of these structures degenerate and only about 10% produce germ cells of 50 to 70 μm in diameter. Oocyte-like structures within these aggregates are in general smaller than their natural counterparts and quite fragile, and isolation of the individual germ cells from accompanying cells by standard procedures such as trituration or enzymatic digestion reduces the viability of these cells dramatically. Some oocyte-like structures with a thin zona pellucida are released during this culture period and can be collected for further analysis from the supernatant of both replated aggregate cultures and at lower yield also from primary cultures (Fig. 7).

Immunohistochemistry on single oocyte structures should show a positive reactivity for ZP2, one of the key components of the zona pellucida. RT-PCR analyses may also detect *figα, ZP2,* and *ZP3* gene expression in the

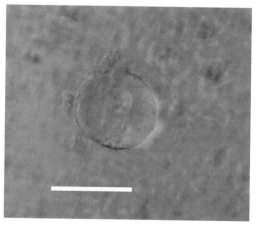

FIG. 7. Phase contrast image of a floating oocyte in the supernatant of a master plate at day 24 of ESC differentiation. Scale bar: 50 μm. Magnification 400×.

primary adherent and in the secondary replated aggregate cultures supportive of female gametogenesis occurring in these cultures. ZP1, the linker protein between ZP2 and ZP3, could not be detected and might explain the fragile nature of the matrix surrounding *in vitro*-derived oocytes.

Oocyte Maturation

Between 21 and 25 days of culture a few of the replated oocyte structures demonstrate resumption of meiosis I by extrusion of the first polar body and formation of a second meiotic spindle. Determination of the exact number of MII oocytes in these cultures proves difficult, as the majority of germ cells are surrounded or embedded by proliferating somatic cells and matrix at this time. Hoechst 33342 staining of the nuclear material within the PB was only possible on oocytes that were exposed on the surface of the cell layer.

Cleavage Stage Embryos

Days 42 to 45. Primary and secondary cultures contain adherent as well as floating embryonic structures that resemble preimplantation embryos (multicellular embryo-like, morula, and blastocyst-like structures) (Fig. 8). These embryos are surrounded by a zona pellucida-like matrix when embedded in neighboring cells or within a follicle-like structure, but frequently appear naked when floating in the supernatant. Some of the blastocyst-like structures show abnormal morphologies, whereas others

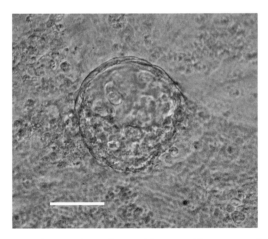

FIG. 8. Phase contrast image of an adherent blastocyst-like structure in a primary culture plate. Scale bar: 50 μm. Magnification 400×.

resemble *in vivo*-matured blastocysts. The yield of these embryonic structures fluctuates greatly from experiment to experiment, and identification and subsequent isolation of these structures require careful monitoring of each individual plate.

Parthenogenetic activation might be responsible for cleavage of *in vitro*-derived oocytes, but the exact mechanisms remain to be determined. Morula and blastocyst-like structures were analyzed by IHC for Oct4 and Troma-1, respectively, and exhibited the typical protein distribution for these embryonic stages. Expression of trophectodermal markers such as *MASH2* by RT-PCR further suggested the formation of trophoblast in these cultures.

Reoccurring Events

Day 35+. A very interesting phenomenon is the detection of reoccurring early germ cell formation in the primary master plates after approximately 35 days of differentiation, similar to the initial germ cell formation observed between days 4 and 7 of culture. A significant increase in the GFP signal in Oct4-GFP ESC cultures indicated a new cycle of germ cell proliferation and was coincident with the sudden cell death of somatic cell types. Undetermined enzymatic activities could be responsible for the breakdown of the extracellular matrix and a subsequent release of previously formed germ cells and/or activation of signaling pathways could stimulate the second cycle of germ cell proliferation. Further development, analogous to initial 8- to 18-day cultures, was observed only in low-density areas of the plates, where aggregate formation was too inefficient for collection and replating. However, single oocytes of varying size and surrounded by ZP were again found floating in the medium between 50 and 60 days of culture.

Schematic Overview of the *In Vitro* Derivation of Oocytes
from Murine ESCs (Fig. 9)

Analytical Methods

Flow Cytometry. Plates of differentiating ESCs are trypsinized as for ESCs, inactivated in differentiation medium, centrifuged at 800g for 6 min at 4°, and resuspended in PBS with 0.4% bovine serum albumin (PBS-BSA). Single cell suspensions are incubated for 30 min on ice in PBS-BSA containing 1 μg/million cells of a monoclonal antibody against the c-Kit receptor (anti-CD117 antibody, clone 2B6) conjugated with allophycocyanin (APC) or with an APC-isotype control (both from BD Pharmingen) and then rinsed twice in PBS-BSA. FACS analyses and cell sorting are performed using Becton Dickinson FACScalibur and FACScan flow cytometers (Becton Dickinson).

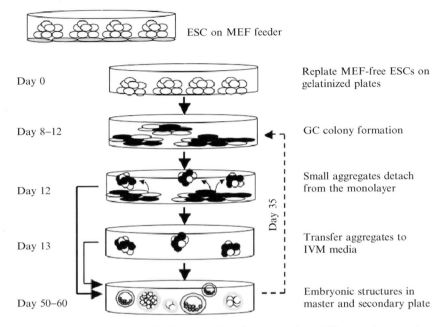

FIG. 9. Overview of the derivation procedure for oocytes from differentiating monolayer cultures of murine ESCs. Conventionally grown undifferentiated ESCs are depleted of MEF feeder cells and LIF and replated for differentiation on gelatinized TC plates. After 8 to 12 days of culture, large GC colonies form from which small aggregates, consisting of GCs and somatic cells, detach. Aggregates are collected and replated at high density and subsequently switched to IVM medium to support maturation of oocytes within follicle-like structures. Between days 13 and 50 various developmental stages of germ cell maturation can be observed in primary as well as secondary cultures. Around day 35 of culture, reoccurring early germ cell formation, similar to the initial germ cell formation between days 4 and 7 of differentiation, can be observed in master plates.

Hormone Measurements

Estradiol concentrations in media are determined using the respective Coat-A-Count tubes and reagents (Diagnostic Products, Los Angeles, CA), according to the manufacturer.

RT-PCR. Total cellular RNA is prepared by the guanidinium thiocyanate–phenol–chloroform extraction procedure using the TôTALLY RNA kit (Ambion). After checking the RNA integrity on an ethidium bromide-stained 1% agarose gel, RNA are treated with DNase I (Amplification grade, Invitrogen) according to the manufacturer's instructions. Reverse transcription is then carried out using the SuperScript II first-strand synthesis kit (Invitrogen). RT-PCR analysis is performed as described (Botquin *et al.*, 1998) with the following modification: amplification is achieved after 35 to 40 cycles of <94°, 30 s; 50 to 55°, 30 s; 72°, 45 s>, depending on the primer sets (Table I).

Quantitative Real-Time RT-PCR. cDNA from the reverse transcriptase (RT) reaction is diluted 1:10 in nuclease-free water, and 1-μl aliquots of each sample are tested in triplicate for each different quantitative PCR reaction as described (Hübner *et al.*, 2003). Primers for analysis of the mouse StAR, GDF-9, cytochrome P450 aromatase (Cyp 19), and cytochrome P450 17-α cDNAs are designed with the Primer Express software package that accompanies the Applied Biosystems Model 7700 and 7900 Sequence Detection Systems as described previously (Hiroi *et al.*, 2004). Primers and conditions used to quantitate StAR cDNA are as described previously (Hiroi *et al.*, 2004). The SYBR green reagent (Applied Biosystems) is used to detect amplicons of the murine cDNAs (Table II).

Primer concentrations for each target cDNA are determined empirically. Agarose gel electrophoresis and analysis of the dissociation curve indicate the presence of a single PCR product for all the amplifications. To account for differences in starting material, rodent glyceraldehyde-3-phosphate dehydrogenase (GAPDH) primers and reagents from Applied Biosystems are used following the manufacturer's instructions. The relative amounts of experimental and GAPDH PCR products are determined by comparison to a standard curve generated by serial dilution of a sample containing high levels of the target amplicons that are also run in triplicate (Jabara *et al.*, 2003). An arbitrary value for the template concentration is assigned to the highest standard, and the corresponding values for the subsequent dilutions (4- to 6-point standard curves were used) are derived from the former. These relative values are plotted against the threshold value for each dilution to generate a standard curve. The relative amount of each target transcript and the endogenous control GAPDH are derived by interpolation on the standard curve. The average of the experimental triplicates is normalized against GAPDH, and the resulting values are used to compare expression across the different cultures (Jabara *et al.*, 2003)(Table II).

Immunocytochemistry

Adherent cells are fixed in 4% paraformaldehyde (PFA) in PBS, pH 7.4; single cells are fixed in 2% PFA for 1 h at ambient temperature. Fixed cells are rinsed in PBS, pH 7.4, quenched for 15 min in 50 mM glycine, pH 7.5, and treated with blocking solution for a minimum of 1 h before stained with primary antibodies for Mvh (Vasa) protein (1:500), ZP2 (IE-3 1:200) or ZP3 (IE-10 1:300) (East *et al.*, 1984, 1985), Oct4 (1:100), or Troma-1 (1:10) for 1 to 24 h. Cells are rinsed three times in PBS or PBS with 0.1% Triton

TABLE I
RT-PCR PRIMERS

Gene	Sense (5'-3')	Antisense (3'-5')	Accession #	Size (bp)
β-actin	GGTCAGAAGGACTCCTATGT	ATGAGGTAGTCTGTCAGGTC	X03672	428
Oct4	ATGGCATACTGTGGACCTCA	CAGAAACATGGTCTCCAGAC	X52437	528
c-Kit	CATCCATCCATCCAGCACAA	CACAACAGGGATAGCCTTGA	NM_021099	536
ZP1	CTGCTCTATCTGCTACCACT	CTACAGAATCCAGCTCAGTC	NM_009580	455
ZP2	CCAGTTCTACCCTCTTCAC	CCAGCAGTCATCTAAGACCA	NM_011775	637
ZP3	TCCTCATCACCTGTGAAGGT	GTGGAAGTCCACGATGAAGT	NM_011776	603
FIGα	TGCTGGAGGAGCTGATTCAG	TCTTCAAGCCACTCGCACAG	NM_012013	429
Vasa	ACTCCTGTGCAGAAGTACAC	GCTTGCCCAACAGCAACAAAC	NM_010029	589
SRY	GGAGGCACAGAGATTGAAGA	TGGTGGTGGTTATGGAACTG	NM_011564	312
DMC1	GTGTGACAGCTCAACTTCCA	AAGGTCATAGTTGCTCCTGG	NM_010059	421
SCP3	GTGGAAGAAAGCATTCTGGG	TTTGCCATCTCTTGTGCTG	NM_011517	661
MASH2	TGAAGGTGCAAACGTCCACT	CTCCACCTTACTCAGCTTCT	NM_008554	376
HAND1	CCAAACGAAAAGGCTCAGGA	CCGTCTTTTTGAGTTCAGCC	NM_008213	231
Pl1	TGGTGTCAAGCCTACTCCTT	TGTTAGCCTGAGACCTGTTG	M35662	438
Tpbp	TCCTAGTCATCCTATGCCTG	TCATCAACAACTGGCTGTGG	NM_009411	295

TABLE II
REAL-TIME RT-PCR PRIMER

cDNA	Forward	Reverse	Accession #
GDF-9	ATCGAGTGCA GTGTCCGTAGGT	TTCACTTGGTTT ATGGCAACGA	L06444
P450 Aromatase	TCATAGCTCCT ATGGTTTGTCATCA	TCACTGGT CCCCAACACAGA	NM_007810
P450 17-α	TGGAGGCCA CTATCCGAGAA	TGTTAGCCTT GTGTGGGATGAG	NM_007809
StAR	GAACAACCCT TGAGCACCTCAG	CCAACCCA CACTCACCTTTCAT	NM_011485.3

X-100 and incubated with the appropriate TRITC or FITC conjugates for 20 min in the dark. All incubations are in PBS, pH 7.4, 5% BSA, except for Mvh and Oct4 analyses when 0.1% Triton X-100 is added to the blocking solution to permeabilize the cell membranes. For nuclear staining, fixed cells are incubated in PBS for 15 min with Hoechst 33342 dye at a concentration of 1 μg/ml. Imaging is performed on a Leica DMIRB fluorescence microscope.

Reagents and Solutions

PFA: 2 or 4% PFA (Applichem) in PBS, pH 7.4

D-PBS: 1× D-PBS, Ca^{2+} and Mg^{2+} free (Invitrogen)

Triton X-100: 10% Triton X-100 (Applichem) in PBS as a 100× stock solution

BSA: IgG free, protease free (Jackson ImmunoResearch Labs)

Blocking solution: PBS, pH 7.4, 5% BSA, with or without 0.1% Triton X-100

Anti-Oct4: mouse monoclonal IgG2b (Santa Cruz)

Anti-Troma-1: mouse antirat monoclonal IgG (Developmental Studies Hybridoma Bank)

Anti-Mvh: kindly provided by Dr. T. Noce, Mitsubishi Kagaku Institute of Life Science

Anti-ZP2 (IE-3) and anti-ZP3 (IE-10): kindly provided by Dr. J. Dean, Laboratory of Cellular and Developmental Biology, NIDDK, National Institutes of Health, Bethesda, Maryland

Fluorescent secondary antibodies: TRITC or FITC conjugates (Jackson ImmunoResearch Labs)

Hoechst dye: Hoechst 33342 (Sigma, 1 μg/ml) in PBS

TABLE III
TROUBLESHOOTING

Unanticipated results	Critical parameters
ESC culture	
ESCs grow slow, colonies stay small	Inadequate serum batch; toxic effect of serum
ESCs grow fast and form flat cell patches rather than colonies	Serum promotes ESC differentiation; poor MEF quality; too high passage number of ESCs
MEF removal	
MEF cells do not reattach efficiently during feeder removal	Poor MEF quality, too high passage number of MEFs
ESC differentiation	
Excessive EB formation upon initial plating of ESCs on day 0	Incomplete homogenization of cell suspension; too high cell density; TC plates were not gelatinized
ESCs differentiate as colonies and not as a monolayer on gelatinized TC plates	Incomplete homogenization of cell suspension before plating
Early stage germ cell formation	
Cultures contain less than 10% of early germ cells by day 7 of differentiation	Initial cell density too low; serum batch does not support germ cell differentiation; presence of MEFs due to insufficient feeder removal; poor ESC quality
Late stage germ cell formation	
Cultures contain only few large germ cell colonies	Initial cell density too low; serum batch does not support germ cell differentiation; presence of MEFs due to insufficient feeder removal; poor ESC quality
Formation of follicle-like structures	
Failure to obtain viable, detaching cell clusters or low yield of aggregates	Initial cell density too low; poor initial ESC quality; small number of GC colonies
Small cell clusters attach immediately to Petri dish after replating and do not form larger aggregates	Detaching cells were collected prematurely; cells replated at too low density; presence of contaminating dead cells and debris
Aggregates disintegrate rapidly in IVM medium or develop into cyst-like structures	Too high aggregate concentration per well; insufficient germ cell number; presence of contaminating dead cells and debris
Oocyte growth	
Failure to detect floating oocyte-like structures in aggregate and/or primary master plates	Detection is difficult due to low number of cells and thin or absent zona pellucida; floating oocytes disintegrate rapidly in supernatant
Oocyte maturation	
Failure to detect MII oocyte-like structures in aggregate and/or primary master plates	High density of surrounding somatic cells; detection requires careful screening of cultures
Cleavage stage embryos	
Failure to detect embryo-like structures in aggregate and/or primary master plates	Embryo-like structures are frequently embedded in surrounding cell material or are attached to the surface of the cell layer; their morphological appearance might differ from wild type

Epifluorescence Microscopy

SCP3/COR1 and SCP1/SYN1 fluorescence microscopy is performed as described previously (De La Fuente and Eppig, 2001) with the following modification: poly-L-lysine-coated slides are dipped in 1% PFA, pH 7.4, 1.5% Triton X-100. Twenty-five to 50 cells are placed in the center of the slide and allowed to attach. Excessive PFA is removed carefully after complete lyses of the cell membranes and slides are air dried briefly. For nuclear staining, slides are incubated at room temperature for 45 min with primary antibodies diluted at 1:1000 for SCP3/COR1 or SCP1/SYN1 (Dobson *et al.*, 1994) in PBS-T, washed three times in PBS-T, and incubated in Cy3 conjugate (1:1000 in PBS). After three washes, slides are mounted in 4 μl of mounting solution.

For cytoplasmic staining, cells are fixed in the absence of Triton X-100 in 2% PFA/PBS, pH 7.4, blocked in PBS, pH 7.4, 10% FCS, and then permeabilized in PBS-T overnight at 4°. Cells are then incubated for 45 min with primary antibodies specific to SCP3/COR1 or SCP1/SYN1 (1:1000) in PBS-T, washed three times, and incubated for 45 min in Cy3 conjugate (1:1000). After three additional washes, cells are transferred to poly-L-lysine-coated slides and mounted in mounting solution. Imaging is performed on a Leica DMIRB fluorescence microscope.

Reagents and Solutions

   PFA: 1 or 2% PFA (Applichem) in PBS, pH 7.4
   D-PBS: 1× D-PBS, Ca^{2+} and Mg^{2+} free (Invitrogen)
   Triton X-100: 10% Triton X-100 (Applichem) in PBS as a 100× stock
      solution
   BSA: IgG free, protease free (Jackson ImmunoResearch Labs)
   FCS: Hyclone
   PBS-T: PBS, pH 7.4, 1% BSA, 0.05% Triton X-100
   Anti-Scp3/Cor1 and Scp1/SYN1: kindly provided by Drs. P. Moens
      and B. Spyropoulos, Department of Biology, York University,
      North York, Ontario, Canada
   Fluorescent secondary antibody: Cy3 conjugate (Jackson Immuno
      Research Labs)
   Mounting solution: Hoechst 33342 (Sigma, 10 μg/ml) in 10% glycerol

Conclusions

In retrospect, the ability of mouse ESCs to differentiate spontaneously into germ cells in culture is not surprising, as they can contribute to the germ line *in vivo* when injected into blastocysts. However, it is the concurrent somatic differentiation supporting the initiation and progression of mammalian oogenesis in culture that is truly surprising. The challenge remains to try and visualize these key transient processes as reproductive development is recapitulated *in vitro*. Initially, the development of new Oct4-GFP transgenes facilitated the detection of germ cells arising in differentiating ESC monolayer cultures and the establishment of the culture techniques presented here. However, this assay has been and can be readily used with wild type or other transgenic ESC lines, as the morphological changes of *in vitro*-derived germ cells become readily apparent as they enter oogenesis. While the ability of these oocyte-like structures to undergo meiotic initiation, maturation, activation, and formation of early embryonic-like structures in monolayer cultures is amazing (Hübner *et al.*, 2003), it will be crucial to demonstrate whether *in vitro*-derived oocytes can support normal embryonic development. Many aspects of the intercommunication between developing oocyte and follicular stromal cells remain to be investigated as well, and ESCs with targeted mutations could also be used in this assay toward that end. It is hoped that by reviewing this new procedure here we will encourage additional laboratories to utilize this *in vitro* assay to dissect the key signaling pathways involved in oogenesis (Wassarman, 1988).

Acknowledgments

The authors acknowledge the contributions discussed in this chapter made by their initial collaborators: M. Boiani, L. Christenson, R. De La Fuente, G. Fuhrmann, R. Reinbold, J. Strauss III, and J. Wood. We thank K. Psathaki, J. Sutter, and L. Gentile for contributions to this chapter. This research was supported in part by the NIH 1RO1HD06274. J. Kehler was supported by the NIH SERCA-NCRR/KO1 (RR019677–01). The Troma-1 hybridoma antibody, developed by P. Brulet and R. Kemler, was obtained from the Developmental Studies Hybridoma Bank developed under the auspices of the National Institute of Child Health and Human Development (of NIH) and maintained by the University of Iowa, Department of Biological Sciences.

References

Adams, I. R., and McLaren, A. (2002). Sexually dimorphic development of mouse primordial germ cells: Switching from oogenesis to spermatogenesis. *Development* **129,** 1155–1164.
Albrecht, K. H., and Eicher, E. M. (2001). Evidence that Sry is expressed in pre-Sertoli cells and Sertoli and granulosa cells have a common precursor. *Dev. Biol.* **240,** 92–107.
Botquin, V., Hess, H., Fuhrmann, G., Anastassiadis, C., Gross, M. K., Vriend, G., and Schöler, H. R. (1998). New POU dimer configuration mediates antagonistic control of an osteopontin preimplantation enhancer by Oct-4 and Sox-2. *Genes Dev.* **12,** 2073–2090.
Chambers, I., Colby, D., Robertson, M., Nichols, J., Lee, S., Tweedie, S., and Smith, A. (2003). Functional expression cloning of Nanog, a pluripotency sustaining factor in embryonic stem cells. *Cell* **113,** 643–655.

De La Fuente, R., and Eppig, J. J. (2001). Transcriptional activity of the mouse oocyte genome: Companion granulosa cells modulate transcription and chromatin remodeling. *Dev. Biol.* **229**, 224–236.

De Miguel, M. P., Cheng, L., Holland, E. C., Federspiel, M. J., and Donovan, P. J. (2002). Dissection of the c-Kit signaling pathway in mouse primordial germ cells by retroviral-mediated gene transfer. *Proc. Natl. Acad. Sci. USA* **99**, 10458–10463.

Dobson, M. J., Pearlman, R. E., Karaiskakis, A., Spyropoulos, B., and Moens, P. B. (1994). Synaptonemal complex proteins: Occurrence, epitope mapping and chromosome disjunction. *J. Cell Sci.* **107**(Pt. 10), 2749–2760.

East, I. J., Gulyas, B. J., and Dean, J. (1985). Monoclonal antibodies to the murine zona pellucida protein with sperm receptor activity: Effects on fertilization and early development. *Dev. Biol.* **109**, 268–273.

East, I. J., Mattison, D. R., and Dean, J. (1984). Monoclonal antibodies to the major protein of the murine zona pellucida: Effects on fertilization and early development. *Dev. Biol.* **104**, 49–56.

Eppig, J., Viveiros, M., Bivens, C., and De La Fuente, R. (2003). Regulation of mammalian oocyte maturation. *In* "The Ovary" (P. Long and E. Adashi, eds.), pp. 113–129. Elsevier, San Diego.

Fan, H. Y., and Sun, Q. Y. (2004). Involvement of mitogen-activated protein kinase cascade during oocyte maturation and fertilization in mammals. *Biol. Reprod.* **70**, 535–547.

Findlay, J. K., Drummond, A. E., Dyson, M. L., Baillie, A. J., Robertson, D. M., and Ethier, J. F. (2002). Recruitment and development of the follicle; the roles of the transforming growth factor-beta superfamily. *Mol. Cell. Endocrinol.* **191**, 35–43.

Geijsen, N., Horoschak, M., Kim, K., Gribnau, J., Eggan, K., and Daley, G. Q. (2004). Derivation of embryonic germ cells and male gametes from embryonic stem cells. *Nature* **427**, 148–154.

Hiroi, H., Christenson, L. K., Chang, L., Sammel, M. D., Berger, S. L., and Strauss, J. F., 3rd. (2004). Temporal and spatial changes in transcription factor binding and histone modifications at the steroidogenic acute regulatory protein (stAR) locus associated with stAR transcription. *Mol. Endocrinol.* **18**, 791–806.

Hübner, K., Fuhrmann, G., Christenson, L. K., Kehler, J., Reinbold, R., De La Fuente, R., Wood, J., Strauss, J. F., 3rd, Boiani, M., and Schöler, H. R. (2003). Derivation of oocytes from mouse embryonic stem cells. *Science* **300**, 1251–1256.

Jabara, S., Christenson, L. K., Wang, C. Y., McAllister, J. M., Javitt, N. B., Dunaif, A., and Strauss, J. F., 3rd (2003). Stromal cells of the human postmenopausal ovary display a distinctive biochemical and molecular phenotype. *J. Clin. Endocrinol. Metab.* **88**, 484–492.

Kehler, J., Hübner, K., Garrett, S., and Schöler, H. R. (2005). Generating oocytes and sperm from embryonic stem cells. *Semin. Reprod. Med.* **23**, 222–233.

Kehler, J., Tolkunova, E., Koschorz, B., Pesce, M., Gentile, L., Boiani, M., Lomeli, H., Nagy, A., McLaughlin, K. J., Schöler, H. R., and Tomilin, A. (2004). Oct4 is required for primordial germ cell survival. *EMBO Rep.* **5**, 1078–1083.

Lacham-Kaplan, O., Chy, H., and Trounson, A. (2005). Testicular cell conditioned medium supports differentiation of embryonic stem (ES) cells into ovarian structures containing oocytes. *Stem Cells* **24**(2), 266–273.

Matsuda, T., Nakamura, T., Nakao, K., Arai, T., Katsuki, M., Heike, T., and Yokota, T. (1999). STAT3 activation is sufficient to maintain an undifferentiated state of mouse embryonic stem cells. *EMBO J.* **18**, 4261–4269.

Matsui, Y., Toksoz, D., Nishikawa, S., Williams, D., Zsebo, K., and Hogan, B. L. (1991). Effect of Steel factor and leukaemia inhibitory factor on murine primordial germ cells in culture. *Nature* **353**, 750–752.

Menke, D. B., Koubova, J., and Page, D. C. (2003). Sexual differentiation of germ cells in XX mouse gonads occurs in an anterior-to-posterior wave. *Dev. Biol.* **262,** 303–312.

Mitsui, K., Tokuzawa, Y., Itoh, H., Segawa, K., Murakami, M., Takahashi, K., Maruyama, M., Maeda, M., and Yamanaka, S. (2003). The homeoprotein Nanog is required for maintenance of pluripotency in mouse epiblast and ES cells. *Cell* **113,** 631–642.

Nagy, A., Gocza, E., Diaz, E. M., Prideaux, V. R., Ivanyi, E., Markkula, M., and Rossant, J. (1990). Embryonic stem cells alone are able to support fetal development in the mouse. *Development* **110,** 815–821.

Nagy, A., and Vintersten, K. (2006). Murine embryonic stem cells. *Methods Enzymol.* **418** (this volume).

Nakamura, T., Arai, T., Takagi, M., Sawada, T., Matsuda, T., Yokota, T., and Heike, T. (1998). A selective switch-on system for self-renewal of embryonic stem cells using chimeric cytokine receptors. *Biochem. Biophys. Res. Commun.* **248,** 22–27.

Nichols, J., Davidson, D., Taga, T., Yoshida, K., Chambers, I., and Smith, A. (1996). Complementary tissue-specific expression of LIF and LIF-receptor mRNAs in early mouse embryogenesis. *Mech. Dev.* **57,** 123–131.

Niwa, H., Burdon, T., Chambers, I., and Smith, A. (1998). Self-renewal of pluripotent embryonic stem cells is mediated via activation of STAT3. *Genes Dev.* **12,** 2048–2060.

Niwa, H., Miyazaki, J., and Smith, A. G. (2000). Quantitative expression of Oct-3/4 defines differentiation, dedifferentiation or self-renewal of ES cells. *Nature Genet.* **24,** 372–376.

Nordhoff, V., Hübner, K., Bauer, A., Orlova, I., Malapetsa, A., and Schöler, H. R. (2001). Comparative analysis of human, bovine, and murine Oct-4 upstream promoter sequences. *Mamm. Genome* **12,** 309–317.

Ohinata, Y., Payer, B., O'Carroll, D., Ancelin, K., Ono, Y., Sano, M., Barton, S. C., Obukhanych, T., Nussenzweig, M., Tarakhovsky, A., Saitou, M., and Surani, M. A. (2005). Blimp1 is a critical determinant of the germ cell lineage in mice. *Nature* **436,** 207–213.

Ohta, H., Wakayama, T., and Nishimune, Y. (2004). Commitment of fetal male germ cells to spermatogonial stem cells during mouse embryonic development. *Biol. Reprod.* **70,** 1286–1291.

Pesce, M., Wang, X., Wolgemuth, D. J., and Schöler, H. (1998). Differential expression of the Oct-4 transcription factor during mouse germ cell differentiation. *Mech. Dev.* **71,** 89–98.

Robertson, E., Bradley, A., Kuehn, M., and Evans, M. (1986). Germ-line transmission of genes introduced into cultured pluripotential cells by retroviral vector. *Nature* **323,** 445–448.

Saitou, M., Barton, S. C., and Surani, M. A. (2002). A molecular programme for the specification of germ cell fate in mice. *Nature* **418,** 293–300.

Smith, A. G., Heath, J. K., Donaldson, D. D., Wong, G. G., Moreau, J., Stahl, M., and Rogers, D. (1988). Inhibition of pluripotential embryonic stem cell differentiation by purified polypeptides. *Nature* **336,** 688–690.

Smith, M. A., Pallister, C. J., and Smith, J. G. (2001). Stem cell factor: Biology and relevance to clinical practice. *Acta Haematol.* **105,** 143–150.

Stahl, N., Boulton, T. G., Farruggella, T., Ip, N. Y., Davis, S., Witthuhn, B. A., Quelle, F. W., Silvennoinen, O., Barbieri, G., Pellegrini, S., *et al.* (1994). Association and activation of Jak-Tyk kinases by CNTF-LIF-OSM-IL-6 beta receptor components. *Science* **263,** 92–95.

Tam, P. P., and Rossant, J. (2003). Mouse embryonic chimeras: Tools for studying mammalian development. *Development* **130,** 6155–6163.

Tilmann, C., and Capel, B. (2002). Cellular and molecular pathways regulating mammalian sex determination. *Recent Prog. Horm. Res.* **57,** 1–18.

Toyooka, Y., Tsunekawa, N., Akasu, R., and Noce, T. (2003). Embryonic stem cells can form germ cells *in vitro*. *Proc. Natl. Acad. Sci. USA* **100,** 11457–11462.

Toyooka, Y., Tsunekawa, N., Takahashi, Y., Matsui, Y., Satoh, M., and Noce, T. (2000). Expression and intracellular localization of mouse Vasa-homologue protein during germ cell development. *Mech. Dev.* **93,** 139–149.

Upadhyay, S., and Zamboni, L. (1982). Ectopic germ cells: Natural model for the study of germ cell sexual differentiation. *Proc. Natl. Acad. Sci. USA* **79,** 6584–6588.

Wassarman, P. M. (1988). The Mammalian Ovum. *In* "The Physiology of Reproduction" (E. Knobil and J. Neill, eds.), pp. 69–102. Raven Press, New York.

Yeom, Y. I., Fuhrmann, G., Ovitt, C. E., Brehm, A., Ohbo, K., Gross, M., Hübner, K., and Schöler, H. R. (1996). Germline regulatory element of Oct-4 specific for the totipotent cycle of embryonal cells. *Development* **122,** 881–894.

Yoshimizu, T., Sugiyama, N., De Felice, M., Yeom, Y. I., Ohbo, K., Masuko, K., Obinata, M., Abe, K., Schöler, H. R., and Matsui, Y. (1999). Germline-specific expression of the Oct-4/green fluorescent protein (GFP) transgene in mice. *Dev. Growth Differ.* **41,** 675–684.

[18]   Male Germ Cells

By Niels Geijsen and George Q. Daley

Abstract

Primordial germ cells, which carry the responsibility for perpetuation of the species, are set aside from their somatic neighbors very early in mammalian embryonic development. The founder population of germ cells is rare and difficult to identify and isolate in quantities suitable for molecular and biochemical analysis, thereby highlighting the importance of an *in vitro* system for deriving germ cells from embryonic stem cells. This chapter details methods for *in vitro* derivation of germ lineage elements and discusses potential applications of these techniques in germ cell research.

Introduction

In many lower species, germ line specification occurs through the inheritance of a preformed complex of RNA and proteins termed "germ plasm" that is asymmetrically localized in the oocyte and inherited by a limited group of cells destined to become the germ cells. In mammals, however, preformed germ line determinants have not been identified. Instead, germ line commitment is achieved through inductive signals that likely emanate from the extraembryonic ectoderm. Mammalian germ line commitment depends on epigenesis rather than preformation and is a regulated event guided by environmental cues (Saitou *et al.*, 2002). Studies using mutant mice or explanted embryos have identified some of the factors controlling primordial germ cell specification, such as Bmp4 and

METHODS IN ENZYMOLOGY, VOL. 418
0076-6879/06 $35.00
DOI: 10.1016/S0076-6879(06)18018-0

Bmp8b (de Sousa Lopes *et al.*, 2004; Fujiwara *et al.*, 2001; Lawson *et al.*, 1999; Ying and Zhao, 2001; Ying *et al.*, 2000, 2001), but considerable gaps still exist in our understanding of the complete array of signals that trigger germ line commitment. A complicating factor in the investigation of these signals is the inaccessibility of the germ cell population in the early embryo. However, several laboratories, including our own, have derived primordial germ cells from embryonic stem cells *in vitro*, thus providing a tractable tool for the study of early germ line commitment and germ cell differentiation (Clark *et al.*, 2004; Geijsen *et al.*, 2004; Hübner *et al.*, 2003; Toyooka *et al.*, 2003).

We use embryoid bodies (EBs) from embryonic stem (ES) cells to induce germ line differentiation. EBs are three-dimensional structures in which early embryonic events, such as gastrulation and the induction of embryonic germ layers, are faithfully recapitulated (Ling and Neben, 1997). In the context of EB differentiation, the fate of numerous cell types is specified in a choreographed, stepwise process and therefore EBs provide a means to investigate otherwise inaccessible cell populations of the early murine embryo. Utilizing the EB differentiation system we were able to demonstrate the *in vitro* development of male gametes that expressed molecular germ cell markers, were able to self-renew in the presence of retinoic acid, and underwent erasure of imprints, a hallmark property of embryonic germ cells (Geijsen *et al.*, 2004). Upon further EB differentiation, molecular markers of male meiosis were observed and a rare haploid cell population could be isolated that, upon injection into recipient oocytes, gave rise to blastocyst embryos.

This chapter describes methods for the (1) generation of ES cells from murine blastocyst embryos, (2) differentiation of ES cells into EBs, (3) characterization of EBs by immunostaining, and (4) isolation and selection of germ cells from EBs.

Methods

ES Cell Derivation and Culture

C57BL/6-TGN(ACTbEGFP) are from Jackson Laboratories (Bar Harbor, ME)(Okabe *et al.*, 1997). 129SvEv are from Taconic (Germantown, NY). Set up timed matings between male C57BL/6-TGN(ACTbEGFP) mice and female 129SvEv. The next morning (day 0.5) check female mice for the presence of a copulation plug. At day 3.5, euthanize fertilized females, and flush blastocyst embryos from the uterine horns using phosphate-buffered saline (PBS). Transfer individual blastocysts to individual wells of a four-well dish (Corning) using a glass transfer pipette. Plate blastocysts in ES cell medium (DME [Invitrogen], 15% fetal bovine serum [FBS; Invitrogen], 0.1 mM nonessential amino acids [Invitrogen], 2 mM L-glutamine [Invitrogen], penicillin/streptomycin [Invitrogen], 0.1 mM β-mercaptoethanol [Sigma]) supplemented with 1000 U/ml leukemia inhibitory factor (LIF, Chemicon) onto murine embryonic feeder cells (MEFs, Chemicon CF-1) in four-well dishes. Pretreat all tissue culture dishes with 0.2 gelatin (from porcine skin, Sigma) in ddH$_2$O for 15 min at 37° prior to seeding the MEFs and/or ES cells. After 4 to 5 days, trypsinize and replate blastocyst outgrowths into a new gelatin-treated four-well dish with MEFs in ES culture medium supplemented with LIF. At this stage, ES cell colonies start to appear, which are propagated by continuous passaging (1:7 to 1:10) on MEFs in gelatinized tissue culture dishes in ES culture medium supplemented with LIF.

Embryoid Body Differentiation

Obtain a single cell suspension of ES cells by washing the cells twice with PBS followed by trypsinization of ES cells with 0.25% trypsin/EDTA (Invitrogen) for 3 min. Dissociate ES cell colonies to single cells in EB medium (IMDM [Invitrogen], 15% FBS [Invitrogen], 0.1 mM nonessential amino acids [Invitrogen], 1 mM sodium pyruvate [Invitrogen], 2 mM L-glutamine [Invitrogen], penicillin/streptomycin [Invitrogen], 200 μg/ml iron-saturated bovine holo-transferrin [Sigma], 4.5 mM monothioglycerol [Sigma], 50 μg/ml ascorbic acid [Sigma]) using a P1000 pipettor (Gilson). Because this cell suspension is a mix of ES cells and fibroblast feeders, the MEFs need to be removed by replating the cell suspension for 30 min at 37° into a new tissue culture dish. The size of the dish to be chosen will vary depending on the number of cells, but ideally the dish should be covered by a thin film of the cell suspension. Because the MEFs are larger than the ES cells and adhere to tissue culture plastic quickly, the MEFs can be removed based on their differential adherence properties. After 30 min the nonadhered cell suspension will contain primarily ES cells. If many contaminating MEFs are still present, the procedure should be repeated. After removal of the MEFs, pellet the ES cells and resuspend in 5 ml EB medium. Filter the ES cell suspension using a 70-μm cell strainer and count the cell density. To generate EBs, resuspend ES cells at 300 cells per 25 μl (6×10^5 cells/ 50 ml) in EB medium. Spot 25-μl drops of cells onto a 15-cm Petri dish (Fig. 1A). It is critical that regular Petri dishes be used in this procedure because tissue culture plastic will cause the ES cells to adhere to the plate and prevent EB formation. Drops can be spotted using a manual multichannel pipette, but for large numbers of EBs we use an automated multichannel pipette (CLP, Electronic Repeat Pipettor, 9620). After spotting the ES cell droplets onto the Petri dish, turn the plate upside down so that the drops hang from the

FIG. 1. Making embryoid bodies. (A) Twenty-five-microliter drops containing approximately 300 ES cells are spotted onto Petri dishes using a multichannel pipette and turned upside down so that the drops hang downward (inset). (B) After 2 to 3 days the hanging drops are collected into smaller Petri dishes and incubated with continuous slow rotation. Place an extra Petri dish with PBS under and on top of the stack to prevent excessive evaporation of medium.

plate (Fig. 1A, *insert*). Incubate the hanging drops at 37°, 5% CO_2 in a humidified incubator for 2 to 3 days. After 2 to 3 days rinse EBs off the plate and collect into 10 ml EB medium in a 10-cm Petri dish. This is necessary to prevent the drops from drying out and also to allow the medium to be renewed. Keep the Petri dishes in the incubator on a slowly rocking or rotating platform to prevent the EBs from aggregating to each other or adhering to the plastic (Fig. 1B). From this point on, feed EBs every other day by exchanging half of their spent medium for fresh EB medium.

Gelatin Embedding and EB Sectioning for Immunohistologic Examination

Collect EBs in an Eppendorf tube. Make sure to use a large enough pipette to avoid disruption of the EBs. Wash the EBs four times with ice-cold PBS before fixing them in 4% PFA (MP Biomedicals) overnight at 4° while rotating. Wash the EBs three times with PBS before substituting them with 15% sucrose (Sigma) in PBS. Add 15% sucrose to EBs for 2 h at 4° while rotating, refresh sucrose, and leave rotating overnight at 4°. Make a 7.5% gelatin (Sigma)/15% sucrose solution in PBS at 37°. Remove as much sucrose from the EBs before adding the prewarmed gelatin/sucrose solution. Place the Eppendorf tubes containing the EBs at 37° for 1 h. Refresh gelatin/sucrose and place the tubes back at 37° for another hour. During the hour wait, pour gelatin in the bottom of a small weighing tray and let it sit at room temperature to solidify. Then place the trays at 4°.

Take the trays out of the refrigerator when the EBs are finished substituting in gelatin/sucrose. Pipette the EBs from the tubes into the trays and let them sit at room temperature for 10 min. Add more gelatin on top to cover the EBs and let them sit at room temperature for another 10 min before putting the trays at 4° to solidify the gelatin.

Take the trays out of the refrigerator when the gelatin is solid and use scissors to cut a square around the EBs. Mark the location of the EBs with a permanent marker. Attach the gelatin to a labeled piece of cardboard with a drop of OCT (Tissue-Tek, OCT compound, Sakura).

Fill a Styrofoam container with liquid nitrogen and a plastic beaker with isopentane (Sigma). Use large forceps to hold the beaker and put in liquid nitrogen until it reaches a temperature between −50° and −60°. Take the beaker from the liquid nitrogen and place it on the bench top. Stir until the temperature is −60°. Use forceps to place the gelatin square in isopentane and leave for 1 min with occasional stirring. Take the frozen gelatin block from the isopentane and move to the cryostat for sectioning.

Make 10-μm-thick sections on slides that are positively charged (Fisher-brand superfrost/plus slides). Dry the sections properly before starting Oct4 immunostaining.

Immune Staining

Wash slides three times with PBS before placing them in a humidified box with blocking solution (2% donkey serum, 1% BSA, 0.05% Tween 20, 0.15% Triton X-100 in PBS) for 30 min at room temperature. Replace the blocking solution with blocking solution containing 0.05 to 0.1 μg antibody per slide (e.g., Oct3/4, Santa Cruz Biotechnology, SSEA1, Hybridoma Bank University of Iowa, MC-480) and cover sections with a cover slide. Incubate for at least 1 h at 37° or overnight at 4°. Wash slides three times with blocking solution or PBS before adding the secondary antibody. An additional 30-min incubation with blocking solution is required when slides are washed in PBS. Add secondary antibody solution (For Oct4 we use donkey antigoat FITC, Jackson ImmunoResearch Laboratories Inc.) diluted 1:200 and incubate for 1 h at 37°. Wash slides three times with PBS and mount in Vectashield (Vector Laboratories Inc., Burlingame, CA) containing 200 ng 4′,6′diamino-2-phenylindole per 100 microliter.

EB Dissociation for Isolation of Embryonic Germ Cells

To isolate cells committed to the germ lineage, collect fresh EBs and wash twice with PBS. Dissociate EBs by collagenase IV digestion for 15 min followed by resuspension in cell dissociation buffer (Invitrogen). Pellet cells for 3 min at 1200 rpm in a Thermo Electron IEC Centra GP8 centrifuge. Resuspend the pellet in 10 ml ice-cold DME + 5% BSA and pass the cells through a 70-μm cell strainer to remove cell clumps. Pellet the cell suspension for 3 min at 1200 rpm and resuspend into ice-cold DME + 5% BSA. Store the cells on ice until further use.

Immunomagnetic Isolation of Germ Cells

Incubate cells harvested from the EBs for 30 min with monoclonal antibody against the cell surface antigen SSEA1 (hybridoma bank, Iowa, MC-480) at 4° in DME/5% BSA. Supply the antibody as a hybridoma supernatant; the antibody concentration will vary with different lots. It is recommended to titrate the antibody on ES cells prior to use in FACS analysis or magnetic bead isolation. Wash cells twice with ice-cold DME/ 0.5% BSA, add immunomagnetic beads (Dynal, rat antimouse IgM), and incubate cells and beads for 1 h at 4° with slow rotation. Perform magnetic separation of SSEA1-positive beads associated with the cells according to the protocol provided by the manufacturer and wash the beads three times with ice-cold DME/0.5% BSA.

Selection of Germ Cells through Retinoic Acid (RA) Culture

Germ cells can be differentially selected and cultured from a mixed population of germ cells and ES cells by the addition of all-*trans* retinoic acid to the tissue culture medium (Geijsen *et al.*, 2004; Koshimizu *et al.*, 1995). While retinoic acid acts to promote growth or primordial germ cells, it is a strong inducer of primarily neuronal differentiation for ES cells. Thus, when purified SSEA1-positive cells from developing embryoid bodies are cultured in the presence of 2 μM RA, the contaminating undifferentiated ES cells are lost from the culture, whereas primordial germ cells form loosely compact germ cell colonies. Plate freshly isolated EB-derived primordial germ cells on STO feeder cells on gelatinized tissue culture plastic in ES cell medium containing 1000 U/ml LIF (Chemicon), 20 ng/ml recombinant human bFGF (Peprotech), 100 ng/ml recombinant murine SCF (Peprotech), and 2 μM all-*trans* RA. After 3 to 4 days in culture or when the culture becomes confluent, passage cells 1:5 onto a new gelatinized dish with STO feeders. At this point the germ cells will form tightly compact EG colonies resembling ES cells, and upon further passaging RA should be omitted from the culture.

Discussion

The germ lineage represents a privileged class of cells that maintains a unique capacity for developmental potency. During embryonic development (Donovan, 1994; Labosky *et al.*, 1994a,b; Matsui *et al.*, 1991, 1992) and postnatal life (Guan *et al.*, 2006; Kanatsu-Shinohara *et al.*, 2004), germ cells retain the capacity to be established in culture as pluripotent cells. How the germ lineage becomes specified apart from somatic tissues and how these cells remain unperturbed by the waves of somatic differentiation throughout the embryo represent central questions that are destined to yield critical insights into the epigenetic regulation of the genome.

The first steps of commitment to the germ lineage appear to be faithfully executed during the *in vitro* differentiation of ES cells. Molecular markers identified in the earliest germ cell elements of the mouse embryo, such as Stella, are expressed in a punctate pattern in EBs (Payer *et al.*, 2006). Stella expression indeed coincides with the ability to detect an array of molecular markers of primordial germ cell development and the capacity to isolate and culture embryonic germ cells from differentiating ES cell cultures. Such an *in vitro* system will be invaluable for determining the role of specific genes in germ cell commitment and in defining the machinery involved in the unique epigenetic remodeling that occurs within the germ lineage at imprinted gene loci. Furthermore, a cell culture system should facilitate obtaining adequate quantities of cellular material to enable microarray and biochemical studies. It remains to be determined whether the EB system can support terminal differentiation of functional gametes, but the foundation exists for asking a range of questions that previously were impossible or impractical.

Acknowledgment

The authors thank Dr. Maureen Eijpe for her critical input in this manuscript.

References

Clark, A. T., Bodnar, M. S., Fox, M., Rodriquez, R. T., Abeyta, M. J., Firpo, M. T., and Pera, R. A. (2004). Spontaneous differentiation of germ cells from human embryonic stem cells *in vitro. Hum. Mol. Genet.* **13,** 727–739.

de Sousa Lopes, S. M., Roelen, B. A., Monteiro, R. M., Emmens, R., Lin, H. Y., Li, E., Lawson, K. A., and Mummery, C. L. (2004). BMP signaling mediated by ALK2 in the visceral endoderm is necessary for the generation of primordial germ cells in the mouse embryo. *Genes Dev.* **18,** 1838–1849.

Donovan, P. J. (1994). Growth factor regulation of mouse primordial germ cell development. *Curr. Top. Dev. Biol.* **29,** 189–225.

Fujiwara, T., Dunn, N. R., and Hogan, B. L. (2001). Bone morphogenetic protein 4 in the extraembryonic mesoderm is required for allantois development and the localization and survival of primordial germ cells in the mouse. *Proc. Natl. Acad. Sci. USA* **98,** 13739–13744.

Geijsen, N., Horoschak, M., Kim, K., Gribnau, J., Eggan, K., and Daley, G. Q. (2004). Derivation of embryonic germ cells and male gametes from embryonic stem cells. *Nature* **427,** 148–154.

Guan, K., Nayernia, K., Maier, L. S., Wagner, S., Dressel, R., Lee, J. H., Nolte, J., Wolf, F., Li, M., Engel, W., and Hasenfuss, G. (2006). Pluripotency of spermatogonial stem cells from adult mouse testis. *Nature* **440**, 1199–1203.

Hübner, K., Fuhrmann, G., Christenson, L. K., Kehler, J., Reinbold, R., De La Fuente, R., Wood, J., Strauss, J. F., 3rd, Boiani, M., and Schöler, H. R. (2003). Derivation of oocytes from mouse embryonic stem cells. *Science* **300**, 1251–1256.

Kanatsu-Shinohara, M., Inoue, K., Lee, J., Yoshimoto, M., Ogonuki, N., Miki, H., Baba, S., Kato, T., Kazuki, Y., Toyokuni, S., Toyoshima, M., Niwa, O., Oshimura, M., Heike, T., Nakahata, T., Ishino, F., Ogura, A., and Shinohara, T. (2004). Generation of pluripotent stem cells from neonatal mouse testis. *Cell* **119**, 1001–1012.

Koshimizu, U., Watanabe, M., and Nakatsuji, N. (1995). Retinoic acid is a potent growth activator of mouse primordial germ cells *in vitro*. *Dev. Biol.* **168**, 683–685.

Labosky, P. A., Barlow, D. P., and Hogan, B. L. (1994a). Embryonic germ cell lines and their derivation from mouse primordial germ cells. *Ciba Found. Symp.* **182**, 157–168; discussion 168–178.

Labosky, P. A., Barlow, D. P., and Hogan, B. L. (1994b). Mouse embryonic germ (EG) cell lines: Transmission through the germline and differences in the methylation imprint of insulin-like growth factor 2 receptor (Igf2r) gene compared with embryonic stem (ES) cell lines. *Development* **120**, 3197–3204.

Lawson, K. A., Dunn, N. R., Roelen, B. A., Zeinstra, L. M., Davis, A. M., Wright, C. V., Korving, J. P., and Hogan, B. L. (1999). Bmp4 is required for the generation of primordial germ cells in the mouse embryo. *Genes Dev.* **13**, 424–436.

Ling, V., and Neben, S. (1997). *In vitro* differentiation of embryonic stem cells: Immunophenotypic analysis of cultured embryoid bodies. *J. Cell. Physiol.* **171**, 104–115.

Matsui, Y., Toksoz, D., Nishikawa, S., Williams, D., Zsebo, K., and Hogan, B. L. (1991). Effect of Steel factor and leukaemia inhibitory factor on murine primordial germ cells in culture. *Nature* **353**, 750–752.

Matsui, Y., Zsebo, K., and Hogan, B. L. (1992). Derivation of pluripotential embryonic stem cells from murine primordial germ cells in culture. *Cell* **70**, 841–847.

Okabe, M., Ikawa, M., Kominami, K., Nakanishi, T., and Nishimune, Y. (1997). 'Green mice' as a source of ubiquitous green cells. *FEBS Lett.* **407**, 313–319.

Payer, B., Chuva de Sousa Lopes, S. M., Barton, S. C., Lee, C., Saitou, M., and Surani, M. A. (2006). Generation of stella-GFP transgenic mice: A novel tool to study germ cell development. *Genesis* **44**, 75–83.

Saitou, M., Barton, S. C., and Surani, M. A. (2002). A molecular programme for the specification of germ cell fate in mice. *Nature* **418**, 293–300.

Toyooka, Y., Tsunekawa, N., Akasu, R., and Noce, T. (2003). Embryonic stem cells can form germ cells *in vitro*. *Proc. Natl. Acad. Sci. USA* **100**, 11457–11462.

Ying, Y., Liu, X. M., Marble, A., Lawson, K. A., and Zhao, G. Q. (2000). Requirement of Bmp8b for the generation of primordial germ cells in the mouse. *Mol. Endocrinol.* **14**, 1053–1063.

Ying, Y., Qi, X., and Zhao, G. Q. (2001). Induction of primordial germ cells from murine epiblasts by synergistic action of BMP4 and BMP8B signaling pathways. *Proc. Natl. Acad. Sci. USA* **98**, 7858–7862.

Ying, Y., and Zhao, G. Q. (2001). Cooperation of endoderm-derived BMP2 and extra-embryonic ectoderm-derived BMP4 in primordial germ cell generation in the mouse. *Dev. Biol.* **232**, 484–492.

[19] Insulin-Producing Cells

By INSA S. SCHROEDER, GABRIELA KANIA, PRZEMYSLAW BLYSZCZUK,
and ANNA M. WOBUS

Abstract

Embryonic stem (ES) cells offer great potential for cell replacement and tissue engineering therapies because of their almost unlimited proliferation capacity and the potential to differentiate into cellular derivatives of all three primary germ layers. This chapter describes a strategy for the *in vitro* differentiation of mouse ES cells into insulin-producing cells. The three-step protocol does not select for nestin-expressing cells as performed in previous differentiation systems. It includes (1) the spontaneous differentiation of ES cells via embryoid bodies and (2) the formation of progenitor cells of all three primary germ layers (multilineage progenitors) followed by (3) directed differentiation into the pancreatic lineage. The application of growth and extracellular matrix factors, including laminin, nicotinamide, and insulin, leads to the development of committed pancreatic progenitors, which subsequently differentiate into islet-like clusters that release insulin in response to glucose. During differentiation, transcript levels of pancreas-specific transcription factors (i.e., Pdx1, Pax4) and of genes specific for early and mature β cells, including insulin, islet amyloid pancreatic peptide, somatostatin, and glucagon, are upregulated. C-peptide/insulin-positive islet-like clusters are formed, which release insulin in response to high glucose concentrations at terminal stages. The differentiated cells reveal functional properties with respect to voltage-activated Na^+ and ATP-modulated K^+ channels and normalize blood glucose levels in streptozotocin-treated diabetic mice. In conclusion, we demonstrate the efficient differentiation of murine ES cells into insulin-producing cells, which may help in the future to establish ES cell-based therapies in diabetes mellitus.

Introduction

Diabetes mellitus is caused by insufficient or abolished insulin release due to autoimmune destruction or malfunction of pancreatic β cells located in the endocrine pancreas in the so-called islets of Langerhans. The lack of insulin and the resulting inadequate control of glycemia lead to a life-threatening metabolic dysfunction that requires insulin injections to alleviate hyperglycemia. However, this does not provide dynamic control of glucose homeostasis, and patients with long-term diabetes suffer from

METHODS IN ENZYMOLOGY, VOL. 418
0076-6879/06 $35.00
DOI: 10.1016/S0076-6879(06)18019-2

complications such as neuropathy, nephropathy, retinopathy, and vascular disorders. Consequently, cell replacement therapies are required to circumvent such adverse side effects. The transplantation of islets of Langerhans has been successfully established, but as the availability of human donor pancreas for islet grafting is limited, *in vitro* β-cell engineering is one promising way to overcome the limitation of donor cells.

Embryonic stem (ES) cells have been studied intensively as potential cellular systems to analyze lineage commitment and differentiation (Wobus and Boheler, 2005). Due to their pluripotent character they are capable of self-renewing and differentiating into practically any cell type of the endo-, ecto-, and mesodermal lineage and therefore may serve as a promising substitute for cell therapy and organ transplantation.

For many years, ES cells of the mouse (mES cells) have represented an excellent experimental system to study basic mechanisms of cell differentiation. Spontaneous differentiation of mES cells results in heterogeneous cell populations with a predominant fraction showing ectodermal characteristics, supporting the idea that ectodermal differentiation is a default pathway and does not require complex extracellular signaling (Ying *et al.*, 2003). In contrast, the yield of endocrine pancreatic cells is relatively low (Kahan *et al.*, 2003). Therefore, the generation of sufficient amounts of insulin-producing cells requires directed differentiation through selection/gating of pancreatic phenotypes (Leon-Quinto *et al.*, 2004; Soria *et al.*, 2000), transgenic expression of pancreatic developmental control genes (Blyszczuk *et al.*, 2003; Ishizaka *et al.*, 2002; Miyazaki *et al.*, 2004; Shiroi *et al.*, 2005; Soria *et al.*, 2000), and/or by applying specific growth and extracellular matrix (ECM) factors (Hori *et al.*, 2002; Lumelsky *et al.*, 2001; Sipione *et al.*, 2004).

Growth and extracellular matrix factors that induce or promote pancreatic differentiation include progesterone, putrescine, insulin, transferrin, sodium selenite, fibronectin (ITSFn), nicotinamide, and laminin, all of which have been used in the protocol described here. Other groups suggested the use of retinoic acid (Micallef *et al.*, 2005) or the use of conditioned medium from fetal pancreatic buds containing soluble factors, which promote pancreatic differentiation of mES cells (Vaca *et al.*, 2005). However, Micallef *et al.* (2005), while showing induction of Pdx1, a marker of pancreatic progenitor cells, could not show differentiation of mES cells into insulin-producing cells, and the use of conditioned medium may be questionable, as repeatable differentiation of ES cells requires a reproducible composition of the conditioned medium, which may be hard to achieve.

Several previously published protocols of pancreatic differentiation used ITSFn/FGF-2 to support proliferation of nestin-positive cells (Lumelsky *et al.*, 2001; Rajagopal *et al.*, 2003; Sipione *et al.*, 2004). However, these protocols not only promoted pancreatic, but also massive neuronal differentiation. It is well known that, during embryogenesis, neuroectodermal and pancreatic differentiation are partially regulated by the same transcription factors, such as Ngn3, Isl-1, or Pax6 in a spatially and temporally distinct manner (Gradwohl *et al.*, 2000; Habener *et al.*, 2005; Lee *et al.*, 2003; Nakagawa and O'Leary, 2001; Schwitzgebel *et al.*, 2000). Likewise, nestin is transiently expressed in ES-derived neuronal (Okabe *et al.*, 1996) and pancreatic (Blyszczuk *et al.*, 2003, 2004; Kania *et al.*, 2004) differentiation, respectively. Therefore, any induction and/or selection of cells expressing these factors leads to a parallel induction of neuronal and pancreatic differentiation, as during *in vitro* differentiation, neuroectodermal and pancreatic progenitor cells are not separated from each other. It has to be mentioned that the differentiation systems that were unsuccessful in demonstrating controlled insulin release and other functional parameters (i.e., insulin-positive secretory granules) used the original protocol (Lumelsky *et al.*, 2001) selecting for nestin-positive cells (Hansson *et al.*, 2004; Paek *et al.*, 2005; see Rajagopal *et al.*, 2003; Sipione *et al.*, 2004).

Consequently, we avoided any selection or support of nestin-expressing cells and found that the selection of nestin-positive cells with ITSFn and FGF-2 is neither obligatory nor profitable for successful pancreatic differentiation and does not promote the generation of specific pancreatic progenitors when applied to ES-derived cells (Blyszczuk *et al.*, 2004; Kania *et al.*, 2004). However, nestin, as well as cytokeratin 19, a marker expressed in pancreatic duct epithelial cells but not in mature islets, is expressed in intermediate stages of differentiation, but is downregulated at terminal stages.

In 2003, Kubo *et al.* found that endoderm could be induced in EBs by limited exposure to serum or treatment with activin A under serum-free conditions (Kubo *et al.*, 2004). Activin A, a member of the transforming growth factor β (TGF-β) superfamily, was also used by Shi *et al.* (2005) in combination with all-*trans* retinoic acid to differentiate mES cells into pancreatic β-like cells. Still, activin A induces the formation of neuronal extensions and neurofilament proteins in PC12 cells (Iwasaki *et al.*, 1996), pointing toward the involvement of this substance in neural differentiation. Our own studies of pancreatic differentiation according to the protocol of Shi *et al.* (2005) as a result led to both, the induction of pancreatic and neuronal differentiation (unpublished data). Evidently, activin A treatment requires a very sophisticated application scheme for successful enrichment of endodermal progenitor cells, as shown by D'Amour *et al.* (2005). However, successful pancreatic differentiation and the emergence of neuronal cells have not been analyzed in this study.

This chapter reports an easy and reproducible *in vitro* differentiation system without any selection for specific cell types to generate insulin-producing cells from mES cells, whose efficacy can be further enhanced by using Pax4-overexpressing cells (Blyszczuk *et al.*, 2003, 2004).

Materials and Methods

Culture of Undifferentiated mES Cells

Mouse R1 ES cells are cultivated on feeder layers of mouse embryonic fibroblasts (MEFs, for preparation, see Wobus *et al.* [2002]) on gelatin-coated (0.1%) Petri dishes (Falcon Becton Dickinson, Heidelberg, Germany) in Dulbecco's modified Eagle's medium (Invitrogen, Karlsruhe, Germany) supplemented with 15% heat-inactivated fetal calf serum (FCS, selected batches, Invitrogen), L-glutamine (Invitrogen, 2 mM), β-mercaptoethanol (Serva, Heidelberg, Germany, final concentration 5 × 10^{-5} M), nonessential amino acids (Invitrogen, 100× stock solution diluted 1: 100), penicillin–streptomycin (Invitrogen, 100× stock solution diluted 1:100), and 10 ng/ml recombinant human leukemia inhibitory factor (LIF) prepared from LIF expression vectors (see Smith and Johnson, 1988; Wobus *et al.*, 2002) or obtained from commercial sources (Chemicon, Hampshire, UK). To maintain their undifferentiated state, the mES cells must be cultured at relatively high density. The use of STO feeder layers is not recommended, as in our experience the supportive capacity for mES cells is dependent on specific sublines, which may not be available commonly. Good-quality, batch-tested FCS is critical for long-term culture of mES cells and for subsequent successful differentiation. As mouse ES cells divide every 12 to 15 h, the culture medium should be replenished daily and the cells passaged every 24 to 48 h onto freshly prepared feeder layers. For passaging, ES cells must be dissociated carefully by treatment with trypsin/EDTA solution. If one or more of these requirements are not complied with, ES cells may differentiate spontaneously during culture and become unsuitable for differentiation studies.

Generation of mES Cell-Derived Multilineage Progenitor Cells

Withdrawal of feeder cells and LIF leads to spontaneous differentiation of mES cells into cells of all three germ layers. This formation of so-called multilineage progenitor cells is the basis of the ES cell differentiation protocol. Controlled production of multilineage progenitors from mES cells is composed of two steps: (i) the formation of three-dimensional aggregates or embryoid bodies (EBs) to promote differentiation into all three germ layers and (ii) the expansion and further differentiation on adhesive substrata. EB formation of mES cells may be induced either by the "hanging drop" method (Wobus *et al.*, 2002) or by "mass culture" in bacteriological-grade dishes (Doetschman *et al.*, 1985). However, the "hanging drop" method has several advantages, including low variation in size of the EBs due to a defined number of mES cells in the starting aggregates, as well as greater reproducibility of differentiation. Therefore, mES cells are cultivated as EBs in "hanging drops" (600 cells/20 μl) for 2 days, transferred into bacteriological plates (Greiner, Germany), and cultured in suspension in Iscove's modified DMEM (IMDM, Invitrogen) supplemented with 20% FCS, L-glutamine, nonessential amino acids (see earlier discussion), and α-monothioglycerol (Sigma, Steinheim, Germany; final concentration 450 μM) instead of β-mercaptoethanol for another 3 days (for differentiation scheme, see Fig. 1). Penicillin and streptomycin may be added to the cultures (see earlier discussion). EBs are seeded onto gelatin-coated (0.1%) dishes (20 to 30 EBs/ 60 mm) and grown in IMDM (see earlier discussion). Medium is changed every second to third day until 9 days after EB plating (= stage 5+9d).

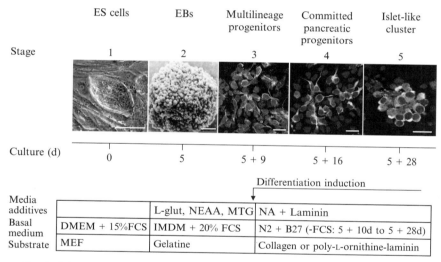

	ES cells	EBs	Multilineage progenitors	Committed pancreatic progenitors	Islet-like cluster
Stage	1	2	3	4	5
Culture (d)	0	5	5 + 9	5 + 16	5 + 28

Differentiation induction

Media additives		L-glut, NEAA, MTG	NA + Laminin	
Basal medium	DMEM + 15%FCS	IMDM + 20% FCS	N2 + B27 (-FCS: 5 + 10d to 5 + 28d)	
Substrate	MEF	Gelatine	Collagen or poly-L-ornithine-laminin	

FIG. 1. Schematic representation of ES cell-derived pancreatic differentiation. Mouse ES cells cultured on fibroblast feeder layers (1) are differentiated via embryoid bodies (EBs, scanning electron microscopy) (2) into multilineage progenitor cells (3) and after differentiation induction by growth factors into committed pancreatic progenitors (4) and insulin-producing cells in islet-like clusters (5). Stages of differentiation with examples of cell morphology and media, additives, and substrates used during *in vitro* differentiation are shown. Cells at stage 5+9d are dissociated and replated onto collagen or laminin I-coated tissue culture plates and cultured in complex differentiation medium with differentiation factors and 10% FCS to support attachment of cells. One day later, the differentiation medium is replenished without serum and cells are differentiated for up to 28 or more days. Immunostaining shows nestin/CK19 (3), nestin/C-peptide (4), and C-peptide/insulin (5) coexpression in cells at different stages. Bars: 20 μm (3, 4, 5), 50 μm (2), and 100 μm (1). (See color insert.)

Induction of Pancreatic Differentiation

The differentiation of mES cells into specific cell types requires the enhancement and subsequent differentiation of lineage-committed progenitor cells via defined growth- and differentiation-inducing factors. However, parameters such as the dissociation of EBs, the choice of suitable adhesive substrata, and the cell density after replating also determine the differentiation efficiency. The way of dissociating EB outgrowths is crucial, as cell-to-cell interactions within the complex and heterogeneous structure of EBs may influence the fate of progenitor cells. Likewise, ECM factors of the adhesive substratum affect adhesion, proliferation, and migration of specific progenitor cells after replating of dissociated EBs. Finally, improper cell density may decrease differentiation efficiency: overgrowth can result in metabolic starvation, necrosis, and cell death, whereas too low a cell density can lead to reduced cell-to-cell contacts and reduced release of essential autocrine factors.

Factors that induce pancreatic differentiation include nicotinamide (Otonkoski *et al.*, 1993) and laminin (Jiang *et al.*, 1999). In addition, factors required for pancreatic cell survival, such as progesterone, putrescine, insulin, sodium selenite, and transferrin, also promote pancreatic differentiation. These factors have been used to direct committed progenitor cells toward pancreatic insulin-producing cells. EB outgrowths generated by day 5+9 (see Fig. 1) are dissociated by 0.1% trypsin (Serva)/0.08% EDTA (Sigma) in phosphate-buffered saline (PBS) (1:1) for 1 min, collected by centrifugation, and replated onto collagen- or laminin-coated tissue culture plates (Nunc, Wiesbaden, Germany) in DMEM/F12 containing 20 nM progesterone, 100 μM putrescine, 1 μg/ml laminin, 10 mM nicotinamide (all from Sigma), B27 medium supplement (Invitrogen), 25 μg/ml insulin, 50 μg/ml transferrin, 5 μg/ml fibronectin, and 30 nM sodium selenite (all from Sigma) supplemented with 10% FCS and penicillin–streptomycin (see earlier discussion). Collagen proved to be a more desirable substrate than laminin, as the latter induced neuronal differentiation (data not shown). For immunofluorescence analysis, cells are plated onto collagen-coated coverslips, for ELISA onto 3-cm culture dishes, and for reverse transcriptase-polymerase chain reaction (RT-PCR) onto 6-cm culture dishes. One day after replating (at day 5 + 10), FCS is removed and the cells are cultivated until days 5 + 16 and 5 + 28 for further analysis (Blyszczuk *et al.*, 2004).

Analysis of Differentiated Phenotypes

For proper characterization of cells in the various stages of differentiation it is crucial to use multiple markers of pancreatic and nonpancreatic cells. It is a precondition to use several phenotypic as well as functional assays to evaluate the extent of differentiation (see currently used analytical methods in Table I). As an example, immunostaining for insulin alone may lead to false-positive results regarding endogenous production of this hormone as it is a compound of most treatment protocols and can be taken up easily by apoptotic cells from the culture medium (Hansson *et al.*, 2004; Rajagopal *et al.*, 2003). Therefore, C-peptide, a by-product of insulin synthesis, is a more reliable marker of insulin production and should always be used for costaining with other pancreatic markers. Characterization should include pancreatic developmental control genes known to be specific for proper β-cell formation, such as HNF3β, Pdx1, Pax4, and Nkx6.1 (see Cerf *et al.*, 2005) in addition to mature pancreatic markers such as insulin 1 and 2, Glut-2, glucagon, islet amyloid polypeptide (IAPP), and pancreatic polypeptide. Moreover, it has to be taken into consideration that both endodermal and ectodermal cells produce insulin: in ectodermal cells insulin acts as a growth factor and is expressed at relatively low levels, whereas in endoderm-derived pancreatic islets insulin is involved in hormonal regulations of glucose homeostasis. Therefore, the *in vitro* generation of ectoderm-derived insulin-producing cells can simulate pancreatic β-cell formation. Rodents possess two insulin genes, insulin 1 and 2. The insulin 1 gene is expressed exclusively in pancreatic tissue, whereas insulin 2 is expressed in pancreatic islets and certain neurons (Melloul *et al.*, 2002).

In the current protocol, the characterization of cells is carried out by RT-PCR, immunohistochemistry, and ELISA. This allows the qualitative and quantitative determination of progenitor- and pancreas-specific markers at the mRNA and protein level, as well as the determination of proper cell function.

Semiquantitative RT-PCR Analysis

ES-derived cells are collected and suspended in lysis buffer composed of 4 *M* guanidinium thiocyanate, 25 m*M* sodium citrate (pH 7), 0.5% (w/v) sarcosyl, and 0.1 *M* β-mercaptoethanol.

Total RNA is isolated by the single-step extraction method according to Chomczynski and Sacchi (1987), including a proteinase K digest for 1 h at 56°. mRNA is reverse transcribed using oligo(dT) and Revert Aid M-MuL-V reverse transcriptase (Fermentas, St. Leon-Rot, Germany). cDNAs are amplified using oligonucleotide primers complementary to transcripts of the analyzed genes (see Table II) and *Taq* polymerase (Fermentas). The PCR reaction is separated electrophoretically on 2% (w/v) agarose gels, visualized using ethidium bromide staining, and analyzed by TINA2.08e software (Raytest Isotopenmessgeräte GmbH, Straubenhardt, Germany).

TABLE I
COMPARISON OF PROTOCOLS AND PARAMETERS OF PANCREATIC DIFFERENTIATION OF MOUSE ES-DERIVED CELLS

Insulin mRNA	C-peptide/ insulin co-expression	In vitro glucose response	In vitro C-peptide secretion	Rescue of diabetes in animal models	Electro-physiological studies	ELMI studies (insulin granules)	Nestin+ cell selection	Transgene expression	References
n.d.	n.d.	+	−	+ (But 40% of animals became hyperglycemic 12 weeks after transplantation)	n.d.	n.d.	−	Gene trapping via human insulin	Soria et al., 2000
Ins. 1: − Ins. 2:+	n.d.	+	n.d.	Survival (no sustained correction of hyperglycemia)	n.d.	n.d.	+	−	Lumelsky et al., 2001
Ins. 1: + Ins. 2: n.d.	+	+	n.d.	+	n.d.	n.d.	+	−	Hori et al., 2002
Ins. 1: + Ins. 2: weak	−	n.d.	n.d.	n.d.	n.d.	−	+	−	Rajagopal et al., 2003
+ (No distinction between Ins. 1+2)	n.d.	+	n.d.	+	n.d.	+	Experiments with and without nestin selec.	wt, Pdx1 and Pax4	Blyszczuk et al., 2003
Ins. 1: − Ins. 2: +	+ (C-peptide/ insulin single staining)	−	n.d.	−	n.d.	n.d.	+	wt and Pdx1	Miyazaki et al., 2004

Insulin expression	− C-peptide+ and insulin+ cells, but no coexpression	− Non-glucose-dependent insulin release		− No improvement of hyperglyc. within 15–25 days			+ In addition: selection for Sox2+ cells	Comments	Reference
Ins. 1: − Ins. 2: +	−	+	n.d.	n.d.	n.d.	n.d.	+	−	Sipione et al., 2004
+ (No distinction between Ins. 1+2[a])	+	+	−	n.d.	n.d.	n.d.	+	−	Hansson et al., 2004
Proins. 1: + Proins. 2: +	n.d.	−	n.d.	+	+	n.d.	−	wt and Pax4	Blyszczuk et al., 2004
Ins. 1: + Ins. 2: n.d.	+	+	n.d.	n.d.	n.d.	n.d.	−	wt and Nkx2.2	Shiroi et al., 2005
	+	+	n.d.	+	n.d.	n.d.	+	−	Shi et al., 2005
+ (No distinction between Ins. 1+2)	+	+	+	+	+	n.d.	−	Gene trapping via human insulin	Vaca et al., 2005
n.d.	Only when cultured with human or bovine insulin	−	n.d.	n.d.	n.d.	n.d.	+	−	Paek et al., 2005

[a] Repetition of RT-PCR with primers specific for insulin 1 and 2 revealed marked induction of insulin 1, while insulin 2 was not expressed (see Fig. 2).

TABLE II

PRIMER SEQUENCES, ANNEALING TEMPERATURE, AND LENGTH OF THE AMPLIFIED FRAGMENT APPLICABLE FOR RT-PCR AMPLIFICATION OF PROGENITOR AND PANCREAS-SPECIFIC GENES

Gene	Primer sequence (forward/reverse)	Annealing temperature	Product size (bp)
Sox 17	5'-CCA TAG CAG AGC TCG GGG TC-3' 5'-GTG CGG AGA CAT CAG CGG AG-3'	62°	627
HNF3β(Foxa2)	5'-ACT GGA GCA GCT ACT ACG-3' 5'-CCC ACA TAG GAT GAC ATG-3'	55°	152
Cytokeratin 19	5'-CTG CAG ATG ACT TCA GAA CC-3' 5'-GGC CAT GAT CTC ATA CTG AC-3'	62°	299
Isl-1	5'-GTT TGT ACG GGA TCA AAT GC-3' 5'-ATG CTG CGT TTC TTG TCC TT-3'	60°	503
Nestin	5'-CTA CCA GGA GCG CGT GGC-3' 5'-TCC ACA GCC AGC TGG AAC TT-3'	60°	219
Ngn3 (MATH4B)	5'-TGG CGC CTC ATC CCT TGG ATG-3' 5'-AGT CAC CCA CTT CTG CTT CG-3'	60°	159
Pax4	5'-ACC AGA GCT TGC ACT GGA CT-3' 5'-CCC ATT TCA GCT TCT CTT GC-3'	60°	300
Pax6	5'-TCA CAG CGG AGT GAA TCA G-3' 5'-CCC AAG CAA AGA TGG AAG-3'	58°	332
Pdx1 (IPF-1)	5'-CTT TCC CGT GGA TGA AAT CC-3' 5'-GTC AAG TTC AAC ATC ACT GCC-3'	60°	205

Insulin 1/ Preproinsulin 1	5'-TAG TGA CCA GCT ATA ATC AGA GAC-3' 5'-CGC CAA GGT CTG AAG GTC-3'	60°	288/406
Insulin 2/ Preproinsulin 2	5'-CCC TGC TGG CCC TGC TCT T-3' 5'-AGG TCT GAA GGT CAC CTG CT-3'	65°	213/701
Glucagon	5'- CAT TCA CAG GGC ACA TTC ACC-3' 5'-CCA GCC CAA GCA ATG AAT TCC-3'	55°	207
Amylase	5'-CAG GCA ATC CTG CAG GAA CAA-3' 5'-CAC TTG CGG ATA ACT GTG CCA-3'	60°	484
Glut-2	5'-TTC GGC TAT GAC ATC GGT GTG-3' 5'-AGC TGA GGC CAG CAA TCT GAC-3'	60°	556
IAPP	5'-TGA TAT TGC TGC CTC GGA CC-3' 5'-GGA GGA CTG GAC CAA GGT TG-3'	65°	233
PP	5'-ACT AGC TCA GCA CAC AGG AT-3' 5'-AGA CAA GAG AGG CTG CAA GT-3'	60°	364
Somatostatin/ Preprosomatostatin	5'-TCG CTG CTG CCT GAG GAC CT-3' 5'-GCC AAG AAG TAC TTG GCC AGT TC-3'	60°	232/897
β5-Tubulin	5'-TCA CTG TGC CTG AAC TTA CC-3' 5'-GGA ACA TAG CCG TAA ACT GC-3'	60°	318

All markers are normalized to the housekeeping gene β5-tubulin (for a detailed description of RT-PCR, see Wobus *et al.* [2002]).

Immunofluorescence Analysis

For immunofluorescence, EB outgrowths of ES cells growing on cover-slips are either fixed with 4% paraformaldehyde (PFA) in PBS at room temperature for 20 min or in methanol:acetone (Met:Ac; 7:3, vol:vol) at –20° for 10 min, depending on the antibody used (see Table III). After rinsing (three times) in PBS, bovine serum albumin (BSA, 1% in PBS) is used to inhibit unspecific labeling (30 min) at room temperature. Cells are incubated with the primary antibodies in specific dilutions (Table III) at 37° for 60 min. Samples are washed (three times) in PBS and incubated with fluorescence-labeled secondary antibodies (diluted in 0.5% BSA in PBS) at 37° for 45 min (Table IV). To label nuclei for a semiquantitative estimation of immunofluorescence signals, cells are incubated in 5 μg/ml Hoechst 33342 in PBS at 37° for 10 min. After washing (three times) in PBS and once in Aqua destillata, specimens are embedded in mounting medium (Vectashield, Vector Laboratories Inc., Burlingame, CA).

TABLE III
SELECTED PRIMARY ANTIBODIES USED TO CHARACTERIZE PROGENITOR AND
PANCREATIC CELL TYPES

Primary antibody	Dilution	Supplier	Fixation[a]
Progenitor cells			
Mouse anti-nestin IgG (clone rat 401)	1:3	Developmental Studies Hybridoma Bank, Iowa	4% PFA
Rabbit anti-desmin IgG	1:100	Dako, Denmark	4% PFA
Mouse anti-cytokeratin 19 IgM	1:100	Cymbus, UK	MeOH:Ac[b] 4% PFA[c]
Rabbit anti-Isl-1 IgG	1:50	Abcam, UK	4% PFA
Rabbit anti-carbonic anhydrase II IgG	1:200	Abcam, UK	4% PFA
Pancreatic markers			
Mouse anti-insulin IgG (clone K36AC10)	1:100	Sigma-Aldrich, Munich, Germany	4% PFA+ 0.1% glutaraldehyde
Sheep anti-C-peptide IgG	1:100	Acris, Germany	4% PFA
Rabbit anti-glucagon IgG	1:40	Abcam, UK	4% PFA
Rabbit anti-somatostatin IgG	1:40	Biomeda, USA	4% PFA
Rabbit anti-PP IgG	1:40	Dako, Denmark	4% PFA

[a] MeOH:Ac: methanol:acetone (7:3, vol:vol) fixation at –20° for 10 min. PFA: 4% paraformaldehyde fixation at room temperature for 20 min.
[b] Filament structures.
[c] Dot-like structures.

TABLE IV
FLUORESCENCE-LABELED SECONDARY ANTIBODIES

Secondary antibody	Dilution	Supplier
Cy^3-conjugated goat anti-mouse IgG	1:600	
Cy^3-conjugated goat anti-rabbit IgG	1:600	Jackson ImmunoResearch Laboratories, USA
Cy^3-conjugated goat anti-mouse IgM	1:600	
ALEXA 488-conjugated donkey anti-sheep IgG	1:100	
ALEXA 488-conjugated goat anti-mouse IgG	1:100	Molecular Probes, Germany

Labeled cells are analyzed by the fluorescence microscope ECLIPSE TE300 (Nikon, Japan) or the confocal laser scanning microscope (CLSM) LSM-410 (Carl Zeiss, Jena, Germany) using the following excitation lines/barrier filters: 364 nm/450–490BP (Hoechst 33342), 488 nm/510–525BP (ALEXA 488), 543 nm/570LP (Cy3).

Semiquantitative Determination of Immunofluorescence Signals

Quantification of immunofluorescence signals is performed by two alternative methods depending on the cell culture status. Cells growing in monolayer may be analyzed by direct determination of immunolabeled cells (percentage values), whereas for cells growing in multilayered clusters, the "labeling index" technique is proposed.

1. Determination of percentage values of Hoechst-labeled cells: Cells are analyzed for immunofluorescence signals, and the percentage number of immunopositive cells relative to a total number of ($n = 1000$) Hoechst 33342-labeled cells is given.

2. Estimation of the "labeling index" (Blyszczuk et al., 2003): For cells growing in clusters, immunofluorescence analysis is performed using the inverted fluorescence microscope ECLIPSE TE300 (Nikon, Japan) equipped with a 3CCD color video camera DXC-9100P (Sony, Japan) and LUCIA M - Version 3.52a software (LIM, Nikon). For each sample, at least 20 randomly but representative selected pictures are analyzed for the "area fraction" value, which is the ratio of the immunopositive signal area to the measured area. To discriminate the immunopositive signal from background fluorescence, the pictures are binarized with the specific threshold fluorescence values.

Insulin ELISA

The analysis of differentiated pancreatic endocrine cells should include the determination of insulin production as a functional assay. The intracellular insulin content can be measured by commercialized specific insulin ELISA. Additionally the glucose responsiveness should be tested. For this purpose, insulin release in the presence of low (5.5 mM, as a control) and high (27.7 mM) glucose concentration is determined. Tolbutamide (10 μM), a sulfonylurea known to stimulate insulin secretion, together with 5.5 mM glucose, can also be used. However, failure of the glucose response may be dependent on insufficient maturation during differentiation. Such effects were already described during pancreatic differentiation of mouse ES cells, where insulin was secreted in response to glucose at an advanced stage of 32 days of differentiation, but not at day 28 (Blyszczuk *et al.*, 2003).

Embryonic stem-derived cells differentiated into the pancreatic lineage are cultured in differentiation medium without insulin for 3 h prior to ELISA. Cells are washed in PBS (five times) and preincubated in freshly prepared KRBH (Krebs' Ringer bicarbonate HEPES) buffer containing 118 mM sodium chloride, 4.7 mM potassium chloride, 1.1 mM potassium dihydrogen phosphate, 25 mM sodium hydrogen carbonate (all from Carl Roth GmbH & Co., Karlsruhe, Germany), 3.4 mM calcium chloride (Sigma), 2.5 mM magnesium sulfate (Merck), 10 mM HEPES, and 2 mg/ml BSA supplemented with 2.5 mM glucose (all from Invitrogen) for 90 min at 37°.

To estimate glucose-induced insulin secretion, the buffer is replaced by 27.7 mM glucose and alternatively with 5.5 mM glucose and 50 μM tolbutamide dissolved in KRBH buffer for 15 min at 37°. The control is incubated in KRBH buffer supplemented with 5.5 mM glucose. The supernatant is collected and stored at –20° for determination of insulin release.

Cells are washed two times with 0.2% trypsin:0.02% EDTA in PBS (1:1), trypsinization is stopped with 1.5 ml DMEM containing 10% FCS, and cells are collected by centrifugation. Proteins are extracted from the cells with 50 μl acid ethanol (1 M hydrochloric acid:absolute ethanol = 1:9), incubated at 4° overnight, sonicated, and stored at –20° for the determination of total cellular insulin and protein content, respectively.

The insulin enzyme-linked immunosorbent assay (Mercodia AB, Sweden) is performed according to manufacturer recommendations.

The total protein content is determined by the protein Bradford assay according to manufacturer recommendations (Bio-Rad Laboratories GmbH, Munich, Germany).

Released insulin levels are presented as a ratio of released insulin per 15 min and intracellular insulin content. The intracellular insulin level is given as nanogram insulin per milligram protein (Blyszczuk *et al.*, 2003, 2004).

Results

By applying a protocol composed of spontaneous differentiation of mES cells via EBs followed by directed differentiation using specific growth and extracellular matrix factors, islet-like clusters were formed (Fig. 1), which expressed Pdx1, Pax4, IAPP, insulin 1, glucagon, amylase, and somatostatin (Fig. 2A and B). Cells at the committed pancreatic progenitor stage (5 + 16 d) showed coexpression of nestin with Isl-1 and C-peptide and of C-peptide with CK19, respectively (Fig. 3A, B, and C). Cells at the terminal differentiation stage (5 + 28 d) did not coexpress C-peptide and nestin in islet-like clusters (Fig. 3D), but nestin-positive cells were found outside the clusters (Fig. 3D). CK 19 showed only a low level of coexpression with C-peptide-positive cells, suggesting that the cells still represent an immature phenotype (Fig. 3E). C-peptide expression in insulin-producing cells (Fig. 3F) and glucose-dependent insulin release (Fig. 2C and D) present evidence that differentiated mES cells indeed produced and released insulin rather than taking it up from the medium.

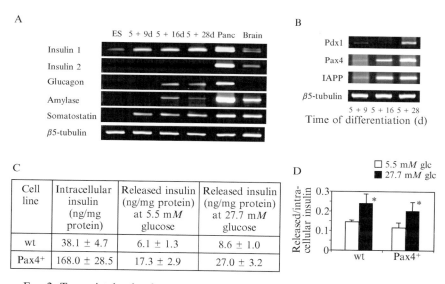

FIG. 2. Transcript levels of pancreas-specific genes and insulin release by ELISA. (A and B) RT-PCR results of mES cells and cells at differentiation stages 5 + 9d, 5 + 16d, and 5 + 28d. Mouse pancreas and brain served as positive controls. (C and D) ELISA data of insulin levels in ES-derived cells after pancreatic differentiation. (C) Levels of intracellular and released insulin in wild type (wt) and Pax4-overexpressing cells (Pax4$^+$). (D) Glucose-dependent insulin release shown as a ratio of secreted and intracellular insulin values. Each value represents the mean ± SEM. Statistical significance was tested by the Student t test: *$P < 0.05$ (B–D according to Blyszczuk et al., 2004).

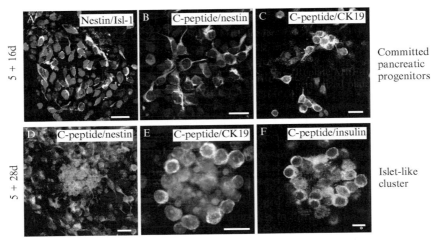

FIG. 3. Double immunofluorescence analysis of ES-derived cells differentiated into the pancreatic lineage. ES-derived cells at intermediate (5 + 16d, A–C) and terminal (5 + 28d, D–F) stages following the three-step pancreatic differentiation protocol are shown. (A and B) Images show immunohistochemical analysis of nestin/ Isl-1 and nestin/ C-peptide coexpression, and (C) coexpression of C-peptide and cytokeratin 19 (CK19). (D–F) Images demonstrate the lack of C-peptide/nestin coexpression in islet-like clusters, a weak coexpression of C-peptide with CK19, but C-peptide/ insulin colabeling in islet-like clusters at the terminal differentiation stage, 5 + 28d. Bar: 20 μm (see Blyszczuk et al., 2004). (See color insert.)

Summary

Our differentiation model represents a reproducible method to generate insulin-producing cells, avoiding all selection steps that might critically affect ES cell differentiation into the pancreatic lineage. Especially, we avoid the induced propagation and specific selection of nestin-expressing cells, as it has been shown by several independent studies that this may result in apoptotic pathways, induction of neural differentiation, and lack of functional pancreatic insulin-producing cells. Moreover, our differentiation protocol allows a further analysis of pancreatic differentiation factors and signaling mechanisms necessary for the generation and maturation of islet-like cells in vitro.

Acknowledgments

We thank Mrs. S. Sommerfeld, O. Weiss, and K. Meier for excellent technical assistance and Dr. A. Rolletschek for helpful comments. The financial support by the German Research Foundation (DFG, WO 503/3-3) and the EU FunGenES program to A.M.W. is gratefully acknowledged. We thank the Int. J. Dev. Biol. for permission to reproduce parts of figures 2 and 3.

References

Blyszczuk, P., Asbrand, C., Rozzo, A., Kania, G., St Onge, L., Rupnik, M., and Wobus, A. M. (2004). Embryonic stem cells differentiate into insulin-producing cells without selection of nestin-expressing cells. *Int. J. Dev. Biol.* **48,** 1095–1104.

Blyszczuk, P., Czyz, J., Kania, G., Wagner, M., Roll, U., St Onge, L., and Wobus, A. M. (2003). Expression of Pax4 in embryonic stem cells promotes differentiation of nestin-positive progenitor and insulin-producing cells. *Proc. Natl. Acad. Sci. USA* **100,** 998–1003.

Cerf, M. E., Muller, C. J., Du Toit, D. F., Louw, J., and Wolfe-Coote, S. A. (2005). Transcription factors, pancreatic development, and beta-cell maintenance. *Biochem. Biophys. Res. Commun.* **326,** 699–702.

Chomczynski, P., and Sacchi, N. (1987). Single-step method of RNA isolation by acid guanidinium thiocyanate-phenol-chloroform extraction. *Anal. Biochem.* **162,** 156–159.

D'Amour, K. A., Agulnick, A. D., Eliazer, S., Kelly, O. G., Kroon, E., and Baetge, E. E. (2005). Efficient differentiation of human embryonic stem cells to definitive endoderm. *Nature Biotechnol.* **23,** 1534–1541.

Doetschman, T. C., Eistetter, H., Katz, M., Schmidt, W., and Kemler, R. (1985). The *in vitro* development of blastocyst-derived embryonic stem cell lines: Formation of visceral yolk sac, blood islands and myocardium. *J. Embryol. Exp. Morphol.* **87,** 27–45.

Gradwohl, G., Dierich, A., LeMeur, M., and Guillemot, F. (2000). Neurogenin3 is required for the development of the four endocrine cell lineages of the pancreas. *Proc. Natl. Acad. Sci. USA* **97,** 1607–1611.

Habener, J. F., Kemp, D. M., and Thomas, M. K. (2005). Minireview: Transcriptional regulation in pancreatic development. *Endocrinology* **146,** 1025–1034.

Hansson, M., Tonning, A., Frandsen, U., Petri, A., Rajagopal, J., Englund, M. C. O., Heller, R. S., Hakansson, J., Fleckner, J., Skold, H. N., Melton, D., Semb, H., and Serup, P. (2004). Artifactual insulin release from differentiated embryonic stem cells. *Diabetes* **53,** 2603–2609.

Hori, Y., Rulifson, I. C., Tsai, B. C., Heit, J. J., Cahoy, J. D., and Kim, S. K. (2002). Growth inhibitors promote differentiation of insulin-producing tissue from embryonic stem cells. *Proc. Natl. Acad. Sci. USA* **99,** 16105–16110.

Ishizaka, S., Shiroi, A., Kanda, S., Yoshikawa, M., Tsujinoue, H., Kuriyama, S., Hasuma, T., Nakatani, K., and Takahashi, K. (2002). Development of hepatocytes from ES cells after transfection with the HNF-3beta gene. *FASEB J.* **16,** 1444–1446.

Iwasaki, S., Hattori, A., Sato, M., Tsujimoto, M., and Kohno, M. (1996). Characterization of the bone morphogenetic protein-2 as a neurotrophic factor: Induction of neuronal differentiation of PC12 cells in the absence of mitogen-activated protein kinase activation. *J. Biol. Chem.* **271,** 17360–17365.

Jiang, F. X., Cram, D. S., DeAizpurua, H. J., and Harrison, L. C. (1999). Laminin-1 promotes differentiation of fetal mouse pancreatic beta-cells. *Diabetes* **48,** 722–730.

Kahan, B. W., Jacobson, L. M., Hullett, D. A., Ochoada, J. M., Oberley, T. D., Lang, K. M., and Odorico, J. S. (2003). Pancreatic precursors and differentiated islet cell types from murine embryonic stem cells: An *in vitro* model to study islet differentiation. *Diabetes* **52,** 2016–2024.

Kania, G., Blyszczuk, P., and Wobus, A. M. (2004). The generation of insulin-producing cells from embryonic stem cells: A discussion of controversial findings. *Int. J. Dev. Biol.* **48,** 1061–1064.

Kubo, A., Shinozaki, K., Shannon, J. M., Kouskoff, V., Kennedy, M., Woo, S., Fehling, H. J., and Keller, G. (2004). Development of definitive endoderm from embryonic stem cells in culture. *Development* **131,** 1651–1662.

Lee, J., Wu, Y., Qi, Y., Xue, H., Liu, Y., Scheel, D., German, M., Qiu, M., Guillemot, F., and Rao, M. (2003). Neurogenin3 participates in gliogenesis in the developing vertebrate spinal cord. *Dev. Biol.* **253,** 84–98.

Leon-Quinto, T., Jones, J., Skoudy, A., Burcin, M., and Soria, B. (2004). *In vitro* directed differentiation of mouse embryonic stem cells into insulin-producing cells. *Diabetologia* **47,** 1442–1451.

Lumelsky, N., Blondel, O., Laeng, P., Velasco, I., Ravin, R., and McKay, R. (2001). Differentiation of embryonic stem cells to insulin-secreting structures similar to pancreatic islets. *Science* **292,** 1389–1394.

Melloul, D., Marshak, S., and Cerasi, E. (2002). Regulation of insulin gene transcription. *Diabetologia* **45,** 309–326.

Micallef, S. J., Janes, M. E., Knezevic, K., Davis, R. P., Elefanty, A. G., and Stanley, E. G. (2005). Retinoic acid induces Pdx1-positive endoderm in differentiating mouse embryonic stem cells. *Diabetes* **54,** 301–305.

Miyazaki, S., Yamato, E., and Miyazaki, J. (2004). Regulated expression of Pdx-1 promotes *in vitro* differentiation of insulin-producing cells from embryonic stem cells. *Diabetes* **53,** 1030–1037.

Nakagawa, Y., and O'Leary, D. D. M. (2001). Combinatorial expression patterns of LIM-homeodomain and other regulatory genes parcellate developing thalamus. *J. Neurosci.* **21,** 2711–2725.

Okabe, S., Forsberg-Nilsson, K., Spiro, A. C., Segal, M., and McKay, R. D. (1996). Development of neuronal precursor cells and functional postmitotic neurons from embryonic stem cells *in vitro. Mech. Dev.* **59,** 89–102.

Otonkoski, T., Beattie, G. M., Mally, M. I., Ricordi, C., and Hayek, A. (1993). Nicotinamide is a potent inducer of endocrine differentiation in cultured human fetal pancreatic cells. *J. Clin. Invest.* **92,** 1459–1466.

Paek, H. J., Moise, L. J., Morgan, J. R., and Lysaght, M. J. (2005). Origin of insulin secreted from islet-like cell clusters derived from murine embryonic stem cells. *Cloning Stem Cells* **7,** 226–231.

Rajagopal, J., Anderson, W. J., Kume, S., Martinez, O. I., and Melton, D. A. (2003). Insulin staining of ES cell progeny from insulin uptake. *Science* **299,** 363.

Schwitzgebel, V. M., Scheel, D. W., Conners, J. R., Kalamaras, J., Lee, J. E., Anderson, D. J., Sussel, L., Johnson, J. D., and German, M. S. (2000). Expression of neurogenin3 reveals an islet cell precursor population in the pancreas. *Development* **127,** 3533–3542.

Shi, Y., Hou, L., Tang, F., Jiang, W., Wang, P., Ding, M., and Deng, H. (2005). Inducing embryonic stem cells to differentiate into pancreatic β cells by a novel three-step approach with activin A and all-*trans* retinoic acid. *Stem Cells* **23,** 656–662.

Shiroi, A., Ueda, S., Ouji, Y., Saito, K., Moriya, K., Sugie, Y., Fukui, H., Ishizaka, S., and Yoshikawa, M. (2005). Differentiation of embryonic stem cells into insulin-producing cells promoted by Nkx2.2 gene transfer. *World J. Gastroenterol.* **11,** 4161–4166.

Sipione, S., Eshpeter, A., Lyon, J. G., Korbutt, G. S., and Bleackley, R. C. (2004). Insulin expressing cells from differentiated embryonic stem cells are not beta cells. *Diabetologia* **47,** 499–508.

Smith, D. B., and Johnson, K. S. (1988). Single-step purification of polypeptides expressed in *Escherichia coli* as fusions with glutathione S-transferase. *Gene* **67,** 31–40.

Soria, B., Roche, E., Berna, G., Leon-Quinto, T., Reig, J. A., and Martin, F. (2000). Insulin-secreting cells derived from embryonic stem cells normalize glycemia in streptozotocin-induced diabetic mice. *Diabetes* **49,** 157–162.

Vaca, P., Martin, F., Vegara-Meseguer, J., Rovira, J., Berna, G., and Soria, B. (2005). Induction of differentiation of embryonic stem cells into insulin secreting cells by fetal soluble factors. *Stem Cells* **24,** 258–265.

Wobus, A. M., and Boheler, K. R. (2005). Embryonic stem cells: Prospects for developmental biology and cell therapy. *Physiol. Rev.* **85,** 635–678.

Wobus, A. M., Guan, K., Yang, H.-T., and Boheler, K. (2002). Embryonic stem cells as a model to study cardiac, skeletal muscle, and vascular smooth muscle cell differentiation. *Methods Mol. Biol.* **185,** 127–156.

Ying, Q. L., Stavridis, M., Griffiths, D., Li, M., and Smith, A. (2003). Conversion of embryonic stem cells into neuroectodermal precursors in adherent monoculture. *Nat. Biotechnol.* **21,** 183–186.

[20] Pulmonary Epithelium

By Anne E. Bishop and Julia M. Polak

Abstract

Repair or regeneration of defective lung epithelium would be of great therapeutic potential. Cellular sources for such repair have long been searched for within the lung, but the identification and characterization of stem or progenitor cells have been hampered by the complexity and cellular heterogeneity of the organ. In recent years, various pulmonary cells have been identified that meet the criteria for stem cells but it remains to be seen how far manipulation of these tissue-specific cell pools can upregulate epithelial repair. The initial excitement that greeted the results of animal experiments showing cells of bone marrow origin in murine lung has been tempered by more recent data suggesting that the cells do not repair pulmonary epithelium. However, there are reports of engraftment of bone marrow-derived cells in human lung, albeit at a low level, so the administration of cell therapy via the circulation, for repair and/or gene delivery, needs further investigation. The potential of human embryonic stem cells to generate any cell, tissue, or organ on demand for tissue repair or replacement is promising to revolutionize the treatment of human disease. Although some headway has been made into making pulmonary epithelium from these stem cells, human embryonic stem cell technology is still in its infancy and many technical, safety, and ethical hurdles must be cleared before clinical trials can begin. This chapter focuses on the potential role of stem cells in future approaches to lung repair and regeneration.

Introduction

Without a doubt, readily available means to repair, replace, and/or regenerate human lung tissue would have an enormous clinical impact. Lung diseases are widespread and debilitating and present a significant biomedical

METHODS IN ENZYMOLOGY, VOL. 418   0076-6879/06 $35.00
DOI: 10.1016/S0076-6879(06)18020-9

problem. As a single example, chronic obstructive pulmonary disease (smoker's lung) is estimated to be the fourth leading cause of death worldwide (2.74 million in 2000), with 12.1 million adults aged 25 and older being diagnosed in the United States alone in 2001, generating healthcare costs in excess of $32.1 billion. However, the structural complexity and cellular diversity of the lung, coupled with the slow cell turnover rates of pulmonary epithelium, make it a particularly difficult target for regenerative medicine. Various advances have been made in recent years that begin to promise some concrete developments in achieving targeted lung regeneration. These include further understanding of the molecular events that take place during lung morphogenesis and the discovery of previously unknown regenerative pathways. Perhaps some of the most pivotal advances have been made in the area of stem cell biology, which has moved to the forefront of medical research in the past decade. Current regenerative medicine strategies are exploring the possibility of using stem cells as key tools to mediate repair *in vivo*, to form pulmonary epithelial cells and lung tissue *in vitro* for implantation, and to create *in vitro* models of lung development for further investigation and manipulation.

Development and Organization of Pulmonary Epithelium

The mammalian lung develops as an outgrowth of the embryonic gut originating in humans from a diverticulum of the ventral wall of the primitive esophagus between 4 and 5 weeks of gestation. From then on, the nascent epithelium undergoes dichotomous branching into the surrounding splanchnic mesenchyme in the highly ordered process called branching morphogenesis (Hogan, 1999). Mammalian lung development is divided into four phases and, in human, the timings of these are embryonic, 0 to 5 weeks; glandular, 5 to 16 weeks; canalicular, 16 to 26 weeks; and saccular, 26 weeks to term (Adamson, 1991). The primordial lining that forms in the embryonic stage develops into pseudostratified epithelium during the early glandular phase and, as branching progresses, columnar epithelium is formed. During the glandular and into the canalicular phase, the initial thick layer of stratified epithelium starts to grow thinner and shows gradation, becoming gradually more and more thin along the length of the tree. Submucosal glands are first seen at around 10 weeks in the trachea but not until 16 weeks in the bronchi. Bronchioles appear during the canalicular stage, marking the initiation of gas exchange unit formation. The final formation of alveoli takes place postnatally. Thus, the mature human lung has distinct anatomical regions lined by different types of epithelial cells. The trachea and major bronchi are lined by pseudostratified epithelium. The major phenotypes in the proximal airways are ciliated and mucous secretory (or goblet) cells, with the more infrequent neuroendocrine cells and the less well-differentiated basal cells lying in a basal position. Ciliated cells also line the bronchioles that possess another phenotype known as Clara cells, which are nonciliated. The alveoli are lined by flattened squamous (type I) and cuboidal (type II) pneumocytes. A further class of epithelial cells that populates the lung comprises the neuroendocrine cells that first appear around 8 weeks of gestation. These contain biogenic amines, usually serotonin (Lauweryns et al., 1986), and/or peptides, mainly bombesin (Wharton et al., 1978) or calcitonin gene-related peptide (Johnson and Wobken, 1987). They are relatively frequent in the developing lung, where they play a major role in airway growth and development, but form <1% of epithelial cells in adult lung, where they are seen as scattered elements in the epithelium or in innervated epithelial cell clusters known as neuroepithelial bodies (Cutz and Orange, 1977; Lauweryns and Cokelaere, 1973).

Lung Stem Cells

    A widely used definition of stem cells is that they are clonogenic and capable of self-renewal and multilineage differentiation (Blau et al., 2001; Fuchs and Segre, 2000; Metcalf and Moore, 1971; Till and McCulloch, 1961; Weissman, 2000). Subsets of cells in the lung have been described that fulfill these criteria, although their identification has been hampered by the difficulty of isolating them coupled with the very low level of regeneration in the lung. The classical view of stem/progenitor cells of the pulmonary epithelium is that they comprise the basal and mucous cells of the proximal airways (Breeze and Wheeldon, 1977; Donnelly et al., 1982; Kauffman, 1980; Reid and Jones, 1979), Clara cells in the bronchioles (Boers et al., 1999; Clara, 1937; Evans et al. 1976, 1978; Plopper and Dungworth; 1987) and type II pneumocytes in the alveoli (Adamson and Bowden, 1974, 1975; Bowden, 1981; Evans et al., 1971, 1975; Kauffman, 1980; Kauffman et al., 1974; Macklin, 1954; Witschi, 1976). More recently, variant Clara cells have been described that express Clara cell secretory protein (CCSP) but are not typical Clara cells as they are resistant to airway pollutants such as naphthalene (Giangreco et al., 2002; Hong et al., 2001; Mahvi et al., 1977; Plopper et al., 1992; Reynolds et al., 2000a,b; Stripp et al., 1995). These variant CCSP-expressing (or vCE) cells show multipotent differentiation and are located in discrete pools in neuroepithelial bodies and at the bronchoalveolar duct junction. Type II pneumocytes also appear to exist in at least two populations, one of which shows proliferation, is relatively resistant to injury, has high telomerase activity, and probably comprises the cells that repopulate damaged alveolar epithelium (Reddy et al., 2004).

In addition, the existence in the lung of a "universal" pluripotent cell has long been speculated upon and now some initial evidence has emerged with the identification of a spore-like cell that can differentiate *in vitro* to bronchiolar tissue. A pluripotent stem cell has been described in the lungs of adult sheep and rats that can generate lung-like tissue *in vitro*, specifically of the alveolar (Cortiella *et al.*, 2000) and bronchiolar regions (Vacanti *et al.*, 2001). The isolated cells were extremely small with a very low oxygen demand, leading the researchers to call them "spore like" (Vacanti *et al.*, 2001). It is thought that these cells lie dormant until activated by injury or disease.

Circulating Stem Cells

Our understanding of the regeneration of the lung is also being revised as a result of the description of pulmonary epithelium derived from blood-borne cells. Adult bone marrow has long been know to contain pluripotent stem cells that have the capacity for self-renewal and can give rise to hematopoietic and mesenchymal cell lineages (Pittenger *et al.*, 1999). These stem cells have gained in importance with the recognition that they can differentiate not only toward multiple mesenchymal lineages, such as adipocytes (Pittenger *et al.*, 1999), osteocytes (Pereira *et al.*, 1998; Prockop, 1997), myocytes (Ferrari *et al.*, 1998), and cardiomyocytes (Orlic *et al.*, 2001), but also toward ectodermal, for example, neurons (Mezey *et al.*, 2000), and endodermal lineages, such as hepatocytes (Alison *et al.*, 2000; Lagasse *et al.*, 2000; Petersen *et al.*, 1999; Theise *et al.*, 2000) and renal parenchymal cells (Poulsom *et al.*, 2001). It has also been reported that multipotent adult progenitor cells from murine bone marrow can differentiate *in vitro* at the single cell level to all three germ layers (Jiang *et al.*, 2002). The exact mechanisms by which cells of bone marrow origin are recruited by, engraft, and differentiate in the various tissues are not known, although engraftment has been reported to be enhanced by tissue injury (Ferrari *et al.*, 1998; Kotton *et al.*, 2001; Okamoto *et al.*, 2002; Ortiz *et al.*, 2003; Theise *et al.*, 2002).

There is some controversy as to whether stem cells are recruited from the circulation and engraft in the lung lining and, if so, whether or not this occurs by fusion with cells *in situ*. A series of publications has described how differing fractions of donor mouse bone marrow cells, labeled or mismatched, have been traced to the pulmonary epithelium of recipient mice with or without overt lung damage (Beckett *et al.,* 2005; Grove *et al.,* 2002; Harris *et al.,* 2004; Ishizawa *et al.,* 2004; Kotton *et al.,* 2001; Krause *et al.,* 2001; Loi *et al.,* 2006; Mattsson *et al.,* 2004; Ortiz *et al.,* 2003; Theise *et al.,* 2002; Yamada *et al.,* 2004). Fusion of bone marrow stem cells with somatic cells has been demonstrated *in vitro* (Alvarez-Dolado *et al.,* 2003;

Terada *et al.*, 2002; Ying *et al.*, 2002) and also *in vivo* in cells known to form heterokaryons in certain pathologies, including hepatocytes and cardio-myocytes (Alvarez-Dolado *et al.*, 2003; Vassilopoulos *et al.*, 2003; Wang *et al.*, 2003). Tests of fusion with pulmonary epithelium specifically gave mixed results; *in vitro* experiments clearly demonstrated fusion with bone marrow stem cells (Spees *et al.*, 2003) but this was not seen *in vivo* (Alvarez-Dolado *et al.*, 2003; Grove *et al.*, 2002; Harris *et al.*, 2004). More recently, however, doubt has been cast as to whether any actual engraft-ment occurs in the pulmonary epithelium. Citing limitations of the histo-logical methods used to assess cell engraftment in previous publications, two studies in animal models have been published that argue that bone marrow stem cells do not contribute to the repair process in the lung (Chang *et al.*, 2005; Kotton *et al.*, 2005).

So far, investigation of human lung has either not detected engraftment of bone marrow cells or shown that it occurs at a low rate (Albera *et al.*, 2005; Bittmann *et al.*, 2001; Kleeberger *et al.*, 2003; Kubit *et al.*, 1994; Suratt *et al.*, 2003; Zander *et al.*, 2005). However, these studies were also fraught with technical problems, mainly because of the paucity and poor quality of most of the lung samples. If circulating cells are indeed recruited by damaged pulmonary epithelium to effect repair, there could be major clinical implications. The observation that bone marrow cells engraft in murine and human heart was translated rapidly to clinic where administra-tion of autologous bone marrow, directly to the cardiac wall or via the vasculature, is now a widely used and successful means to treat myocardial infarction and heart failure (for review see Wollert and Dexter, 2005). The complex structure and cellular heterogeneity of the lung may make similar cell therapy a more difficult goal for respiratory diseases, but the potential benefits make this a possibility that needs to be explored thoroughly; in addition to repair, delivery of transfected stem cells and/or their progeny could provide a novel means for gene therapy.

In Vitro Generation of Pulmonary Epithelium from Stem Cells

Stem cells can be isolated from embryos, fetuses, or adult tissue, but the range of cell types to which they can differentiate varies according to origin. For regenerative medicine purposes, stem cells can provide a virtually inexhaustible cell source, and current research is focused on promoting their differentiation to required lineages, purification of consequent cells, and implantation in a form that will replace, or augment the function of, diseased or injured tissues (for review see Polak and Bishop, 2006). An initial step is the selection of the most appropriate stem cell to form the required tissue. However, as mentioned previously, the complexity of the lung precludes isolation of stem cells in sufficient numbers for study of their biology, let alone for potential therapeutic applications. For this reason, attempts have been made to produce pulmonary epithelium from exogenous stem cells. The first choice has been embryonic stem cells, in view of their relative availability, known provenance, and proliferative capacity. Early work established the possibility of deriving alveolar airway epithelium, specifically type II pneumocytes, from murine embryonic stem cells using medium supplementation (Ali *et al.*, 2002; Rippon *et al.*, 2004, 2006) (Fig. 1). The following provides a three-step strategy for the generation of distal lung epithelial progenitors cells, thought to be representative of those present in the early branching lung at approximately E10–11 of murine development (Rippon *et al.*, 2006).

Derivation of Distal Lung Epithelial Cells from Murine Embryonic Stem Cells (mESC) Using Medium Supplementation

The derivation of lung epithelium, like all cell types arising from the endodermal embryonic germ layer, is comparatively inefficient from mESC. This optimized protocol yields highly mixed cultures in which the majority of cells are not lung epithelial cells. This method reliably obtains an average of at least 3 to 5% differentiation efficiency. However, it should be noted that under control conditions spontaneous differentiation of mESC to lung epithelial phenotypes is never detectable.

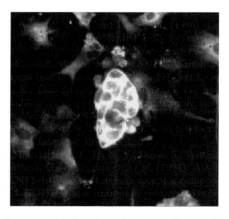

FIG. 1. A clump of differentiated embryonic stem cells in culture immunostained for surfactant protein C, a specific marker for type II pneumocytes. The clump of pneumocytes is surrounded by nonimmunoreactive cells showing that the resultant cell populations were heterogeneous (indirect immunofluorescence method). (See color insert.)

Cell culture reagents are obtained from Invitrogen (Paisley, UK) unless otherwise stated.

Ten-Day Suspension Culture of Embryoid Bodies (EBs). Embryoid bodies are treated with high levels of activin A early in differentiation to enhance the formation of definitive endoderm, the germ layer from which lung epithelium is derived.

> Day 0: Form embryoid bodies from a healthy culture of mESC by limited trypsin digestion. We use one T25 flask of undifferentiated mESC (24 h after passage) to generate one 90-mm Petri dish of EBs. Embryoid bodies are formed in medium consisting of high glucose Dulbecco's modified Eagle's medium (DMEM), 10% fetal bovine serum (FBS), and 2 mM L-glutamine.
>
> Day 2.5: Switch embryoid bodies into serum-free medium containing saturating levels of recombinant activin A for 4.5 days. The activin A-containing medium consists of high glucose DMEM, 10% knockout Serum Replacement, 2 mM L-glutamine, and 100 ng/ml activin A (R&D systems, Abingdon, UK). Refresh the medium on day 5.
>
> Day 7: Remove activin A from the culture medium. Continue to culture EBs in suspension for a further 3 days in high glucose DMEM, 10% knockout Serum Replacement, and 2 mM L-glutamine.

Eleven-Day Adherent Culture of Embryoid Bodies. Embryoid bodies are then plated onto gelatinized tissue culture-treated plastic to allow outgrowth and maturation of the early differentiated mESC. It is very important that EBs are not dissociated at this stage. Dissociation of embryoid bodies abolishes all lung epithelial differentiation, suggesting that the presence of a three-dimensional structure is critical to the specification of lung epithelial cell types.

> Day 10: Distribute each dish of EBs between two gelatinized-well plates in fresh medium (high glucose DMEM, 10% knockout Serum Replacement, and 2 mM L-glutamine). The majority of embryoid bodies should adhere in the next 3 to 4 days. After most embryoid bodies have adhered, refresh the medium twice weekly, removing any EBs still floating.

Selection of Lung Epithelial Progenitors with Lung-Specific Medium. The final step in the differentiation protocol selects for the lung epithelial progenitors that have formed using a serum-free commercial medium, small airway basal medium (SABM; Cambrex Corp., NJ). This step causes widespread cell death of nonlung epithelial cells, thus enriching the population for the desired cell type.

Day 21: Remove medium and wash cells with PBS. Replace medium with SABM and culture for 4 to 10 days, replacing the medium at least twice weekly to remove dead cells. At the end of the culture time, harvest the remaining live cells for analysis (patent: preparation of type II pneumocytes from stem cells. GB 0218332; WO2004015091A3; PCT/GB2003/03407; EU03784254.9. Novathera Ltd., UK).

Subsequently, other means have been used to drive the formation of pulmonary epithelium from embryonic stem cells. It has been reported that a combination of medium supplementation and growth at the air interface can induce the formation of fully differentiated tracheobronchial airway epithelium from murine embryonic stem cells (Coraux et al., 2005). A method originally used to convert fibroblasts into T cells using T-cell extracts (Håkelien et al., 2002) was adapted and applied to murine embryonic stem cells. The stem cells were permeabilized and exposed for 1 h to extracts of transformed murine pneumocytes. Following membrane resealing and culture in unsupplemented medium, the cells were found to differentiate to type II pneumocytes (Qin et al., 2005). The protocol for this method is as follows.

Derivation of Distal Lung Epithelial Cells from Murine Embryonic Stem Cells Using Cell Extracts

Cell Culture. Murine ES cells are grown in an undifferentiated state, without feeder layers, on tissue culture plates with 1000 U/ml of leukemia inhibitory factor (LIF, Chemicon, Temecula, CA). Cells are maintained in ESC medium, comprising DMEM, supplemented with 10% FBS, 2 mM L-glutamine, 100 U/μg penicillin/streptomycin, and 0.1 mM β-mercaptoethanol. Following removal of LIF to allow differentiation of the ESC, cells are maintained in the same ESC culture medium. MLE-12 cells (15: American Type Culture Collection, USA) are grown on plates in HITES medium, comprising 50% Ham's F12K medium and 50% DMEM, supplemented with 2% FBS, 10 mM HEPES, 2 mM L-glutamine, 100 U/μg penicillin/streptomycin, 0.005 mg/ml insulin, 0.01 mg/ml apo-transferrin, 30 nM sodium selenite, 10 nM hydrocortisone, and 10 nM β-estradiol. All cultures are maintained in an incubator at 37° in a humidified atmosphere of 5% CO_2. For the undifferentiated ESC cultures, medium is changed every day and, for other cells, on alternate days.

Transfection. A 4.8-kb murine SP-C promoter/GFP construct is transfected into undifferentiated E14tg2a using Lipofectamine 2000, according to the manufacturer's instructions (Invitrogen). The transfected ESC are selected in ES culture medium containing 300 ng/ml geneticin (Invitrogen) for 2 weeks.

Cell Extract Preparation. The MLE-12 extract is prepared from 80% confluent MLE-12 cells. MLE-12 cells are trypsin digested, washed in ice-cold phosphate-buffered saline (PBS), and resuspended in one volume of lysis buffer (10 mM HEPES, pH 8.2, 50 mM NaCl, 5 mM MgCl$_2$, 1 mM dithiothreitol, a cocktail of protease inhibitors, and 0.1 mM phenylmethyl-sulfonyl fluoride; Sigma, Poole, UK). The cells are snap frozen in liquid nitrogen for 2 min, thawed, and disrupted by vortexing. After centrifugation (15,000g for 15 min at 4°), the supernatant (extract) is used fresh or can be frozen in liquid nitrogen and stored at –80°. The protein concentration of the extract should be 30 mg/ml (Bradford).

Cell Reprogramming. To initiate the differentiation, transfected ES cells are cultured in suspension for 10 days to allow EB formation, and the resulting 10-day EBs are cultured in tissue culture T75 flasks for 3 days. These cells are then permeabilized at 5×10^5 cells/500 μl with 800 ng/ml streptolysin O (SLO) in Ca^{2+}, Mg^{2+}-free Hank's balanced salt solution (GIBCO-BRL) for 50 min at 37°. SLO is replaced with 100 μl of MLE extract containing an ATP-generating system (1 mM ATP, 1 mM GTP, 1 mM NTP, 10 mM phosphocreatine, and 25 μg/ml creatine kinase, Sigma) and incubated for 60 min at 37°. To reseal plasma membranes, 2 mM CaCl$_2$ is added to ES culture medium and cells are cultured overnight at 37°. Extract-treated cells are cultured in ES medium, and the type II pneumocyte phenotype (surfactant protein C expression, lamellar body formation, ability to form type I pneumocytes) begins to emerge after 3 days (patent: cell extract-based derivation of type 2 pneumocytes. GB 0510319.7. Novathera Ltd., UK).

In addition, the possibility that distal pulmonary mesenchyme from embryonic lung could stimulate alveolar airway patterning, as occurs *in vivo*, was examined by wrapping partially differentiated embryonic stem cells in microdissected mesenchyme. After only 5 days, mesenchyme and embryonic stem cells had coalesced and small channels had formed that were lined by cells expressing markers of alveolar epithelium (Van Vranken *et al.*, 2005)(Fig. 2). Clearly, this latter approach is too limited to provide cells or tissue for implantation, but it provides an *in vitro* model of distal lung development for the investigation of mesodermal–endodermal interactions.

Embryonic Stem Cell and Pulmonary Mesenchyme Coculture

ES Cell Culture. Murine embryonic stem cells are grown undifferentiated in DMEM medium (Invitrogen, UK) supplemented with 10% (v/v) fetal calf serum (FCS), 2 mM L-glutamine, 1% (w/v) penicillin/streptomycin, 0.1 mM β-mercaptoethanol, and LIF (500 U/ml) (complete DMEM)

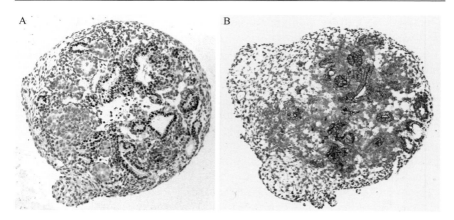

FIG. 2. Serial sections through a preparation of murine embryonic (E11.5) pulmonary mesenchyme wrapped around an embryoid body (day 8) following culture for 5 days in basic medium. The mesenchyme and embryoid body have coalesced and channels have formed lined by cells showing immunoreactivity for (A) thyroid transcription factor-1, a marker for immature pulmonary epithelium, and (B) cytokeratin, a general marker for epithelium (avidin biotin peroxidase complex method). (See color insert.)

on feeder layers comprising murine embryonic (E16) fibroblasts (SNL) inactivated with mitomycin C (Sigma, UK) according to a standard protocol. After weaning ES cells from the feeder layer for at least one passage, differentiation is induced via the formation of embryoid bodies from clusters of approximately 15 cells obtained by partial trypsin digestion (0.05% [v/v] trypsin/0.53 mM EDTA in 0.1 M PBS, Ca or Mg; 2% [v/v] chicken serum). EBs are cultured in nonadherent bacterial-grade Petri dishes in complete DMEM without LIF for 8 days to allow for formation of endoderm with medium being replenished every 36 to 48 h.

   Embryonic Lung Mesenchyme. Lungs are dissected from murine embryos at E11.5 and E13.5 and placed in ice-cold Hank's Balanced salt solution (HBSS) supplemented with 1% (v/v) antibiotic/antimycotic (1% A/A) solution (HBSS/1% A/A). The technique of mesenchyme isolation and formation of cocultures was adapted from the original protocol by Shannon and co-workers (1998). Briefly, the tips of distal lung buds are dissected intact using Moria microsurgery knives (Fine Science Tools Inc., Germany) and digested in dispase for 10 min at 37°. They are then washed three times in HBSS and separated into epithelial and mesenchymal components using tungsten needles.

   Direct Contact EB/Lung Mesenchyme Cocultures. For each coculture, one 8-day-old EB and three to four pieces of distal lung mesenchyme are used. EBs are washed twice in 5% (v/v) FCS DMEM. Portions of mesenchyme are arranged around EBs on 0.5% (v/v) agarose in a 35-mm Petri dish, and a small drop of 5 or 10% (v/v) FCS DMEM is added. Cocultures are incubated at 37° in a humidified 5% (v/v) CO_2 atmosphere for 5 or 12 days, as arbitrary short and long periods, and 200 to 300 μl of fresh medium is added daily to each dish.

Type II pneumocytes have been differentiated successfully from human embryonic stem cells (hESC) (Samadikuchaksaraei *et al.*, 2006)(Fig. 3), the first crucial step in the development of clinical applications.

Derivation of Distal Lung Epithelial Cells from hESC Using Medium Supplementation

Basic Human Embryonic Stem Cell Culture. Human embryonic stem cells are propagated in an undifferentiated state on mitotically inactivated primary mouse embryonic fibroblasts (MEFs). MEFs are inactivated by a 2-h incubation in culture medium containing 8 μg/ml mitomycin C (Sigma-Aldrich, Dorset, UK). Undifferentiated hESC are grown in "undifferentiated ES cell medium" (unESCM) consisting of DMEM with 2 mM L-glutamine and 15 mM HEPES, supplemented with 20% knockout Serum Replacement, 1% MEM nonessential amino acids (all from GIBCO Invitrogen Corp., Paisley, UK), and 4 ng/ml recombinant human basic fibroblast growth factor (bFGF) (157 amino acids) (R&D Systems, Oxon, UK). Cells are split using 0.1% collagenase IV (GIBCO Invitrogen Corp.) for 20 min followed by scraping.

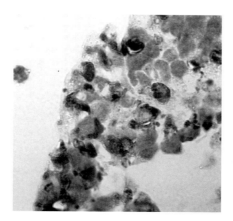

FIG. 3. Human embryonic stem cells following differentiation in SAGM showing immunoreactivity for surfactant protein A, found in the distal airway epithelium (avidin biotin peroxidase complex method). (See color insert.)

Induction of Differentiation. Undifferentiated cells are treated with collagenase for 5 min at 37° and collected in clumps. Cell clumps are cultured in suspension in polymethylpentene Petri dishes (Nalge Nunc International Corp., Rochester, NY) in embryoid body medium consisting of knockout D-MEM supplemented with 20% (v/v) heat-inactivated FBS, 2 mM L-glutamine, and 1% (v/v) MEM nonessential amino acids (all from GIBCO Invitrogen Corp.). After 7 days, the embryoid bodies are transferred to tissue culture test plates (Orange Scientific, Braine-l'Alleud, Belgium). The cells, which should adhere to the culture plates, are fed with the embryoid body medium for 2 days. Then, the medium is changed to the differentiating ES cell medium (difESCM) consisting of unESCM without bFGF. After 10 days of feeding with difESCM, cells are transferred to small airway growth medium (SAGM; Biowhittaker, Watersville, MD) consisting of a basal medium (SABM) plus the following factors: 0.5 mg/ml bovine serum albumin, 5 μg/ml insulin, 10 μg/ml transferrin, 30 μg/ml bovine pituitary extract, 0.5 μg/ml epinephrine, 6.5 ng/ml triiodothyronine, 0.1 ng/ml retinoic acid, 0.5 μg/ml hydrocortisone, and 0.5 ng/ml human epidermal growth factor. Cells are fed with SAGM for 28 days, after which time type II pneumocytes expressing surfactant protein C can be identified readily.

Whether embryonic stem cell-derived pulmonary epithelium can be used in the clinic remains to be seen. The implantation of mature phenotypes obtained from embryonic stem cells into human beings has yet to happen, not least because of the potential danger of implanting poorly or completely undifferentiated cells. However, although direct clinical applications remain questionable, there is no doubt that embryonic stem cell-derived pulmonary epithelium can have a range of indirect therapeutic applications, such as in disease modeling, drug discovery, toxicological screening, and possibly as part of a biohybrid gas-exchange device.

References

Adamson, I. Y., and Bowden, D. H. (1974). The type 2 cell as progenitor of alveolar epithelial regeneration: A cytodynamic study in mice after exposure to oxygen. *Lab. Invest.* **30,** 35–42.

Adamson, I. Y., and Bowden, D. H. (1975). Derivation of type 1 epithelium from type 2 cells in the developing rat lung. *Lab. Invest.* **32,** 736–745.

Adamson, I. Y. (1991). Development of lung structure. *In* "The Lung: Scientific Foundation." (R. G. Crystal and J. B. West, eds.), pp. 663–666. Raven Press, New York.

Albera, C., Polak, J. M., Janes, S., Griffiths, M. J. D., Alison, M. R., Wright, N. A., Navaratnarasah, S., Poulsom, R., Jeffery, R., Fisher, C., Burke, M., and Bishop, A. E. (2005). Repopulation of human pulmonary epithelium by bone marrow cells: A potential means to promote repair. *Tissue Engin.* **11,** 1115–1121.

Ali, N. N., Edgar, A. J., Samadikuchaksaraei, A., Timson, C. M., Romanska, H. M., Polak, J. M., and Bishop, A. E. (2002). Derivation of type II alveolar epithelial cells from murine embryonic stem cells. *Tissue Eng.* **8,** 541–550.

Alison, M. R., Poulsom, R., Jeffrey, R., Dhillon, A. P., Quaglia, A., Jacob, J., Novelli, M., Prentice, G., Williamson, J., and Wright, N. A. (2000). Hepatocytes from non-hepatic adult stem cells. *Nature* **406,** 257.

Alvarez-Dolado, M., Pardal, R., Garcia-Verdugo, J. M., Fike, J. R., Lee, H. O., Pfeffer, K., Lois, C., Morrison, S. J., and Alvarez-Buylla, A. (2003). Fusion of bone marrow derived cells with Purkinje neurons, cardiomyocytes and hepatocytes. *Nature* **425,** 968–973.

Beckett, T., Loi, R., Prenovitz, R., Poynter, M., Goncz, K. K., Suratt, B. T., and Weiss, D. J. (2005). Acute lung injury with endotoxin or NO2 does not enhance development of airway epithelium from bone marrow. *Mol. Ther.* **12,** 680–686.

Bittmann, I., Dose, T., Baretton, G. B., Muller, C., Schwaiblmair, M., Kur, F., and Lohrs, U. (2001). Cellular chimerism of the lung after transplantation: An interphase cytogenetic study. *Am. J. Clin. Pathol.* **115,** 525–533.

Blau, H. M., Brezelton, T. R., and Weimann, J. M. (2001). The evolving concept of a stem cell: Entity or function? *Cell* **105,** 829–841.

Boers, J. E., Ambergen, A. W., and Thunnissen, F. B. (1999). Number and proliferation of clara cells in normal human airway epithelium. *Am. J. Respir. Crit. Care Med.* **159,** 1585–1591.

Bowden, D. H. (1981). Alveolar response to injury. *Thorax* **36,** 801–804.

Breeze, R. G., and Wheeldon, E. B. (1977). The cells of the pulmonary airways. *Am. Rev. Respir. Dis.* **116,** 705–777.

Chang, J. C., Summer, R., Sun, X., Fitzsimmons, K., and Fine, A. (2005). Evidence that bone marrow cells do not contribute to the alveolar epithelium. *Am. J. Respir. Cell Mol. Biol.* **33,** 335–342.

Clara, M. (1937). Zur Histobiologie des Bronchialepithels. *Z. Microsk Anat. Forsch.* **41,** 321.

Coraux, C., Nawrocki-Raby, B., Hinnrasky, J., Kileztky, C., Gaillard, D., Dani, C., and Puchelle, E. (2005). Embryonic stem cells generate airway epithelial tissue. *Am. J. Respir. Cell Mol. Biol.* **32,** 87–92.

Cortiella, J., Kojima, K., Bonassar, L. J., Hendricks, G., Vacanti, C. A., and Vacanti, M. P. (2000). Tissue engineered lung. *Tissue Eng.* **6,** 661.

Cutz, E., and Orange, R. P. (1977). Mast cells and endocrine (APUD) cells of the lung. *In* "Asthma: Physiology, Immunopharmacology and Treatment" (L. M. Lichtenstein and K. F. Austen, eds.), p. 51. Academic Press, New York.

Donnelly, G. M., Haack, D. G., and Heird, C. S. (1982). Tracheal epithelium: Cell kinetics and differentiation in normal rat tissue. *Cell Tissue Kinet.* **15,** 119–130.

Evans, M. J., Cabral Anderson, L. J., and Freeman, G. (1978). Role of the Clara cell in renewal of the bronchiolar epithelium. *Lab. Invest.* **38,** 648–653.

Evans, M. J., Johnson, L. V., Stephens, R. J., and Freeman, G. (1976). Renewal of the terminal bronchiolar epithelium in the rat following exposure to NO$_2$ or O$_3$. *Lab. Invest.* **35,** 246–257.

Evans, M. J., Stephens, R. J., and Freeman, G. (1971). Effects of nitrogen dioxide on cell renewal in the rat lung. *Arch. Intern. Med.* **128,** 57–60.

Ferrari, G., Cusella-De Angelis, G., Coletta, M., Paolucci, E., Stornaiuolo, A., Cossu, G., and Mavilio, F. (1998). Muscle regeneration by bone marrow-derived myogenic progenitors. *Science* **279,** 1528–1530.

Fuchs, E., and Segre, J. A. (2000). Stem cells: A new lease on life. *Cell* **100,** 143–155.

Giangreco, A., Reynolds, S. D., and Stripp, B. R. (2002). Terminal bronchioles harbor a unique airway stem cell population that localizes to the bronchoalveolar duct junction. *Am. J. Pathol.* **161,** 173–182.

Grove, J. E., Lutzko, C., Priller, J., Henegariu, O., Thiese, N. D., Kohn, D. B., and Krause, D. S. (2002). Marrow-derived cells as vehicles for delivery of gene therapy to pulmonary epithelium. *Am. J. Respir. Cell Mol. Biol.* **27,** 645–651.

Håkelien, A. M., Landsverk, H. B., Robl, J. M., Skalhegg, B. S., and Collas, P. (2002). Reprogramming fibroblasts to express T-cell functions using cell extracts. *Nature Biotechnol.* **20,** 460–466.

Harris, R. G., Herzog, E. L., Bruscia, E. M., Grove, J. E., Van Arnam, J. S., and Krause, D. S. (2004). Lack of a fusion requirement of development of bone marrow-derived epithelia. *Science* **305,** 90–93.

Hogan, B. L. M. (1999). Morphogenesis. *Cell* **96,** 225–233.

Hong, K. U., Reynolds, S. D., Giangreco, A., Hurley, C. M., and Stripp, B. R. (2001). Clara cell secretory protein-expressing cells of the airway neuroepithelial body microenvironment include a label-reatining subset and are critical for epithelial renewal after progenitor cell depletion. *Am. J. Respir. Cell Mol. Biol.* **24,** 671–681.

Ishizawa, K., Kubo, H., Yamada, M., Kobayashi, S., Numasaki, M., Ueda, S., Suzuki, T., and Sasaki, H. (2004). Bone marrow-derived cells contribute to lung regeneration after elastase-induced pulmonary emphysema. *FEBS Lett.* **556,** 249–252.

Jiang, Y., Jahagirdar, B. N., Reinhardt, R. L., Schwartz, R. E., Keene, C. D., Ortiz-Gonzalez, X. R., Reyes, M., Lenvik, T., Lund, T., Blackstad, M., Du, J., Aldrich, S., Lisberg, A., Low, W. C., Largaespada, D. A., and Verfaillie, C. M. (2002). Pluripotency of mesenchymal stem cells derived from adult marrow. *Nature* **418,** 41–49.

Johnson, D. E., and Wobken, J. D. (1987). Calcitonin gene-related immunoreactivity in airway epithelial cells of the human fetus and infant. *Cell Tissue Res.* **250,** 579–583.

Kauffman, S. L. (1980). Cell proliferation in the mammalian lung. *Int. Rev. Pathol.* **22,** 131–191.

Kauffman, S. L., Burri, P. H., and Weibel, E. R. (1974). The postnatal growth of the rat lung. II. Autoradiography. *Anat. Rec.* **180,** 63–76.

Kleeberger, W., Versmold, A., Rothamel, T., Glockner, S., Bredt, M., Haverich, A., Lehmann, U., and Kreipe, H. (2003). Increased chimerism of bronchial and alveolar epithelium in human lung allografts undergoing chronic injury. *Am. J. Pathol.* **162,** 1487–1494.

Kotton, D. N., Fabian, A. J., and Mulligan, R. C. (2005). Failure of bone marrow to reconstitute lung epithelium. *Am. J. Respir. Cell Mol. Biol.* **33,** 328–334.

Kotton, D. N., Ma, B. Y., Cardoso, W. V., Sanderson, E. A., Summer, R. S., Williams, M. C., and Fine, A. (2001). Bone marrow-derived cells as progenitors of lung alveolar epithelium. *Development* **128,** 5181–5188.

Krause, D. S., Theise, N. D., Collector, M. I., Henegariu, O., Hwang, S., Gardner, R., Neutzel, S., and Sharkis, S. J. (2001). Multi-organ, multi-lineage engraftment by a single bone marrow-derived stem cell. *Cell* **105,** 369–377.

Kubit, V., Sonmez-Alpan, E., Zeevi, A., Paradis, I., Dauber, J. H., Iacono, A., Keenan, R., Griffith, B. P., and Yousem, S. A. (1994). Mixed allogeneic chimerism in lung allograft recipients. *Hum. Pathol.* **25,** 408–412.

Lagasse, E., Connors, H., Al-Dhalimy, M., Reitsma, M., Dohse, M., Osborne, L., Wang, X., Finegold, M., Weissman, I. L., and Grompe, M. (2000). Purified hematopoietic stem cells can differentiate into hepatocytes *in vivo. Nature Med.* **6,** 1229–1234.

Lauweryns, J. M., and Cokelaere, M. (1973). Hypoxia-sensitive neuro-epithelial bodies: Intrapulmonary secretory neuroreceptors modulated by the CNS. *Z. Zellforsch. Milrosk. Anat.* **145,** 521–540.

Lauweryns, J. M., Van Ranst, L., and Verhofstad, A. A. J. (1986). Ultrastructural localization of serotonin in the intrapulmonary neuroepithelial bodies of neonatal rabbits by use of immunoelectron microscopy. *Cell Tissue Res.* **243**, 455–459.

Loi, R., Beckett, T., Goncz, K. K., Suratt, B. T., and Weiss, D. J. (2006). Limited restoration of cystic fibrosis lung epithelium *in vivo* with adult bone marrow-derived cells. *Am. J. Respir. Crit. Care Med.* **173**, 171–179.

Macklin, C. C. (1954). The pulmonary alveolar mucoid film and the pneumocytes. *Lancet* **29**, 1099–1104.

Mahvi, D., Bank, H., and Harley, R. (1977). Morphology of a naphthalene-induced bronchiolar lesion. *Am. J. Pathol.* **86**, 559–572.

Mattsson, J., Jansson, M., Wernerson, A., and Hassan, M. (2004). Lung epithelial cells and type II pneumocytes of donor origin after allogeneic hematopoietic stem cell transplantation. *Transplantation* **78**, 154–157.

Metcalf, D., and Moore, M. A. S. (1971). "Haematopoietic Cells." North Holland, Amsterdam.

Mezey, E., Chandross, K. J., Harta, G., Maki, R. A., and McKercher, S. R. (2000). Turning blood into brain: Cells bearing neuronal antigens generated *in vivo* from bone marrow. *Science* **290**, 1779–1782.

Okamoto, R., Yajima, T., Yamazaki, M., Kanai, T., Mukai, M., Okamoto, S., Ikeda, Y., Hibi, T., Inazawa, J., and Watanabe, M. (2002). Damaged epithelia regenerated by bone marrow-derived cells in the human gastrointestinal tract. *Nature Med.* **8**, 1011–1017.

Orlic, D., Kajstura, J., Chimenti, S., Bodine, D. M., Leri, A., and Anversa, P. (2001). Bone marrow cells generate infarcted myocardium. *Nature* **410**, 701–705.

Ortiz, L. A., Gambelli, F., McBride, C., Gaupp, D., Baddoo, M., Kaminski, N., and Phinney, D. G. (2003). Mesenchymal stem cell engraftment in lung is enhanced in response to bleomycin exposure and ameliorates its fibrotic effects. *Proc. Natl. Acad. Sci. USA* **100**, 8407–8411.

Pereira, R. F., O'Hara, M. D., Laptev, A. V., Halford, K. W., Pollard, M. D., Class, R., Simon, D., Livezey, K., and Prockop, D. J. (1998). Marrow stromal cells as a source of progenitor cells for nonhemapoietic tissues in transgenic mice with a phenotype of osteogenesis imperfecta. *Proc. Natl. Acad. Sci. USA* **95**, 1142–1147.

Petersen, B. E., Bowen, W. C., Patrene, K. D., Mars, W. M., Sullivan, A. K., Murase, N., Boggs, S. S., and Greenberger, J. S. (1999). Bone marrow as a potential source of hepatic oval cells. *Science* **284**, 1168–1170.

Pittenger, M. F., MacKay, A. M., Beck, S. C., Jaiswal, R. K., Douglas, R., Mosca, J. D., Moorman, M. A., Simonetti, D. W., Craig, S., and Marshak, D. R. (1999). Multilineage potential of adult human mesenchymal stem cells. *Science* **284**, 143–147.

Plopper, C. G., and Dungworth, D. L. (1987). Structure, function, cell injury and cell renewal of bronchiolar and alveolar epithelium. *In* "Lung Carcinoma" (E. M. McDowell, ed.), p. 29. Churchill Livingstone, London.

Plopper, C. G., Suverkropp, C., Morin, D., Nishio, S., and Buckpitt, A. (1992). Relationship of cytochrome P-450 activity to Clara cell cytotoxicity. I. Histopathologic comparison of the respiratory tract of mice, rats and hamsters after parenteral administration of naphthalene. *J. Pharmacol. Exp. Ther.* **26**, 353–363.

Polak, J. M., and Bishop, A. E. (2006). Stem cells and tissue engineering: Past, present and future. *Ann. N.Y. Acad. Sci.* **1068**, 352–366.

Poulsom, R., Forbes, S. J., Hodivala-Dilke, K., Ryan, E., Wyles, S., Navaratnarasah, S., Jeffery, R., Hunt, T., Alison, M., Cook, T., Pusey, C., and Wright, N. A. (2001). Bone marrow contributes to renal parenchymal turnover and regeneration. *J. Pathol.* **195**, 229–235.

Prockop, D. J. (1997). Marrow stromal cells for non hemapoietic tissues. *Science* **276**, 71–74.

Qin, M. D., Tai, G. P., Collas, P., Polak, J. M., and Bishop, A. E. (2005). Cell extract-derived differentiation of embryonic stem cells. *Stem Cells* **23**, 712–718.

Reddy, R., Buckley, S., Doerken, M., Barsky, L., Weinberg, K., Anderson, K. D., Warburton, D., and Driscoll, B. (2004). Isolation of a putative progenitor subpopulation of alveolar epithelial type 2 cells. *Am. J. Physiol. Lung Cell Mol. Physiol.* **286**, L658–L667.

Reid, L., and Jones, R. (1979). Bronchial mucosal cells. *Fed. Proc.* **38**, 191–196.

Reynolds, S. D., Giangreco, A., Power, J. H. T., and Stripp, B. R. (2000a). Neuroepithelial bodies of pulmonary airways serve as a reservoir of progenitor cells capable of epithelial regeneration. *Am. J. Pathol.* **156**, 269–278.

Reynolds, S. D., Hong, K. U., Giangreco, A., Mago, G. W., Guron, C., Morimoto, Y., and Stripp, B. R. (2000b). Conditional Clara cell ablation reveals a self-renewing progenitor function of pulmonary neuroendocrine cells. *Am. J. Physiol. Lung Cell Mol. Physiol.* **278**, L1256–L1263.

Rippon, H. J., Ali, N. N., Polak, J. M., and Bishop, A. E. (2004). Initial observations on the effect of medium composition on the differentiation of murine embryonic stem cells to alveolar type II cells. *Cloning Stem Cells.* **6**, 49–56.

Rippon, H. J., Polak, J. M., Qin, M., and Bishop, A. E. (2006). Derivation of distal lung epithelial progenitors from murine embryonic stem cells using a novel 3-step differentiation protocol. *Stem Cells* **24**(5), 1389–1398.

Samadikuchaksaraei, A., Cohen, S., Polak, J. M., Bielby, R. C., and Bishop, A. E. (2006). Derivation of type II pneumocytes from human embryonic stem cells. *Tissue Eng.* **8**(4), 541–550.

Shannon, J. M., Nielsen, L. D., Gebb, S. A., and Randell, S. H. (1998). Mesenchyme specifies epithelial differentiation in reciprocal recombinants of embryonic lung and trachea. *Dev. Dyn.* **212**, 482–494.

Spees, J. L., Olson, S. D., Ylostalo, J., Lynch, P. J., Smith, J., Perry, A., Peister, A., Wang, M. Y., and Prockop, D. J. (2003). Differentiation, cell fusion, and nuclear fusion during *ex vivo* repair of epithelium by human adult stem cells from bone marrow stroma. *Proc. Natl. Acad. Sci. USA* **100**, 2397–2402.

Stripp, B. R., Maxson, K., Mera, R., and Singh, G. (1995). Plasticity of airway cell proliferation and gene expression after acute naphthalene injury. *Am. J. Physiol.* **269**, L791–L799.

Suratt, B. T., Cool, C. D., Serls, A. E., Chen, L., Varella-Garcia, M., Shpall, E. J., Brown, K. K., and Worthen, G. S. (2003). Human pulmonary chimerism after hematopoietic stem cell transplantation. *Am. J. Respir. Crit. Care Med.* **168**, 318–322.

Terada, N., Hamazaki, T., Oka, M., Hoki, M., Mastalerz, D. M., Nakano, Y., Meyer, E. M., Morel, L., Petersen, B. E., and Scott, E. W. (2002). Bone marrow cells adopt the phenotype of other cells by spontaneous cell fusion. *Nature* **416**, 542–545.

Theise, N. D., Henegariu, O., grove, J., Jagirdar, J., Kao, P. N., Crawford, J. M., Badve, S., Saxena, R., and Krause, D. S. (2002). Radiation pneumonitis in mice: A severe injury model for pneumocyte engraftment from bone marrow. *Exp. Hematol.* **30**, 1333–1338.

Theise, N. D., Nimmakayalu, M., Gardner, R., Illei, P. B., Morgan, G., Teperman, L., Henegariu, O., and Krause, D. S. (2000). Liver from bone marrow in humans. *Hepatology* **32**, 11–16.

Till, J. E., and McCulloch, E. A. (1961). A direct measurement of the radiation sensitivity of normal mouse bone marrow cells. *Radiat. Res.* **14**, 1419–1422.

Vacanti, M. P., Roy, A., Cortiella, J., Bonassar, L., and Vacanti, C. (2001). Identification and initial characterization of spore-like cells in adult mammals. *J. Cell. Biochem.* **80**, 455–460.

Van Vranken, B., Romanska, H. M., Polak, J. M., Rippon, H. J., Shannon, J. M., and Bishop, A. E. (2005). Co-culture of embryonic stem cells with pulmonary mesenchyme: A microenvironment that promotes differentiation of pulmonary epithelium. *Tissue Eng.* **11**, 1177–1187.

Vassilopoulos, G., Wang, P. R., and Russell, D. W. (2003). Transplanted bone marrow regenerates liver by cell fusion. *Nature* **422**, 901–904.

Wang, X., Willenbring, H., Akkari, Y., Torimaru, Y., Foster, M., Al-Dhalimy, M., Lagasse, E., Finegold, M., Olson, S., and Grompe, M. (2003). Cell fusion is the principal source of bone marrow derived hepatocytes. *Nature* **422**, 897–901.

Weissman, I. L. (2000). Stem cells: Units of development, units of regeneration, and units in evolution. *Cell* **100**, 157–168.

Wharton, J., Polak, J. M., Bloom, S. R., Ghatei, M. A., Solcia, E., Brown, M. R., and Pearse, A. G. E. (1978). Bombesin-like immunoreactivity in the lung. *Nature* **273**, 769–770.

Witschi, H. (1976). Proliferation of type II alveolar cells: A review of common responses in toxic lung injury. *Toxicology* **5**, 267–277.

Wollert, K. C., and Drexler, H. (2005). Clinical applications of stem cells for the heart. *Circ. Res.* **96**, 151–163.

Yamada, M., Kubo, H., Kobayashi, S., Ishizawa, K., Numasaki, M., Ueda, S., Suzuki, T., and Sasaki, H. (2004). Bone marrow-derived progenitor cells are important for lung repair after lipopolysaccharide-induced lung injury. *J. Immunol.* **172**, 1266–1272.

Ying, Q. L., Nichols, J., Evans, E. P., and Smith, A. G. (2002). Changing potency by spontaneous fusion. *Nature* **416**, 545–548.

Zander, D. S., Baz, M. A., Cogle, C. R., Visner, G. A., Theise, N. D., and Crawford, J. M. (2005). Bone marrow-derived stem-cell repopulation contributes minimally to the type II pneumocyte pool in transplanted human lungs. *Transplantation* **80**, 206–212.

Author Index

A

Abe, K., 285
Abe, M., 147
Abeyta, M. J., 85, 308
Ablonczy, Z., 182
Abramczuk, J., 148
Abramow-Newerly, W., 4, 7
Abramsson, A., 259
Abri, A., 169, 170
Acker, H., 272
Adams, C. E., 117
Adams, I. R., 286
Adamson, I. Y., 334, 335
Adamson, P., 169, 170
Addicks, K., 278
Aden, D. P., 18, 93
Adenot, P. G., 118
Afrikanova, I., 210
Ageyama, N., 244
Aglietta, M., 222, 227
Agulnick, A. D., 317
Ahmad, S., 80
Ahn, H. J., 81
Ährlund-Richter, L., 80, 89, 157
Ailhaud, G., 197, 198
Akasu, R., 284, 288, 308
Akkari, Y., 337
Akkina, R., 232
Akutsu, H., 6, 78, 177, 220
Albanese, J., 54, 62
Albera, C., 337
Albrecht, K. H., 286
Al-Dhalimy, M., 336, 337
Aldrich, S., 336
Alestrom, A., 65, 66, 68
Alestrom, P., 65, 66, 68
Alexander, M. R., 218
Ali, N. N., 338
Alison, M. R., 336, 337
Allen, N. D., 7, 124, 125, 172
Allsopp, T. E., 78
Alvarez-Buylla, A., 159, 336, 337

Alvarez-Dolado, M., 336, 337
Amariglio, N., 255, 256
Ambergen, A. W., 335
Amemiya, K., 170
American Heart Association, 267
Amit, M., 80, 85, 88, 89, 92, 200, 252, 253, 255, 262, 268, 269, 278
Amsterdam, A., 65
Anastassiadis, C., 299
Ancelin, K., 286
Andang, M., 80, 89, 157
Anderson, D. J., 317
Anderson, J. S., 232
Anderson, K. D., 335
Anderson, W. F., 105
Anderson, W. J., 316, 317, 321, 322
Ansell, J. D., 214
Anthony, T. E., 159
Antke, C., 197
Antonelli, M., 65
Anversa, P., 336
Aoki, N., 243
Arai, T., 285
Araki, K., 213
Araki, R., 146
Aramant, R. B., 170
Arechaga, J., 121
Arenas, E., 5, 6, 18
Arentson, E., 210
Ares, X., 88, 92, 108
Arii, S., 169
Arman-Kalcek, G., 219
Armant, D. R., 117
Armstrong, L., 80, 89
Artzt, K., 65
Asahina, K., 169
Asashima, M., 244
Asbrand, C., 317, 318, 328, 329
Asensi, A., 80
Ashtiani, S. K., 80
Assad, J. A., 197
Assadoulina, A., 169, 170
Atala, A., 171, 173, 175, 182, 183

351

H

Subject Index

A

Adipocyte, differentiation from human embryonic stem cells, 202

Alkaline phosphatase
 bovine embryonic stem cell detection, 33–34
 human embryonic stem cell detection, 97–99

Avian embryonic stem cell
 advantages as model, 38
 chimera production
  chimerism maximization, 51–53
  efficiency, 61–62
  embryonic stem cell preparation and injection, 54–57
  recipient embryos
   incubation, 57–59
   storage, 57
  surrogate cell culture
   overview, 53
   shell preparation and use, 53–54
  timeline, 60–61
 cryopreservation, 49
 egg acquisition, storage, and quality control, 59–60
 electroporation, 49–51
 establishment
  blastodermal cell isolation, 42, 44–45
  breed selection, 51–52
  culture, 45
  feeder cells, 41
  media, 39–41
 history of study, 39
 maintenance and passaging, 45–46
 morphology, 46, 48–49

B

B-cell, differentiation from embryonic stem cells, 212, 232

Beta-cells, see Pancreatic cells

Blastocyst cloning, see Somatic cell nuclear transfer

Blastomere, see Mouse embryonic stem cell

BMPs, see Bone morphogenetic proteins

Bone morphogenetic proteins, embryonic mesoderm regulation, 244

Bovine embryonic stem cell
 characterization
  alkaline phosphatase detection, 33–34
  immunocytochemistry, 35–36
  karyotyping, 36
  surface marker immunostaining, 34–35
 establishment
  expansion, 31
  fertilization in vitro, 28–29
  inner cell mass isolation, 30
  materials, 24–25
  media and solutions, 24, 26–27
  mouse embryonic fibroblast feeder layer preparation, 27–28
  oocyte collection and maturation, 28
  parthenogenic activation, 29
  somatic cell nuclear transfer, 29–30
 maintenance, 31, 33
 markers, 22–23

C

Cardiomyocyte
 loss in heart disease, 267–268
 mouse embryonic stem cell differentiation
  bioreactor production of embryoid bodies
   autoclave method, 275–278
   overview, 272
   rotating cell culture system, 273, 275
  embryoid body generation approaches, 268–269
  enrichment, 279–281
  factors affecting, 278
  hanging drop culture of embryoid bodies
   embryonic stem cell dissociation, 269, 271

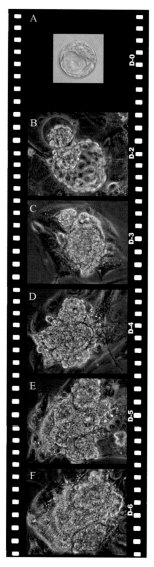

NAGY AND VINTERSTEN, CHAPTER 1, FIG. 1. Phases of blastocyst outgrowth development during preparation for ES cell establishment. All the pictures are of the same scale. (A) High-quality blastocyst ready to be plated on MEFs for ES cell derivation (day 0). (B) Embryo at the final stage of hatching (day 2). (C) Attaching embryo (day 3). Attached trophoblast cells are clearly visible under the outgrowth. (D–F) The outgrowth is increasing in size (days 4 to 6). Areas with ES cell-like cells become visible (E) and grow larger. This outgrowth is now ready for disaggregation (F).

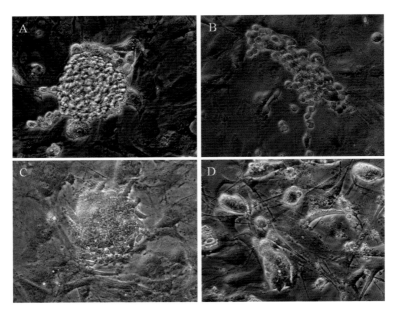

NAGY AND VINTERSTEN, CHAPTER 1, FIG. 2. Cell colonies in the early phase of mES cell establishment. (A) Colony of mixed cell types as a result of improper disaggregation of the initial outgrowth. (B) Differentiated non-ES-like cell. (C) Three days after disaggregation of the outgrowth (Fig. 1F). ES cell-like colonies can be recognized easily by the characteristic morphology. The colonies may, at this stage, still contain a few differentiated cells, but these usually diminish after a few passages. (D) Small colonies of pure mouse ES cells.

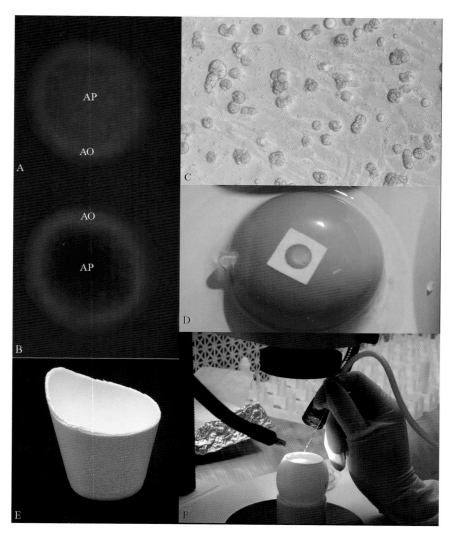

VAN DE LAVOIR AND MATHER-LOVE, CHAPTER 3, FIG. 1. (A and B) Stage X (EG&K) embryos showing the peripheral area opaca (AO) and the central area pellucida (AP). (B) After injection of 2 μl of medium. Note the expansion of the medium throughout the subgerminal cavity. Injections were done and pictures taken using a blue light source. (C) Blastodermal cells collected from the AP from a stage X (EG&K) embryo. (D) Recovery of stage X (EG&K) embryo using a paper filter square. (E) Flexible pouring cup for transferring contents of fertile egg into first surrogate shell. (F) Injection of a cell suspension into a stage X (EG&K) embryo.

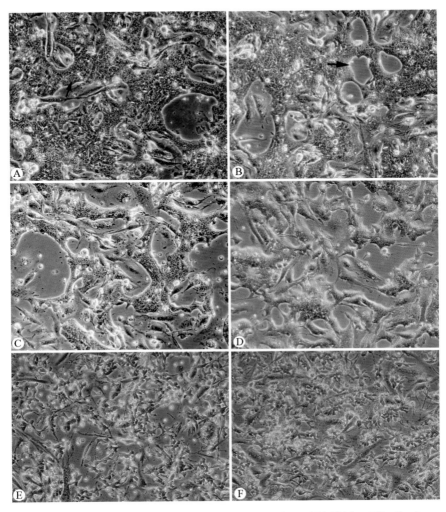

VAN DE LAVOIR AND MATHER-LOVE, CHAPTER 3, FIG. 3. (A and B) Chicken ES cell cultures with desired morphology. Note the single layer of cells. The cells have a large nucleus and a pronounced nucleolus (arrow in A). These panels represent the cell density minimally necessary for successful passaging. (B) The large openings between the cells are evident (arrow), which is indicative of good growing cES cells. These plates were seeded with STO cells, which have been pushed aside by the cES cells and are not visible in the openings. (C and D) Representative of good cultures that are too sparse for passaging. Medium changes on these cultures will have a detrimental effect. Be aware that these cultures can grow very fast and therefore need to be checked first thing the next morning. (E and F) cES cultures that have taken on a pronounced fibroblast morphology. Although these cells do not have the distinct characteristics of cES cells, they can still contribute to the somatic tissues when injected into a recipient embryo.

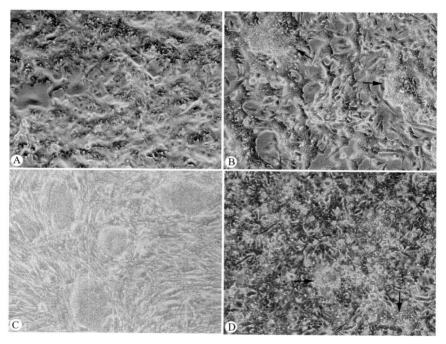

VAN DE LAVOIR AND MATHER-LOVE, CHAPTER 3, FIG. 4. (A) A completely differentiated cES culture. The culture in B displays areas of differentiation (indicated by arrows) that can be removed by passaging gently in Ca^{2+}/Mg^{2+}-free PBS in a 1:1 ratio. The differentiated areas will be left behind and a nice culture can be obtained. Generally a few passages are necessary to eliminate all the differentiation. (C) cES cells that have compacted into colonies reminiscent of mouse ES cells. The STO cells are dense and elongated due to their exposure to cES medium. (D) A culture that has become too dense. Areas of differentiation are apparent (arrows).

VAN DE LAVOIR AND MATHER-LOVE, CHAPTER 3, FIG. 5. (A) System II cultures showing the window sealed with plastic film held in place by plastic rings and rubber bands. (B) System II eggs inside an Octagon 250 incubator. Note their horizontal position in the tray. (C) A Stage 17 (H&H) embryo immediately before transfer from system II to system III. (D) A tray of system III cultures containing 8-day-old embryos. Note how the surrogate shell system provides easy accessibility and visibility of the embryo and the extraembryonic vasculature. (E) Group of two BR and two chimeric chicks. Chimeras are indistinguishable from BR chicks, indicating high levels of chimerism. Reproduced from van de Lavoir *et al.* (2006).

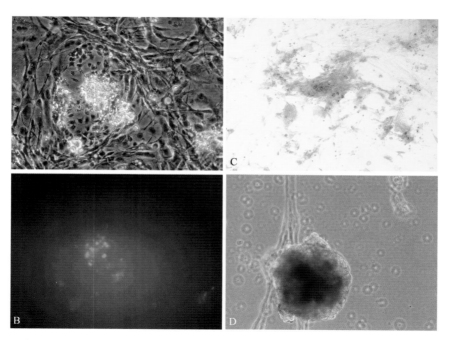

FAN AND COLLODI, CHAPTER 4, FIG. 2. Zebrafish ES cell cultures exhibit *in vitro* characteristics of pluripotency, including (A and B) expression of the SSEA-1 antigen, (C) alkaline phosphatase activity, and (D) formation of differentiated embryoid bodies in suspension culture. The same culture is shown in A and B before and after immunostaining using anti-SSEA-1 antiserum.

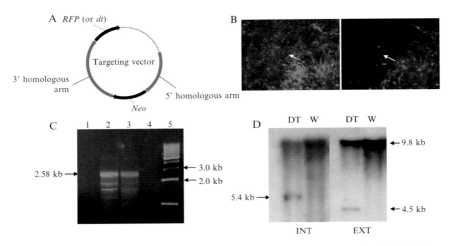

Fan and Collodi, Chapter 4, Fig. 3. Zebrafish ES cells incorporate plasmid DNA in a targeted fashion by homologous recombination. (A) A targeting vector contains neo flanked by arms that are homologous to the targeted gene and either RFP or dt located outside of the homologous region. (B) When RFP is used, following G418 selection, the potential homologous recombinants are identified by the loss of RFP expression (arrow). The targeting vector was introduced into ES cells that constitutively express the green fluorescent protein and the same two G418-resistant colonies are shown by fluorescence microscopy using a green (left) and red (right) filter. The RFP-negative colony (arrow) was removed from the dish, expanded, and confirmed to have undergone homologous recombination by PCR and Southern blot analysis. (C) Following electroporation of a targeting vector that contained dt located outside of the homologous arms, three surviving colonies were isolated and examined by PCR (lanes 1–3) for the presence of a 2.58-kb junction fragment created by targeted insertion of the plasmid. Two of the colonies were found to be homologous recombinants. (D) Southern blot analysis of one dt-resistant colony using probes that hybridized to sequences of the targeted gene that were either internal (INT) or external (EXT) to the homologous arms on the vector. Both probes hybridized to a 9.8-kb fragment corresponding to the nontargeted allele. Each probe also hybridized to a smaller restriction fragment (5.4-kb INT or 4.5-kb EXT) that corresponded to the targeted allele [Fan et al. (2006), B, C, and D reprinted with permission].

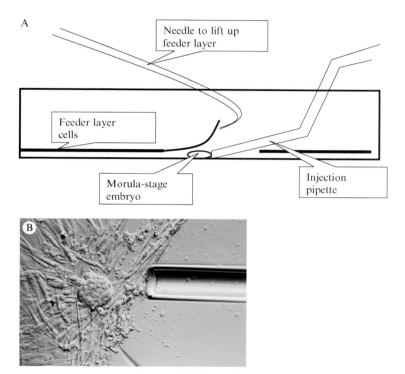

A

Needle to lift up
feeder layer

Feeder layer
cells

Morula-stage
embryo

Injection
pipette

B

STRELCHENKO AND VERLINSKY, CHAPTER 6, FIG. 1. (A) Chart of microtool disposition shows
how to place naked morula under feeder layer. On left side is needle used to lift up cell layer.
On right side is injection pipette. Position tool in front of view. (B) Microtool position after
injection of naked human morula stage embryo under feeder layer. The bottom shows end of
needle to hold feeder layer. Right side shows injection pipette.

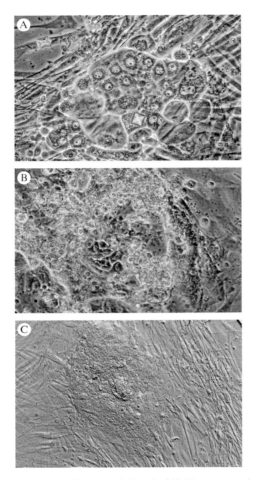

STRELCHENKO AND VERLINSKY, CHAPTER 6, FIG. 2. (A) Human morula the next day after placing under feeder layer. Some cells probably have committed. There are cells with a small amount of cytoplasts, probably ancestors of ES cells, and with a large amount of cytoplasts targeted to trophoblast cells. (B) Cluster of cells derived from human morula on day 5 of culture. Different types of cells can be observed. The center shows small clusters of hESC. Phase contrast. (C) Cluster of cells derived from human morula on day 7 of culture. Differential interference contrast (Hoffman modulation contrast).

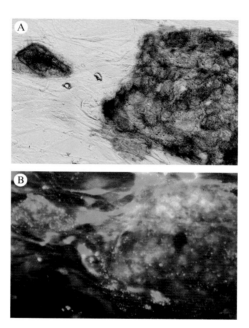

STRELCHENKO AND VERLINSKY, CHAPTER 6, FIG. 3. (A) Human morula-derived stem cells stained for alkaline phosphatase (AP) general enzyme reaction (SK-5300). (B) The same cluster of morula-derived hESC stained with specific monoclonal antibody TRA-2-39. Both approaches in detection of L-AP match each other.

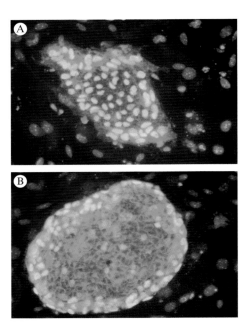

Strelchenko and Verlinsky, Chapter 6, Fig. 5. (A) Simultaneous presence of two markers, TRA-2-39 and Oct-4, in a colony of morula-derived hESC with green fluorescence (FITC) TRA-2-39, red fluorescence (TRITC) Oct-4, and blue fluorescence nonspecific nuclei (DAPI). (B) Large colony of morula-derived hESC where nuclei of hESC are nonpermeable for antibody Oct-4.

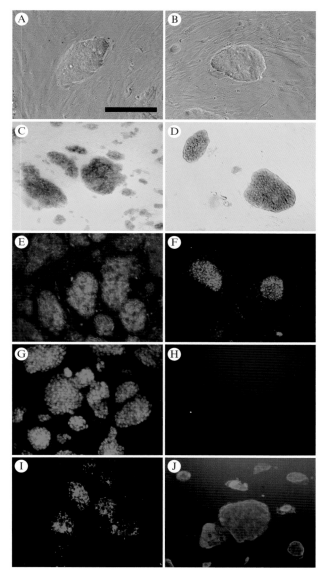

BECKER AND CHUNG, CHAPTER 7, FIG. 2. Comparison of putative ES (left column) and TS (right column) cell lines derived from single blastomeres. (A and B) Phase-contrast photograph of typical colonies. (C and D) Lac-Z-stained colonies showing their single blastomere origin. (E and F) Alkaline phosphatase staining. (G and H) Indirect immunofluorescence with antibodies to Oct-4. (I) SSEA-1 staining of putative ES cells. (J) TROMA-1 staining of putative TS cells (same field as H). Scale bar: 200 μm. From Chung *et al.* (2006), reproduced with permission of Nature Publishing Group.

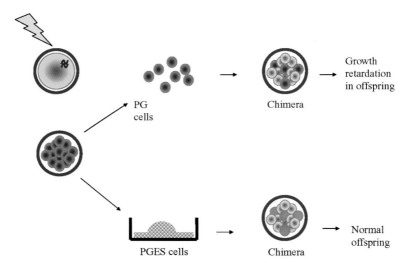

C IBELLI *ET AL.*, C HAPTER 8, F IG. 3. Parthenogenetic blastomeres (PG) when used to make chimeras, generate offspring with growth retardation; however, when a parthenogenetic embryo is used to make ESCs and then those ESCs are used for chimeras, the offspring is normal. *Adapted from Allen et al. Development 1994.*

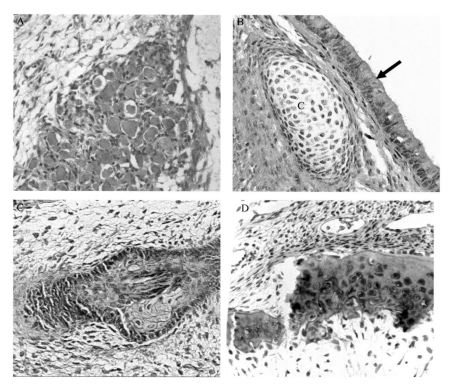

CIBELLI *ET AL.*, CHAPTER 8, FIG. 5. Teratoma produced from Cyno1 cells 15 weeks after injection into SCID mice. (A) ganglion; (B) Cartilage (c) and Respiratory epithelium (arrow); (C) Hair follicle; and (D) bone.

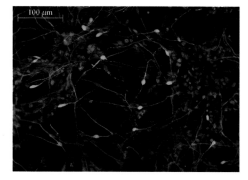

CIBELLI *ET AL.*, CHAPTER 8, FIG. 6. Double ICC for MAP-2 and Serotonin after differentiation of Cyno1. Nuclei are stained with Hoechest dye.

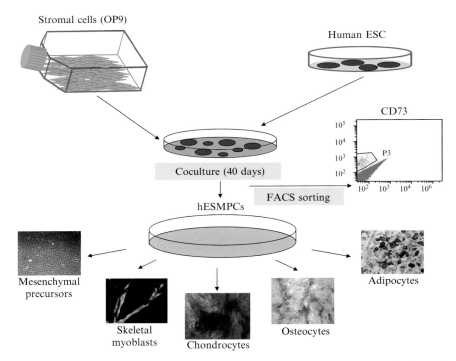

BARBERI AND STUDER, CHAPTER 12, FIG. 1. Schematic illustration of the derivation and differentiation of hESC-derived mesenchymal precursors (hESMPCs). hESCs are dissociated and replated onto confluent OP9 cells. After 40 days of coculture, hESMPCs are isolated and purified using FACS for CD73. hESMPCs can be further expanded and maintained as mesenchymal precursors or selectively differentiated into specific mesenchymal derivatives, including fat, cartilage, bone, and skeletal muscle cells.

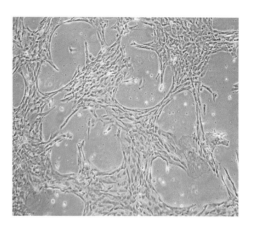

GOLDBERG-COHEN *ET AL.*, CHAPTER 15, FIG. 2. Following filtration of human ES cells cultured on type IV collagen, separated cells were recultured on type IV collagen in the presence of VEGF. Recultured cells are organized in ring-like structures on the collagen plating.

	ES cells	EBs	Multilineage progenitors	Committed pancreatic progenitors	Islet-like cluster
Stage	1	2	3	4	5

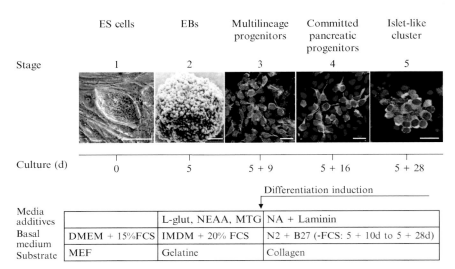

| Culture (d) | 0 | 5 | 5 + 9 | 5 + 16 | 5 + 28 |

Differentiation induction

Media additives		L-glut, NEAA, MTG	NA + Laminin		
Basal medium	DMEM + 15%FCS	IMDM + 20% FCS	N2 + B27 (-FCS: 5 + 10d to 5 + 28d)		
Substrate	MEF	Gelatine	Collagen		

SCHROEDER *ET AL.*, CHAPTER 19, FIG. 1. Schematic representation of ES cell-derived pancreatic differentiation. Mouse ES cells cultured on fibroblast feeder layers (1) are differentiated via embryoid bodies (EBs, scanning electron microscopy) (2) into multilineage progenitor cells (3) and after differentiation induction by growth factors into committed pancreatic progenitors (4) and insulin-producing cells in islet-like clusters (5). Stages of differentiation with examples of cell morphology and media, additives, and substrates used during *in vitro* differentiation are shown. Cells at stage 5 + 9d are dissociated and replated onto collagen I-coated tissue culture plates and cultured in complex differentiation medium with differentiation factors and 10% FCS to support attachment of cells. One day later, the differentiation medium is replenished without serum and cells are differentiated for up to 28 or more days. Immunostaining shows nestin/CK19 (3), nestin/C-peptide (4), and C-peptide/insulin (5) coexpression in cells at different stages. Bars: 20 μm (3, 4, 5), 50 μm (2), and 100 μm (1).

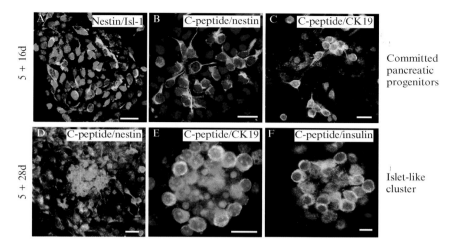

SCHROEDER *ET AL.*, CHAPTER 19, FIG. 3.   Double immunofluorescence analysis of ES-derived cells differentiated into the pancreatic lineage. ES-derived cells at intermediate (5 + 16d, A–C) and terminal (5 + 28d, D–F) stages following the three-step pancreatic differentiation protocol are shown. (A and B) Images show immunohistochemical analysis of nestin/Isl-1 and nestin/C-peptide coexpression, and (C) coexpression of C-peptide and cytokeratin 19 (CK19). (D–F) Images demonstrate the lack of C-peptide/nestin coexpression in islet-like clusters, a weak coexpression of C-peptide with CK19, but C-peptide/insulin colabeling in islet-like clusters at the terminal differentiation stage, 5 + 28d. Bar: 20 μm.

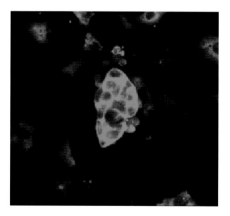

BISHOP AND POLAK, CHAPTER 20, FIG. 1.   A clump of differentiated embryonic stem cells in culture immunostained for surfactant protein C, a specific marker for type II pneumocytes. The clump of pneumocytes is surrounded by nonimmunoreactive cells showing that the resultant cell populations were heterogeneous (indirect immunofluorescence method).

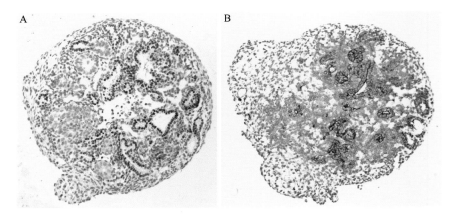

BISHOP AND POLAK, CHAPTER 20, FIG. 2. Serial sections through a preparation of murine embryonic (E11.5) pulmonary mesenchyme wrapped around an embryoid body (day 8) following culture for 5 days in basic medium. The mesenchyme and embryoid body have coalesced and channels have formed lined by cells showing immunoreactivity for (A) thyroid transcription factor-1, a marker for immature pulmonary epithelium, and (B) cytokeratin, a general marker for epithelium (avidin biotin peroxidase complex method).

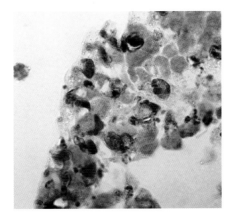

BISHOP AND POLAK, CHAPTER 20, FIG. 3. Human embryonic stem cells following differentiation in SAGM showing immunoreactivity for surfactant protein A, found in the distal airway epithelium (avidin biotin peroxidase complex method).